Phytochemical Methods

A GUIDE TO MODERN TECHNIQUES OF PLANT ANALYSIS

J. B. Harborne

Professor of Botany
University of Reading

Second Edition

LONDON   NEW YORK
CHAPMAN AND HALL

First published 1973 by
Chapman and Hall Ltd
11 New Fetter Lane, London EC4P 4EE
Reprinted 1976
Second edition 1984
Second edition issued as a paperback 1988

Published in the USA by
Chapman and Hall
29 West 35th Street, New York NY 10001

© 1973, 1984 J. B. Harborne

Printed in Great Britain by
Richard Clay Ltd, Bungay, Suffolk

ISBN 0 412 34330 4

British Library Cataloguing in Publication Data

Harborne, J. B. (Jeffrey Barry), 1928
  Phytochemical methods.—2nd ed.
  1. Plants. organic compounds.
  Qualitative analysis. Laboratory techniques
  I. Title
  581.19'24
  ISBN 0 412 34330 4

Library of Congress Cataloguing in Publication Data

Harborne, J. B. (Jeffrey B.)
  Phytochemical methods.

  Includes bibliographies and index.
  I. Plants—Analysis I. Title.
QK865.H27 *1984* 581.19'285   *84-7750*
ISBN 0 412 34330 4

Phytochemical Methods

A GUIDE TO MODERN TECHNIQUES
OF PLANT ANALYSIS

Contents

Preface to First Edition

While there are many books available on methods of organic and biochemical analysis, the majority are either primarily concerned with the application of a particular technique (e.g. paper chromatography) or have been written for an audience of chemists or for biochemists working mainly with animal tissues. Thus, no simple guide to modern methods of plant analysis exists and the purpose of the present volume is to fill this gap. It is primarily intended for students in the plant sciences, who have a botanical or a general biological background. It should also be of value to students in biochemistry, pharmacognosy, food science and 'natural products' organic chemistry.

Most books on chromatography, while admirably covering the needs of research workers, tend to overwhelm the student with long lists of solvent systems and spray reagents that can be applied to each class of organic constituent. The intention here is to simplify the situation by listing only a few specially recommended techniques that have wide currency in phytochemical laboratories. Sufficient details are provided to allow the student to use the techniques for themselves and most sections contain some introductory practical experiments which can be used in classwork.

After a general introduction to phytochemical techniques, the book contains individual chapters describing methods of identifying phenolic compounds, terpenoids, fatty acids and related compounds, nitrogen compounds, sugars and their derivatives and macromolecules. The attempt has been made to cover practically every class of organic plant constituent, although in some cases, the account is necessarily brief because of space limitations. Special attention has, however, been given to detection of endogeneous plant growth regulators and to methods of screening plants for substances of pharmacological interest. Each chapter concludes with a general reference section, which is a bibliographic guide to more advanced texts.

While the enormous chemical variation of secondary metabolism in plants has long been appreciated, variation in primary metabolism (e.g. in the enzymes of respiration and of photosynthetic pathways) has only become apparent quite recently. With the realization of such chemical variation in both the small and large molecules of the plant kingdom, systematists have become interested in phytochemistry for shedding new light on plant relation-ships. A new discipline of biochemical systematics has developed and, since the present book has been written with some emphasis on comparative aspects, it should be a useful implement to research workers in this field.

In preparing this book for publication, the author has been given advice and suggestions from many colleagues. He would particularly like to thank Dr E. C. Bate-Smith, Dr T. Swain, Dr T. A. Smith and Miss Christine Williams for their valuable assistance. He is also grateful to the staff of Chapman and Hall for expeditiously seeing the book through the press.

Reading,   J. B. H.
June 1973

Preface to Second Edition

Since the preparation of the first edition, there have been several major developments in phytochemical techniques. The introduction of carbon-13 NMR spectroscopy now provides much more detailed structural information on complex molecules, while HPLC adds a powerful and highly sensitive analytical tool to the armoury of the chromatographer. With HPLC, it is possible to achieve for involatile compounds the type of separations that GLC produces for volatile substances. Spectacular developments have also occurred in the techniques of mass spectrometry; for example, the availability of a fast-atom bombardment source makes it possible in FAB–MS to determine the molecular weight of both very labile and involatile plant compounds. These techniques are described briefly in the appropriate sections of this new edition.

During the last decade, the number of new structures reported from plant sources has increased enormously and, among some classes of natural constituent, the number of known substances has doubled within this short time-span. The problems of keeping up with the phytochemical literature are, as a result of all this activity, quite considerable, although computerized searches through *Chemical Abstracts* have eased the burden for those scientists able to afford these facilities. In order to aid at least the student reader, literature references in this second edition have been extensively updated to take into account the most recent developments. Additionally, some new practical experiments have been added to aid the student in developing expertise in studying, for example, phytoalexin induction in plants and allelopathic interactions between plants.

Since the first edition, phytochemical techniques have become of increasing value in ecological research, following the realization that secondary constituents have a significant role in determining the food choice of those animals that feed on plants. Much effort has been expended on analysing plant populations for their toxins or feeding deterrents. Most such compounds were included in the first edition, except for the plant tannins. A new section has therefore been included on tannin analysis in Chapter 2.

With this new edition, the opportunity has been taken to add two appendices – a checklist of TLC procedures for all classes of plant substance and a list of useful addresses for phytochemists. Some errors in the first edition have been corrected, but others may remain and the author would welcome suggestions for further improvements.

As with the first edition, the author has benefited considerably from help and advice of many colleagues. He would particularly like to thank his co-workers in the phytochemical unit and his students, who have been willing guinea pigs in the development of new phytochemical procedures.

Jeffrey B. Harborne

Reading,
December 1983

Glossary

GENERAL ABBREVIATIONS

TLC	= thin layer chromatography	MS	= mass spectroscopy
GLC	= gas liquid chromatography	R_F	= mobility relative to front
PC	= paper chromatography	RR_t	= relative retention time
UV	= ultraviolet	nm	= nanometres
IR	= infrared	mol. wt.	= molecular weight
NMR	= nuclear magnetic resonance	M	= molar
HPLC	= high performance liquid chromatography		

CHEMICALS

BSA	= N,O-bis(trimethylsilyl)acetamide
PVP	= Polyvinylpyrrolidone (for removing phenols)
BHT	= butylated hydroxytoluene (anti-oxidant)
EDTA	= ethylenediaminetetracetic acid (chelating agent)
Tris	= tris(hydroxymethyl)methylamine (buffer)
PMSF	= phenylmethane sulphonyl fluoride
SDS	= sodium dodecyl sulphate

CHROMATOGRAPHIC SUPPORTS

Kieselguhr = diatomaceous earth
Decalso = sodium aluminium silicate
DEAE–cellulose = diethylaminoethyl-treated
PEI–cellulose = polyethyleneimine-treated
ECTEOLA–cellulose = epichlorohydrin-triethanolamine-alkali treated

PPE = polyphenyl ether	OV = methyl siloxane polymer
TXP = trixylenylphosphate	SE = silicone oil
DEGS = diethyleneglycol succinate	XE = nitrile silicone
Apiezon L = stop-cock grease	Embacel = acid-washed celite support
Carbowax = polyethylene glycol	
Chromosorb = firebrick support	Poropak = styrene polymer support
ODS = octadecylsilane	

Glossary

CHROMATOGRAPHIC SOLVENTS

MeOH	= methanol	HCO_2H	= formic acid
EtOH	= ethanol	HOAc	= acetic acid
iso-PrOH	= iso-propanol	$CHCl_3$	= chloroform
n-BuOH	= n-butanol	CH_2Cl_2	= methylene dichloride
iso-BuOH	= iso-butanol	EtOAc	= ethyl acetate
PhOH	= phenol	Me_2CO	= acetone
$NHEt_2$	= diethylamine	MeCOEt	= methyl ethyl ketone
Et_2O	= diethyl ether	C_6H_6	= benzene

Methods of Plant Analysis

1.1 INTRODUCTION

The subject of phytochemistry, or plant chemistry, has developed in recent years as a distinct discipline, somewhere in between natural product organic chemistry and plant biochemistry and is closely related to both. It is concerned with the enormous variety of organic substances that are elaborated and accumulated by plants and deals with the chemical structures of these substances, their biosynthesis, turnover and metabolism, their natural distribution and their biological function.

In all these operations, methods are needed for separation, purification and identification of the many different constituents present in plants. Thus, advances in our understanding of phytochemistry are directly related to the successful exploitation of known techniques, and the continuing development of new techniques to solve outstanding problems as they appear. One of the challenges of phytochemistry is to carry out all the above operations on vanishingly small amounts of material. Frequently, the solution of a biological problem in, say, plant growth regulation, in the biochemistry of plant–animal interactions, or in understanding the origin of fossil plants depends on identifying a range of complex chemical structures which may only be available for study in microgram amounts.

It is the purpose of this book to provide, for the first time, an introduction to present available methods for the analysis of plant substances and to provide a key to the literature on the subject. No novelty is claimed for the methods described here. Indeed, the purpose is to outline those methods which have been most widely used; the student or research worker can then most rapidly develop his own techniques for solving his own problems.

Some training in simple chemistry laboratory techniques is assumed as a background. However, it is possible for botanists and other plant scientists, with very little chemistry, to do phytochemistry, since many of the techniques are simple and straightforward. As in other practical subjects, the student must develop his own expertise. No recipe, however precisely written down, can substitute in the laboratory for common-sense and the ability to think things out from first principles. Examples of practical experiments which can be worked through to gain experience are provided in most sections of the following chapters. These can readily be adapted for laboratory courses and many have already been used for this purpose.

The range and number of discrete molecular structures produced by plants is huge and such is the present rate of advance of our knowledge of them that a major problem in phytochemical research is the collation of existing data on each particular class of compound. It has been estimated, for example, that there are now over 5500 known plant alkaloids and such is the pharmacological interest in novel alkaloids that new ones are being discovered and described, possibly at the rate of one a day.

Because the number of known substances is so large, special introductions have been written in each chapter of the book, indicating the structural variation existing within each class of compound, outlining those compounds which are commonly occurring and illustrating the chemical variation with representative formulae. References are given, wherever possible, to the most recent listings of known compounds in each class. Tables are included, showing the R_F values, colour reactions and spectral properties of most of the more common plant constituents. These tables are given mainly for illustrative or comparative purposes and are not meant to be exhaustive.

Phytochemical progress has been aided enormously by the development of rapid and accurate methods of screening plants for particular chemicals and the emphasis in this book is inevitably on chromatographic techniques. These procedures have shown that many substances originally thought to be rather rare in occurrence are of almost universal distribution in the plant kingdom. The importance of continuing surveys of plants for biologically active substances needs no stressing. Certainly, methods of preliminary detection of particular classes of compound are discussed in some detail in the following chapters.

Although the term 'plant' is used here to refer to the plant kingdom as a whole, there is some emphasis on higher plants and methods of analysis for micro-organisms are not dealt with in any special detail. As a general rule, methods used with higher plants for identifying alkaloids, amino acids, quinones and terpenoids can be applied directly to microbial systems. In many cases, isolation is much easier, since contaminating substances such as the tannins and the chlorophylls are usually absent. In a few cases, it may be more difficult, due to the resilience of the microbial cell wall and the need to use mechanical disruption to free some of the substances present.

There are a number of organic compounds, such as the penicillin and tetracycline antibiotics (Turner, 1971; Turner and Aldridge, 1983), which are specifically found in micro-organisms and their identification is not covered here, because of limitations on space. Lichens also make a range of special pigments, including the depsidones and depsides. These are analysed by special microchemical methods, based on colour reactions, chromatographic and spectral techniques. A comprehensive account of the chemistry of lichens is given by Culberson (1969). The analysis of lichen pigments is mentioned briefly here in Chapter 2 (p. 94).

The chemical constituents of plants can be classified in a number of different ways; in this book, classification is based on biosynthetic origin, solubility properties and the presence of certain key functional groups. Chapter 2 covers the phenolic compounds, substances which are readily recognized by their hydrophilic nature and by their common origin from the aromatic precursor shikimic acid. Chapter 3 deals with the terpenoids, which all share lipid properties and a biosynthetic origin from isopentenyl pyrophosphate. Chapter 4 is devoted to organic acids, lipids and other classes of compound derived biosynthetically from acetate. Chapter 5 is on the nitrogen compounds of plants, basic substances recognized by their positive responses to either ninhydrin or the Dragendorff reagent. Chapter 6 deals with the water-soluble carbohydrates and their derivatives. Finally, Chapter 7 briefly covers the macromolecules of plants, nucleic acids, proteins and polysaccharides, which are easily separated from other constituents by their high molecular weights.

In the remainder of this introductory chapter, it is proposed to discuss, in general terms, methods of extraction, separation and identification. A final section will include some examples of the application of phytochemical methods in different areas of plant science.

The only major reference available on methods of plant analysis is the seven volume treatise, edited by Paech and Tracey (1956–64) with chapters written in both German and English. This work is now becoming out-dated, but nevertheless, it provides essential background reading in the subject. Many other texts deal *inter alia* with phytochemical methods and these are listed in the references at the end of this and subsequent chapters. Among current journals that can be consulted for the very latest techniques are *Journal of Chromatography, Phytochemistry, Analytical Biochemistry, Journal of Chromatographic Science* and *Planta*.

1.2 METHODS OF EXTRACTION AND ISOLATION

1.2.1 The plant material

Ideally, fresh plant tissues should be used for phytochemical analysis and the material should be plunged into boiling alcohol within minutes of its collection. Sometimes, the plant under study is not at hand and material may have to be supplied by a collector living in another continent. In such cases, freshly picked tissue, stored dry in a plastic bag, will usually remain in good condition for analysis during the several days required for transport by airmail.

Alternatively, plants may be dried before extraction. If this is done, it is essential that the drying operation is carried out under controlled conditions to avoid too many chemical changes occurring. It should be dried as quickly as possible, without using high temperatures, preferably in a good air draft. Once thoroughly dried, plants can be stored before analysis for long periods of time. Indeed, analyses for flavonoids, alkaloids, quinones and terpenoids have been successfully carried out on herbarium plant tissue dating back many years.

One example of the use of herbarium material is the essential oil analysis that was carried out on type specimens of *Mentha* leaf, the material being obtained from the original collection of Linneaus made before 1800 (Harley and Bell, 1967). Quantitative changes in essential oil content may occur in both leaf and fruit tissue with time and this possibility must be taken into account. For example, Sanford and Heinz (1971) found that the myristicin content of nutmeg, *Myristica fragrans* fruits increased slowly on storage, while the more volatile β-pinene content decreased with time. On the other hand, flavonoids and alkaloids in herbarium specimens are remarkably stable with time; thus, a leaf sample of *Strychnos nuxvomica* originally collected in 1675 still contained 1–2% by weight of alkaloid (Phillipson, 1982).

The freeing of the plant tissue under study from contamination with other plants is an obvious point to watch for at this stage. It is essential, for example, to employ plants which are free from disease, i.e. which are not affected by viral, bacterial or fungal infection. Not only may products of microbial synthesis be detected in such plants, but also infection may seriously alter plant metabolism and unexpected products could be formed, possibly in large amounts.

Contamination may also occur when collecting lower plant material for analysis. When fungi growing parasitically on trees are collected, it is important to remove all tree tissue from the samples. Earlier reports (Paris *et al.*, 1960) of chlorogenic acid, a typical higher plant product, in two fungi are almost certainly incorrect because of contamination; repeat analyses on carefully cleaned material showed no evidence of this compound being present (Harborne J.B. and Hora F.B., unpublished results). Again, mosses often grow in close association with higher plants and it is sometimes difficult to obtain them free from such litter. Finally, in the case of higher plants, mixtures of plants may sometimes be gathered in error. Two closely similar grass species growing side by side in the field may be incorrectly assumed to be the same, or a plant may be collected without the realization that it has a parasite (such as the dodder, *Cuscuta epithymum*) intertwined with it.

In phytochemical analysis, the botanical identity of the plants studied must be authenticated by an acknowledged authority at some stage in the investigation. So many mistakes over plant identity have occurred in the past that it is essential to authenticate the material whenever reporting new substances from plants or even known substances from new plant sources. The identity of the material should either be beyond question (e.g. a common species collected in the expected habitat by a field botanist) or it should be possible for the identity to be established by a taxonomic expert. For these reasons, it is now common practice in phytochemical research to deposit a voucher specimen of a plant examined in a recognized herbarium, so that future reference can be made to the plant studied if this becomes necessary.

1.2.2 Extraction

The precise mode of extraction naturally depends on the texture and water content of the plant material being extracted and on the type of substance that is being isolated. In general, it is desirable to 'kill' the plant tissue, i.e. prevent enzymic oxidation or hydrolysis occurring, and plunging fresh leaf or flower tissue, suitably cut up where necessary, into boiling ethanol is a good way of achieving this end. Alcohol, in any case, is a good all-purpose solvent for preliminary extraction. Subsequently, the material can be macerated in a blender and filtered but this is only really necessary if exhaustive extraction is being attempted. When isolating substances from green tissue, the success of the extraction with alcohol is directly related to the extent that chlorophyll is removed into the solvent and when the tissue debris, on repeated extraction, is completely free of green colour, it can be assumed that all the low molecular weight compounds have been extracted.

The classical chemical procedure for obtaining organic constituents from dried plant tissue (heartwood, dried seeds, root, leaf) is to continuously extract powdered material in a Soxhlet apparatus with a range of solvents, starting in turn with ether, petroleum and chloroform (to separate lipids and terpenoids) and then using alcohol and ethyl acetate (for more polar compounds). This method is useful when working on the gram scale. However, one rarely achieves complete separation of constituents and the same compounds may be recovered (in varying proportions) in several fractions.

The extract obtained is clarified by filtration through celite on a water pump and is then concentrated *in vacuo*. This is now usually carried out in a rotary evaporator, which will concentrate bulky solutions down to small volumes, without bumping, at temperatures between 30 and 40°C. Extraction of volatile components from plants needs special precautions and these are discussed later, in Chapter 3, p. 107.

There are short-cuts in extraction procedures which one learns with practice. For example, when isolating water-soluble components from leaf tissue, the lipids should strictly speaking be removed at an early stage, before concentration, by washing the extract repeatedly with petroleum. In fact, when a direct ethanolic extract is concentrated on a rotatory evaporator, almost all the chlorophyll and lipid is deposited on the side of the flask, and, with skill, the concentration can be taken just to the right point when the aqueous concentrate can be pipetted off, almost completely free of lipid impurities.

The concentrated extract may deposit crystals on standing. If so, these should be collected by filtration and their homogeneity tested for by chromatography in several solvents (see next section).

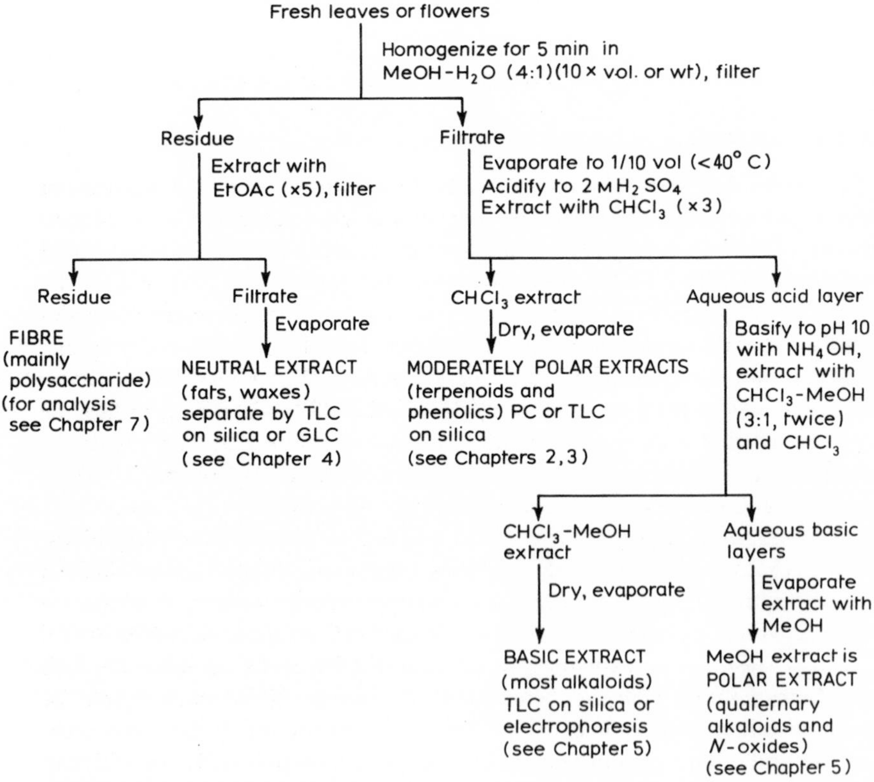

Fresh leaves or flowers

Homogenize for 5 min in MeOH–H_2O (4:1)(10 × vol. or wt), filter

Residue
Extract with EtOAc (×5), filter

Filtrate
Evaporate to 1/10 vol (<40° C)
Acidify to 2 M H_2SO_4
Extract with $CHCl_3$ (×3)

Residue
FIBRE (mainly polysaccharide) (for analysis see Chapter 7)

Filtrate
Evaporate
NEUTRAL EXTRACT (fats, waxes) separate by TLC on silica or GLC (see Chapter 4)

$CHCl_3$ extract
Dry, evaporate
MODERATELY POLAR EXTRACTS (terpenoids and phenolics) PC or TLC on silica (see Chapters 2,3)

Aqueous acid layer
Basify to pH 10 with NH_4OH, extract with $CHCl_3$–MeOH (3:1, twice) and $CHCl_3$

$CHCl_3$–MeOH extract
Dry, evaporate
BASIC EXTRACT (most alkaloids) TLC on silica or electrophoresis (see Chapter 5)

Aqueous basic layers
Evaporate extract with MeOH
MeOH extract is POLAR EXTRACT (quaternary alkaloids and N-oxides) (see Chapter 5)

Fig. 1.1 A general procedure for extracting fresh plant tissues and fractionating into different classes according to polarity.

If a single substance is present, the crystals can be purified by recrystallization and then the material is available for further analysis. In most cases,
mixtures of substances will be present in the crystals and it will then be
necessary to redissolve them in a suitable solvent and separate the constituents
by chromatography. Many compounds also remain in the mother liquor and
these will also be subjected to chromatographic fractionation. As a standard
precaution against loss of material, concentrated extracts should be stored in
the refrigerator and a trace of toluene added to prevent fungal growth.

When investigating the complete phytochemical profile of a given plant
species, fractionation of a crude extract is desirable in order to separate the
main classes of constituent from each other, prior to chromatographic
analysis. One procedure based on varying polarity that might be employed on
an alkaloid-containing plant is indicated in Fig. 1.1. The amounts and type of
compounds separated into the different fractions will, of course, vary from
plant to plant. Also, such a procedure may have to be modified when labile
substances are under investigation.

1.3   METHODS OF SEPARATION

1.3.1   General

The separation and purification of plant constituents is mainly carried out
using one or other, or a combination, of four chromatographic techniques:
paper chromatography (PC), thin layer chromatography (TLC), gas liquid
chromatography (GLC) and high performance liquid chromatography
(HPLC). The choice of technique depends largely on the solubility properties
and volatilities of the compounds to be separated. PC is particularly applicable to water-soluble plant constituents, namely the carbohydrates, amino
acids, nucleic acid bases, organic acids and phenolic compounds. TLC is the
method of choice for separating all lipid-soluble components, i.e. the lipids,
steroids, carotenoids, simple quinones and chlorophylls. By contrast, the third
technique GLC finds its main application with volatile compounds, fatty
acids, mono- and sesquiterpenes, hydrocarbons and sulphur compounds.
However, the volatility of higher boiling plant constituents can be enhanced
by converting them to esters and/or trimethylsilyl ethers so that there are few
classes which are completely unsuitable for GLC separation. Alternatively,
the less volatile constituents can be separated by HPLC, a method which
combines column efficiency with speed of analysis. Additionally, it may be
pointed out that there is considerable overlap in the use of the above techniques and often a combination of PC and TLC, TLC and HPLC or TLC and
GLC may be the best approach for separating a particular class of plant
compound.

All the above techniques can be used both on a micro- and a macro-scale.

For preparative work, TLC is carried out on thick layers of adsorbent and PC on thick sheets of filter paper. For isolation on an even larger scale than this, it is usual to use column chromatography coupled with automatic fraction collecting. This procedure will yield purified components in gram amounts.

One further technique which has fairly wide application in phytochemistry is electrophoresis. In the first instance, this technique is only applicable to compounds which carry a charge, i.e. amino acids, some alkaloids, amines, organic acids and proteins. However, in addition, certain classes of neutral compounds (sugars, phenols) can be made to move in an electric field by converting them into metal complexes (e.g. by use of sodium borate). Sargent (1969) has provided a simple introduction to electrophoretic techniques.

Besides the techniques so far mentioned, a few others are used occasionally in phytochemical research. Separation by simple liquid–liquid extraction is still of some value in the carotenoid field (see Chapter 3, p. 132). The means for automatic liquid–liquid extraction, as embodied in the Craig counter-current distribution apparatus, has been available for some time but it tends only to be used as a last resort when other techniques fail. A more convenient apparatus for liquid–liquid extraction has been developed recently, called droplet counter-current chromatography (DCCC), which works on a preparative scale mainly for separating water-soluble constituents (Hostettmann, 1981). Separation of plant proteins and nucleic acids often requires special techniques not yet mentioned, such as filtration through Sephadex gels, affinity chromatography and differential ultracentrifugation.

Since so much has been written elsewhere on chromatography (see e.g. Heftmann, 1983), it is only necessary here to discuss the main separation techniques as they are applied in phytochemical research and to give a few leading references to other available texts.

1.3.2   Paper chromatography

One of the main advantages of PC is the great convenience of carrying out separations simply on sheets of filter paper, which serve both as the medium for separation and as the support. Another advantage is the considerable reproducibility of R_F values determined on paper, so that such measurements are valuable parameters for use in describing new plant compounds. Indeed, for substances such as the anthocyanins, which do not have other clearly defined physical properties, the R_F is the most important means of describing and distinguishing the different pigments (Harborne, 1967).

Chromatography on paper usually involves either partition or adsorption chromatography. In partition, the compounds are partitioned between a largely water-immiscible alcoholic solvent (e.g. *n*-butanol) and water. The classic solvent mixture, *n*-butanol-acetic acid-water (4:1:5, top layer) (abbreviated as BAW) was indeed devised as a means of increasing the water content of the *n*-butanol layer and thus improving the utility of the solvent mixture. Indeed, BAW is still widely applicable as a general solvent for many classes of plant constituent. By contrast, adsorption forces are one of the main features of PC in aqueous solvent. Pure water is a remarkably versatile chromatographic solvent and it can be used to separate the common purines and pyrimidines and is also applicable to phenolic compounds and to plant glycosides in general.

The choice of apparatus for PC depends to some extent on the amount of laboratory space available. Horizontal or circular PC, for example, takes up little more space than a standard TLC tank. It has remarkably good resolution and is used, for example, for separating carotenoids. In most laboratories, PC is carried out by descent, in tanks which will accommodate Whatman papers of the size 46×57 cm. Descending PC is most useful since the solvent can be more easily over-run if this is desired; it is also slightly more convenient for two-dimensional separations.

A considerable range of 'modified' filter papers are available commercially for achieving particular chromatographic separations. For example, the polar properties of cellulose can be reduced by incorporating silicic acid or alumina into the papers, making them more suitable for separating lipids. Papers can likewise be modified in the laboratory, for example, by soaking them in paraffin or silicone oil in order to carry out 'reversed-phase' chromatography, again for lipids. For large-scale separations, thick sheets of chromatography filter paper are available (Whatman no. 3 or 3 MM) and these will cope with several milligrams of material per sheet.

In PC, compounds are usually detected as coloured or UV-fluorescent spots, after reaction with a chromogenic reagent, used either as a spray or as a dip. For large sheets, dipping is usually easier but the solvent content of the spray should be modified in order to facilitate quick drying and thus avoid diffusion during the dipping. The paper may then be heated in order to develop the colours.

The R_F value is the distance a compound moves in chromatography relative to the solvent front. It is obtained by measuring the distance from the origin to the centre of the spot produced by the substance, and this is divided by the distance between the origin and the solvent front (i.e. the distance the solvent travels). This always appears as a fraction and lies between 0·01 and 0·99. It is convenient to multiply this by 100 and R_Fs are quoted in this book as R_Fs (×100). Elsewhere, R_F (×100) is sometimes referred to as the hR_F value.

When comparing R_F values of a series of structurally related compounds, it is useful to refer to another chromatographic constant, the R_M value. This is related to R_F by the expression:

$$R_M = \log \left(\frac{1}{R_F} - 1 \right)$$

It is valuable for relating chromatographic mobility to chemical structure, since ΔR_M values in a homologous series are usually constants. Thus, with the flavonoid compounds. ΔR_Ms are constant for the number of hydroxyl and glycosyl substitutions present in the molecule (Bate-Smith and Westall, 1956). The procedure can be used to calculate the R_F value of an unknown member of a series of compounds, in order to facilitate the search for the particular compound in plant extracts. Such a procedure was employed, for example, in characterizing a new methyl ether of kaempferol, where the predicted and actual R_F values were in very good agreement (Table 1.1) (Harborne, 1969).

Table 1.1 R_F data of flavonol 5-methyl ethers: comparison of actual R_F and R_F calculated from ΔR_M

Flavonol	R_F (×100)*			
	Forestal	50% HOAc	PhOH	BAW
Kaempferol	62	44	58	91
Quercetin	45	31	28	76
Myricetin	29	21	10	41
Kaempferol 5-methyl ether				
Actual value	70	43	78	82
Value predicted from ΔR_M	70	41	76	89
Quercetin 5-methyl ether (azaleatin)	53	29	42	55
Myricetin 5-methyl ether	37	21	23	27

*Measured on Whatman no. 1 paper.

Within the vast literature on PC, a useful introductory account written for the novice is that of Pereeboom (1971). Books on PC which are particularly valuable as sources of R_F data are Lederer and Lederer (1957), Linskens (1959) and Sherma and Zweig (1971).

1.3.3 Thin layer chromatography

The special advantages of TLC compared to PC include versatility, speed and sensitivity. Versatility is due to the fact that a number of different adsorbents besides cellulose may be spread on to a glass plate or other support and employed for chromatography. Although silica gel is most widely used, layers may be made up from aluminium oxide, celite, calcium hydroxide, ion exchange resin, magnesium phosphate, polyamide, Sephadex, polyvinyl-pyrrolidone, cellulose and from mixtures of two or more of the above materials. The greater speed of TLC is due to the more compact nature of the adsorbent when spread on a plate and is an advantage when working with labile compounds. Finally, the sensitivity of TLC is such that separations on less than μg amounts of material can be achieved if necessary.

One of the original disadvantages of TLC was the labour of spreading glass plates with adsorbent, a labour somewhat eased by the later introduction of automatic spreading devices. Nevertheless, even with these, certain precautions are necessary. The glass plates have to be carefully cleaned with acetone to remove grease. Then the slurry of silica gel (or other adsorbent) in water has to be vigorously shaken for a set time interval (e.g. 90 s) before spreading. Depending on the particle size of the adsorbent, calcium sulphate hemihydrate (15%) may have to be added to help bind the adsorbent on to the glass. Finally, plates after spreading have to be air dried and then activated by heating in an oven at 100–110°C for 30 min. In some separations, it is advantageous to modify the properties of the adsorbent by adding an inorganic salt (e.g. silver nitrate for argentation TLC) and this is best done when the plate is being spread. Another reason for still using plates coated in the laboratory is that the moisture content of the silica gel can be controlled, a factor which is critical for some separations.

Nowadays, however, it is usual to employ precoated plates of commercial manufacture in most work, since these are more uniform and provide more reproducible results. There are a range of such plates available with different adsorbents, coated on glass, aluminium sheets or plastic. These may be with or without a fluorescent indicator, the addition of which allows the detection of all compounds which quench the fluorescence, when the plate is observed in UV light of 254 nm wavelength. The most recent type of TLC plate is that coated with the same fine microparticles of silica that are used in the columns for HPLC. Such chromatography is termed HPTLC and it usually gives more efficient and rapid separations than on conventional silica layers.

A wider range of solvents have been applied to TLC than to PC and in general, there is more latitude in the exact proportions of different solvents used in a solvent system. R_F values are considerably less reproducible than on paper and it is therefore essential to include one or more reference compounds as markers. It is possible to standardize conditions for accurate measurement of R_F in TLC, but this is a very tedious process. TLC is usually carried out by ascent, in a tank which is paper-lined so that the atmosphere inside is saturated with the solvent phase. Horizontal TLC is employed, either when plates need to be over-run with solvent or when TLC is used in combination with electrophoresis.

Detection of compounds on TLC plates is normally carried out by spraying, the smaller area of the plate (20×20 cm) making this a relatively simple procedure. One advantage over PC is that glass plates may be sprayed with conc. H_2SO_4, a useful detection reagent for steroids and lipids.

Preparative TLC is carried out using thick (up to 1 mm) instead of thin (0·10–0·25 mm) layers of adsorbent. Manufactured plates are available for this. Separated constituents are recovered by scraping off the adsorbent at the appropriate places on the developed plate, eluting the powder with a solvent such as ether and finally centrifuging to remove the adsorbent.

Such is the strength of the adsorbent layers on glass that it is possible to repeatedly develop a plate with one or several different solvent systems, drying the plate in between developments. Alternatively, a multiple elimination TLC system, devised by van Sumere (1969), can be used. This involves cutting a long rectangular glass plate spread with adsorbent with a glass cutter at appropriate steps in a complex separation and even spraying fresh adsorbent on to the plate in between separations. Many other modifications of the basic TLC procedure are described in the books on TLC mentioned below.

The literature on TLC is enormous. The most comprehensive book on the topic is probably that edited by Stahl (1969). A simple introduction is the book by Truter (1963). Other important contributions are the works of Bobbitt (1963), Kirchner (1978) and Touchstone and Dobbins (1978).

1.3.4   Gas liquid chromatography

The apparatus required for GLC is sophisticated and expensive, relative to that required for PC or TLC. In principle, however, GLC is no more complicated than other chromatographic procedures.

Apparatus for GLC has four main components as follows:

(1) *The column* is a long narrow tube (e.g. 3 m × 1 mm) usually of metal made in the form of a coil to conserve space. It is packed with a stationary phase (e.g. 5–15% silicone oil) on an inert powder (Chromosorb W, celite, etc.). The packing is not essential and alternatively an open silica column is used in which the stationary phase is spread as a film on the inside surface (capillary GLC).

(2) *The heater* is provided to heat the column progressively from 50 to 350°C at a standard rate and to hold the temperature at the higher limit if necessary. The temperature of the column inlet is separately controlled so that the sample can be rapidly vaporized as it is passed on to the column. The sample dissolved in ether or hexane is injected by hypodermic syringe into the inlet port through a rubber septum.

(3) *Gas flow* consists of an inert carrier gas such as nitrogen or argon. Separation of the compounds on the column depends on passing this gas through at a controlled rate.

(4) *A detection device* is needed to measure the compounds as they are swept through the column. Detection is frequently based on either flame ionization or electron capture. The former method requires hydrogen gas to be added to the gas mixture and to be burnt off in the actual detector. The detection device is linked to a potentiometric recorder, which produces the results of the separation as a series of peaks of varying intensity (see Fig. 1.2).

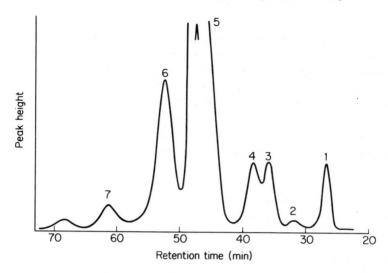

Fig. 1.2 GLC trace of the separation of the mixture of sterol acetates present in oat seed. Key: 1, cholesterol; 2, brassicasterol; 3, campesterol; 4, stigmasterol; 5, sitosterol; 6, Δ^5-avenasterol; and 7, Δ^7-avenasterol. Stationary phase is 1% cyclohexanedimethanol succinate +2% polyvinyl-pyrrolidone on acid-washed Gas chrome P 225 (from Knights, 1965).

The results of GLC can be expressed in terms of retention volume R_V, which is the volume of carrier gas required to elute a component from the column, or in terms of retention time R_t, which is the time required for elution of the sample. These parameters are nearly always expressed in terms relative to a standard compound (as RR_V or RR_t), which may be added to the sample extract or which could take the form of the solvent used for dissolving that sample.

The main variables in GLC are the nature of the stationary phase of the column and the temperature of operation. These are varied according to the polarity and volatility of the compounds being separated. Many classes of substances are routinely converted to derivatives (especially to trimethylsilyl ethers) before being subjected to GLC, since this allows their separation at a lower temperature.

GLC provides both quantitative and qualitative data on plant substances, since measurements of the area under the peaks shown on the GLC trace (Fig. 1.2) are directly related to the concentrations of the different components in the original mixture. There are two general formulae for measuring these areas: (*a*) peak height×peak width at half the height = 94% of the peak area (this only applies to symmetrical peaks), and (*b*) peak area is equivalent to that of a triangle produced by drawing tangents through the points of inflection. Peak areas can be determined automatically, e.g. by use of an electronic integrator.

The GLC apparatus can be set up in such a way that the separated components are further subjected to spectral or other analysis. Most frequently, GLC is automatically linked to mass spectrometry (MS) and the combined GC–MS apparatus has emerged in recent years as one of the most important of all techniques for phytochemical analysis.

Although there are numerous books and reviews on GLC, few have been written with a phytochemical audience in mind. A useful introductory text, from the practical point of view, is that of Simpson (1970). A text dealing more specifically with biochemical applications of GLC is that of Burchfield and Storrs (1962).

1.3.5   High performance liquid chromatography (HPLC)

HPLC is analogous to GLC in its sensitivity and ability to provide both quantitative and qualitative data in the single operation. It differs in that the stationary phase bonded to a porous polymer is held in a narrow-bore stainless steel column and the liquid mobile phase is forced through under considerable pressure. The apparatus for HPLC is more expensive than GLC, mainly because a suitable pumping system is required and all connections have to be screw-jointed to withstand the pressures involved. The mobile phase is a miscible solvent mixture, which either remains constant (isocratic separation) or, may be changed continuously in its proportions, by including a mixing chamber into the set-up (gradient elution). The compounds are monitored as they elute off the column by means of a detector, usually measuring in the ultraviolet. A computing integrator may be added to handle the data as they emerge and the whole operation can be controlled through a microprocessor.

A major difference between HPLC and GLC is that the former procedure normally operates at ambient temperature, so that the compounds are not subjected to the possibility of thermal re-arrangement during separation. Temperature control of the HPLC column may, however, be advantageous for critical separations so that a thermostatically controlled jacket may be needed. The column, which is usually packed with very small spherical particles of silica coated or bonded with stationary phase, is particularly sensitive to poisoning by impurities, so that it is essential to purify and filter plant extracts before injecting them at the head of the column.

HPLC is mainly used for those classes of compound which are non-volatile, e.g. higher terpenoids, phenolics of all types, alkaloids, lipids and sugars. It works best for compounds which can be detected in the ultraviolet or visible regions of the spectrum. An example of the separation of flavonoids by HPLC is shown in Fig. 1.3. For sugars, which do not show any UV absorption, it is possible to use a refractive index detector, but this is a less sensitive procedure.

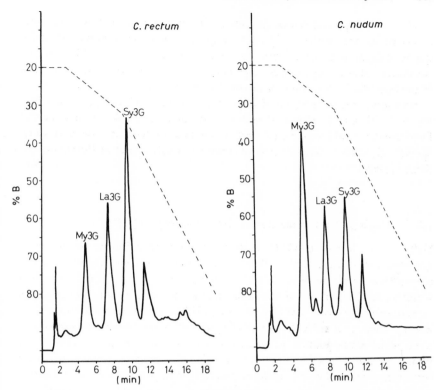

Fig. 1.3 HPLC traces of the flavonoids of two species of *Chondropetalum*, where the same compounds are present but in different amount. Key: Species A, *C. rectum*; species B, *C. nudum*; My3G = myricetin 3-galactoside; La3G = laricytrin 3-galactoside; Sy3G = syringetin 3-galactoside; 4th peak = syringetin 3-diglycoside. Separation on Spherisorb 5 μm C_8 column (25 cm × 4·6 mm), gradient elution with MeOH–H_2O–HOAc (1:18:1) and MeOH–H_2O–HOAc (18:1:1) and detection at 365 nm.

Proteins have been separated by HPLC on columns of modified Sephadex, silica gels or ion exchangers.

In most modern HPLC separations, prepacked columns are employed and many types are available from the manufacturers. However, it is possible to carry out most separations using either a silica microporous particle column (for non-polar compounds) or a reverse-phase C_{18} bonded phase column (for polar compounds) (Hamilton and Sewell, 1982). One final practical detail needs mentioning; the solvents have to be ultrapure and need to be degassed before use.

HPLC is the latest chromatographic technique to be added to the phytochemist's armoury. Apart from the expense of the apparatus and the solvents, it promises to be a most important and versatile method of quantitative plant analysis but it has yet to prove itself for separations on a preparative scale.

1.4 METHODS OF IDENTIFICATION

1.4.1 General

In identifying a plant constituent, once it has been isolated and purified, it is necessary first to determine the class of compound and then to find out which particular substance it is within that class. Its homogeneity must be checked carefully beforehand, i.e. it should travel as a single spot in several TLC and/or PC systems. The class of compound is usually clear from its response to colour tests, its solubility and R_F properties and its UV spectral characteristics. Biochemical tests may also be invaluable: presence of a glucoside can be confirmed by hydrolysis with β-glucosidase, of a mustard oil glycoside by hydrolysis with myrosinase and so on. For growth regulators, a bioassay is an essential part of identification.

Complete identification within the class depends on measuring other properties and then comparing these data with those in the literature. These properties include melting point (for solids), boiling point (for liquids), optical rotation (for optically active compounds) and R_F or RR_t (under standard conditions). However, equally informative data on a plant substance are its spectral characteristics: these include ultraviolet (UV), infrared (IR), nuclear magnetic resonance (NMR) and mass spectral (MS) measurements. A known plant compound can usually be identified on the above basis. Direct comparison with authentic material (if available), should be carried out as final confirmation. If authentic material is not available, careful comparison with literature data may suffice for its identification. If a new compound is present, all the above data should be sufficient to characterize it. With new compounds, however, it is preferable to confirm the identification through chemical degradation or by preparing the compound by laboratory synthesis.

Identification of new plant compounds by X-ray crystallography is now becoming routine, and can be applied whenever the substance is obtained in sufficient amount and in crystalline form. It is particularly valuable in the case of complex terpenoids, since it provides both structure and stereochemistry in the same operation.

Brief comments will now be given on the different spectral techniques and on their comparative importance for phytochemical identification. For a more detailed treatment of spectral methods, the reader is referred to one of the many books available on the application of spectroscopic methods to organic chemistry (Brand and Eglinton, 1965; Williams and Fleming, 1966; Scheinmann, 1970, 1974).

1.4.2 Ultraviolet and visible spectroscopy

The absorption spectra of plant constituents can be measured in very dilute solution against a solvent blank using an automatic recording spectro-photometer. For colourless compounds, measurements are made in the range 200 to 400 nanometres (nm); for coloured compounds, the range is 200 to 700 nm. The wavelengths of the maxima and minima of the absorption spectrum so obtained are recorded (in nm) and also the intensity of the absorbance (or optical density) at the particular maxima and minima (Fig. 1.4). Only traces of material are required, since the standard spectro-photometric cell (1×1 cm) only holds 3 ml of solution and, using special cells, only one tenth of this volume need be available for spectrophotometry. Such spectral measurements are important in the identification of many plant constituents, for monitoring the eluates of chromatographic columns during

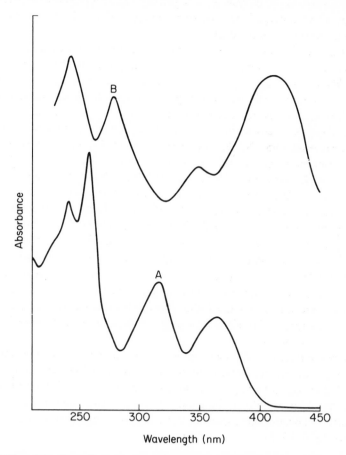

Fig. 1.4 Ultraviolet absorption spectrum of the xanthone mangiferin. Curve A, solvent is 95% EtOH. Wavelength maxima are at 240, 258, 316 and 364 nm; minima are at 215, 248, 285 and 338 nm. Relative intensities of absorbance at the maxima are 82, 100, 50 and 37% respectively. Curve B, 95% EtOH + 2 drops 2N NaOH.

purification of plant products and for screening crude plant extracts for the presence of particular classes of compound such as polyacetylenes.

A solvent widely used for UV spectroscopy is 95% ethanol since most classes of compound show some solubility in it. Commercial absolute alcohol should be avoided, since it contains residual benzene which absorbs in the short UV. Other solvents frequently employed are water, methanol, hexane, petroleum and ether. Solvents such as chloroform and pyridine are generally to be avoided since they absorb strongly in the 200–260 nm region; they are, however, quite suitable for making measurements in the visible region of the spectrum with plant pigments such as the carotenoids.

When substances are isolated as crystalline compounds and their molecular weights are known or can be determined, then the intensities of the wavelength maxima are normally recorded in terms of $\log \epsilon$, where $\epsilon = A/Cl$ (A = absorbance, C = concentration in gm moles/litre, l = cell path length in cm, usually 1). With compounds where neither the concentration nor the molecular weight are known, absorbance values have to be used. In these cases, the heights of the various maxima may be compared by considering absorbances as a percentage of that at the most intense peak.

Purification is an essential preliminary to spectral studies and plant constituents which exhibit characteristic absorption properties should be repeatedly purified till these properties are constant. In chromatographic purifications, allowances for UV-absorbing impurities present in the filter paper can be made by using an eluate of a paper blank, prepared at the same time as the sample, for balancing in the spectrophotometer against the eluate containing that sample. A similar procedure should be adopted when purification is being done on TLC plates.

The utility of spectral measurements for identification purposes can be greatly enhanced by repeating measurements made in neutral solution, either at a range of different pH values or in the presence of particular inorganic salts. For example, when alkali is added to alcoholic solutions of phenolic compounds, the spectra are characteristically shifted towards longer wavelengths (they undergo bathochromic shifts) with increases in absorbance (Fig. 1.4). By contrast, when alkali is added to neutral solutions of aromatic carboxylic acids, the shift is in the opposite direction towards shorter wavelengths (hypsochromic shifts). Reactions such as chemical reduction (with sodium borohydride) or enzymic hydrolysis can equally well be followed in the cell cuvette of a recording spectrophotometer and absorption measurements made at regular time intervals will indicate whether reduction or hydrolysis has taken place.

The value of UV and visible spectra in identifying unknown constituents is obviously related to the relative complexity of the spectrum and to the general position of the wavelength maxima. If a substance shows a single absorption band between 250 and 260 nm, it could be any one of a considerable number of compounds (e.g. a simple phenol, a purine or pyrimidine, an aromatic amino acid and so on). If, however, it shows three distinct peaks in the 400–500 nm region, with little absorption elsewhere, it is almost certainly a carotenoid. Furthermore, spectral measurements in two or three other solvents and comparison with literature data might even indicate which particular carotenoid it is.

The above statements suggest that absorption spectra are of especial value in plant pigment studies and this is certainly true for both water- and lipid-soluble plant colouring matters (Table 1.2). Other classes which show characteristic absorption properties include unsaturated compounds (particularly the polyacetylenes), aromatic compounds in general (e.g. hydroxy-cinnamic acids) and ketones. The complete absence of UV absorption also provides some useful structural information. It is indicative of the presence of saturated lipids or alkanes in lipid fractions of plant extracts, or of organic acids, aliphatic amino acids or sugars in the water-soluble fractions.

Table 1.2 Spectral properties of the different classes of plant pigment

Pigment class	Visible spectral range (nm)*	Ultraviolet range (nm)
Chlorophylls (green)	640–660 and 430–470	
Phycobilins (red and blue)	615–650 or 540–570	intense short UV absorption
Cytochromes (yellow)	545–605 (minor band sometimes at 415–440)	due to protein attachment
Anthocyanins (mauve or red)	475–550	ca. 275
Betacyanins (mauve)	530–554	250–270
Carotenoids (yellow to orange)	400–500 (a major peak with two minor peaks or inflections)	—
Anthraquinones (yellow)	420–460	3–4 intense peaks between 220 and 290
Chalcones and Aurones (yellow)	365–430	240–260
Yellow flavonols (yellow)	365–390	250–270

*All values are approximate; actual values vary according to the solvent used, the pH and the physical state of the pigment.

Because of space limitations, the spectral properties of only a very limited number of plant constituents can be given in this book. These are mainly recorded in the form of tables of spectral maxima but a few illustrations of spectral curves are included in later pages. For more comprehensive tables of spectral data, the reader is referred to one or other of the many compilations of such data, e.g. Hershenson (1956, 1961, 1966) or Lang (1959). Useful introductory texts to absorption spectroscopy are those of Gillam and Stern (1957) and Williams and Fleming (1966). A more advanced text dealing specifically with the spectra of natural products is that of Scott (1964). There is also a comprehensive account of biochemical spectroscopy by Morton (1975).

1.4.3   Infrared (IR) spectroscopy

IR spectra may be measured on plant substances in an automatic recording IR spectrophotometer either in solution (in chloroform or carbon tetrachloride (1–5%)), as a mull with nujol oil or in the solid state, mixed with potassium bromide. In the latter case, a thin disc is prepared under anhydrous conditions from a powder containing about 1 mg of material and 10 to 100 mg KBr, using

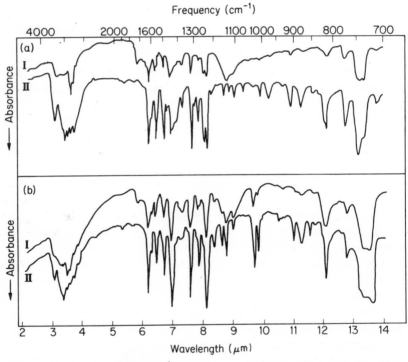

Fig. 1.5 Infrared spectra of two alkaloids from tobacco smoke. Key: (a) harmane, natural (I) and synthetic (II); (b) norharmane, natural (I) and synthetic (II). Note that IR spectra are traditionally recorded upside down compared to the UV and visible spectra (Fig. 1.4). Thus, the absorbance bands here point downwards.

a mould and press. The range of measurement is from 4000 to 667 cm⁻¹ (or 2·5 to 15 μm) and the spectrum takes about three minutes to be recorded. Typical IR spectra obtained in this way are shown in Fig. 1.5.

Table 1.3 Characteristic infrared frequencies of some classes of natural products

Class of compound	*Approximate positions of characteristic bands* above* 1200 cm⁻¹
Alkanes	2940 (S), 2860 (M), 1455 (S), 1380 (M)
Alkenes	3050 (W−M), 1850 (W), 1650 (W−M), 1410 (W).
Aromatics	3050 (W−M), 2100−1700 (W), 1600, 1580, 1500 (W−M)
Acetylenes	3310 (M), 2225 (W), 2150 (W−M), 1300 (W).
Alcohols and Phenols	3610 (W−M), 3600−2400 (broad), 1410 (M)
Aldehydes and Ketones	2750 (W), 2680 (W), 1820−1650 (S), 1420 (W−M).
Esters and Lactones	1820−1680 (S)
Carboxylic acids	3520 (W), 3400−2500 (broad, M), 1760 (S), 1710 (S).
Amines	3500 (M), 3400 (M), 3400−3100 (variable), 1610 (M)
Cyanides	2225 (W−S).
Isocyanates	2270 (VS)

*Bands in 'fingerprint' region are omitted for simplicity. Data adapted from Brand and Eglinton (1965).

The region in the IR spectrum above 1200 cm⁻¹ shows spectral bands or peaks due to the vibrations of individual bonds or functional groups in the molecule under examination (Table 1.3). The region below 1200 cm⁻¹ shows bands due to the vibrations of the whole molecule and, because of its complexity, is known as the 'fingerprint' region. Intensities of the various bands are recorded subjectively on a simple scale as being either strong (S), medium (M) or weak (W).

The fact that many functional groups can be identified by their characteristic vibration frequencies (Table 1.3) makes the IR spectrum the simplest and often the most reliable method of assigning a compound to its class. In spite of this, IR spectroscopy is most frequently used in phytochemical studies as a 'fingerprinting' device, for comparing a natural with a synthetic sample (see Fig. 1.5). The complexity of the IR spectrum lends itself particularly well to this purpose and such comparisons are very important in the complete identification of many types of plant constituent. IR spectra, for example, have been used extensively for identifying known essential oil components as they are separated by GLC on a preparative scale.

An illustration of the use of IR spectra for 'finger-printing' alkaloids is given in Fig. 1.5. Two trace components of tobacco smoke are identified as the bases harmane and norharmane, using the KBr disc procedure (Poindexter and Carpenter, 1962). It may be noted that some of the detail in the fingerprint region of both alkaloids is absent from the natural samples, probably due to the presence of traces of impurity. It may also be observed that although the alkaloids are closely similar in structure (they differ only in that harmane is the *C*-methyl derivative of norharmane), they can be readily distinguished by their IR spectra.

IR spectroscopy can also usefully contribute to structural elucidation, when new compounds are encountered in plants. Although there are many listed correlations between chemical structure and IR absorption peaks, the actual interpretation of a complex spectrum can be difficult and the operation requires much experience. With some classes of compounds, however, interpretation can be a relatively simple matter. Measurement of the carbonyl frequencies between 1800 and 1650 cm^{-1} in quinones can show at once whether the carbonyl group is chelated with an adjacent hydroxyl or not. With the anthraquinones, for example, non-chelated quinones have a band between 1678 and 1653 cm^{-1}; a quinone with one α-OH shows two bands, at 1675–1647 cm^{-1} and at 1637–1621 cm^{-1}; and a quinone with two α-OH groups has bands at 1675–1661 cm^{-1} and at 1645–1608 cm^{-1}.

For sources of IR data, various catalogues can be consulted, e.g. Hershenson (1959, 1964). For a good introductory review on IR spectroscopy, see Eglinton (1970). A popular introductory practical textbook, now in its third edition, is that of Cross and Jones (1969).

1.4.4   Mass spectroscopy (MS)

MS, since its relatively recent introduction (about 1960), has revolutionized biochemical research on natural products and has eased the task of the phytochemist in many ways. The value of the technique is that it requires only microgram amounts of material, that it can provide an accurate molecular weight and that it may yield a complex fragmentation pattern which is often characteristic of (and may identify) that particular compound.

MS, in essence, consists of degrading trace amounts of an organic compound and recording the fragmentation pattern according to mass. The sample vapour diffuses into the low pressure system of the mass spectrometer where it is ionized with sufficient energy to cause fragmentation of the chemical bonds. The resulting positively charged ions are accelerated in a magnetic field which disperses and permits relative abundance measurements of ions of given mass-to-charge ratio. The resulting record of ion abundance versus mass constitutes the mass spectral graph, which thus consists of a series of lines of varying intensity at different mass units (see Fig. 1.6).

In many cases, some of the parent compound will survive the vaporization process and will be recorded as a molecular ion peak. Very accurate mass measurements (to 0·0001 mass units) can then be performed on this (and any other) particular ion. The accuracy is such as to indicate the exact molecular

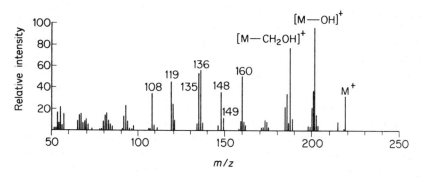

Fig. 1.6 Mass spectrum of the growth regulator zeatin.

formula of the substance, so that conventional elemental analysis (which normally requires several milligrams of substance) is no longer necessary.

Unlike the UV and IR spectrophotometers which are operated by the phytochemist himself, instruments for mass spectral and NMR determinations are more expensive and much more sophisticated, so that they are normally operated by trained personnel. The phytochemist therefore, hands his sample over for analysis and receives back the results in the form of the graph shown in Fig. 1.6. The technique works successfully with almost every type of low-molecular weight plant constituent and it has even been used for peptide analysis. Those compounds which are too involatile to vaporize in the MS instrument are converted to trimethylsilyl ethers, methyl esters or similar derivatives. This is true of the gibberellins (see p. 118). MS is frequently used in conjunction with GLC, and the combined operation provides at one go a qualitative and quantitative identification of the many structurally complex components that may be present together in a particular plant extract.

One example must suffice to illustrate the value of MS measurements in phytochemical research. This is the case of zeatin, the first naturally occurring cytokinin growth regulator to be isolated from higher plants. Its structure as 6-(4-hydroxymethyl-*trans*-2-butenylamino)purine was determined by Shannon and Letham (1966); the results of the MS (Fig. 1.6) considerably helped in this identification. Thus, there was a prominent molecular ion at 219, confirming the molecular formula $C_{10}H_{13}ON_5$. The presence of a primary alcohol was revealed by fragments at m/z 202 (M—OH) and m/z 188 (M—CH_2OH). The location of the alkyl group attachment at N- was indicated by the fragment at m/z 148. Finally, confirmation of the adenine nucleus was obtained from the characteristic fragments (shown by most adenine derivatives) at m/z 136, 135 and 108.

New technical developments continue to emerge in mass spectroscopy and modern spectrometers may be provided with a Fast Atom Bombardment (FAB) source and are then capable of analysing fragile or involatile organic compounds, including salts and higher-molecular weight materials. Previously, when using MS in the analysis of plant glycosides, the *O*-glycosidic sugars were lost in the process and escaped detection but it is now possible with FAB-MS to obtain molecular ions for the original glycoside.

Some of the many applications of MS data to plant biochemical research are covered in two recently published treatises (Waller, 1972; Walker and Dermer, 1980). The reader is also referred to accounts of MS in the general spectroscopic books already mentioned earlier in this section.

1.4.5   Nuclear magnetic resonance spectroscopy (NMR)

Proton NMR spectroscopy essentially provides a means of determining the structure of an organic compound by measuring the magnetic moments of its hydrogen atoms. In most compounds, hydrogen atoms are attached to different groups (as —CH_2—, —CH_3, —CHO, —NH_2, —CHOH—, etc.) and the proton NMR spectrum provides a record of the number of hydrogen atoms in these different situations. It cannot, however, give any direct information on the nature of the carbon skeleton of the molecule; this can only be obtained by carbon 13 NMR spectroscopy, which is described later.

In practice, the sample of the substance is placed in solution, in an inert solvent, between the poles of a powerful magnet and the protons undergo different chemical shifts according to their molecular environments within the molecule. These are measured in the NMR apparatus relative to a standard, usually tetramethylsilane (TMS), which is an inert compound which can be added to the sample solution without the likelihood of chemical reaction occurring.

Chemical shifts are measured in either δ (delta) or τ (tau) units; where $\tau = 10\,\delta$ and $\delta = \Delta\nu \times 10^6/$radio frequency, $\Delta\nu$ being the difference in absorption frequency of the sample and the reference compound TMS in Hertz units. Since total radio frequency is usually 60 MegaHertz (60 million Hertz) and shifts are measured in Hertz units, they are often referred to as p.p.m. Also, the intensity of the signals may be integrated to show the number of protons resonating at any one frequency.

The solvent for NMR measurements has to be inert and without protons. One is, therefore, limited to using carbon tetrachloride, deuterochloroform ($CDCl_3$), deuterium oxide (D_2O), deuteroacetone (CD_3COCD_3) or deuterated dimethylsulphoxide. Polar compounds are often sparingly soluble or insoluble in the available solvents and they have to be converted to the trimethylsilyl ethers for measurement (see Fig. 1.7). At least 5–10 mg of sample is needed and this limits the use of NMR spectroscopy in many phytochemical experiments. However, instruments which only require mg samples for analysis are becoming available. One advantage of NMR spectroscopy over MS is that the sample can at least be recovered unchanged after the operation and used for other determinations.

As with other spectral techniques, proton NMR spectroscopy can be used by the phytochemist as a fingerprinting technique. It has to be remembered, however, that the complexity of the spectrum is directly related to the number of different types of proton present, so that a highly substituted complex alkaloid may in fact give fewer signals than a simple aliphatic hydrocarbon. The major use of proton NMR is for structural determination, in combination with other spectral techniques. Its use for determining the class of compound is quite considerable; some examples of chemical shifts which are typical of certain classes of natural products are listed in Table 1.4. Aromatic protons (either in benzene derivatives or in heterocyclic compounds) are clearly distinct from aliphatic protons. Within a class of compound, too, NMR measurements may often provide the means of identifying individual structures.

Table 1.4 Some proton nuclear magnetic resonance chemical shifts characteristic of different classes of plant products

Class	Type of proton	Range of shift, δ (p.p.m.)
Alkanes and	CH_3—R	0·85–0·95
fatty acids	R—CH_2—R	1·20–1·35
Alkenes	CH_3—C=C	1·60–1·69
	—CH=C	5·20–5·70
Acetylenes	HC≡C	2·45–2·65
Aromatic compounds	Ar—H	6·60–8·00
	Ar—CH_3	2·25–2·50
	Ar—CHO	9·70–10·00
Nitrogen compounds	N—CH_3	2·10–3·00
	N—CHO	7·90–8·10
	N—H	variable

Proton NMR spectra can be quite complex. For example, due to the interaction between protons attached to adjacent carbon atoms, the spectral signals may appear as 'doublets' or 'triplets' instead of as single peaks. Interpretation, therefore, requires much skill. In spite of this, useful phytochemical information can be obtained without necessarily analysing the spectrum in close detail. Two examples will show this. First, while determining the structure of sterculic acid, chemists could not at first accept that this unique fatty acid had an apparently strained cyclopropene ring in its structure and alternative formulae in which the double bond was placed in positions adjacent to a cyclopropane ring were preferred. The proton NMR spectrum, however, showed clearly that it must have a cyclopropene ring, since there was no signal for protons in the olefinic region (5·2–5·7 δ) and the alternative formulae were thus impossible.

A second example is from a structural study of a yellow flavonoid pigment, tambuletin. This was described, by the original Indian workers who isolated it, as a flavonol aglycone but the proton NMR spectrum at once revealed the presence of an unsuspected sugar unit, since there were signals at 5·15 and 5·25 δ due to sugar protons well separated from signals shown by protons attached to the flavone nucleus (between 6 and 7·5 δ) (see Fig. 1.7). Thus, the pigment was clearly a glycoside, not an aglycone (Harborne *et al.*, 1971).

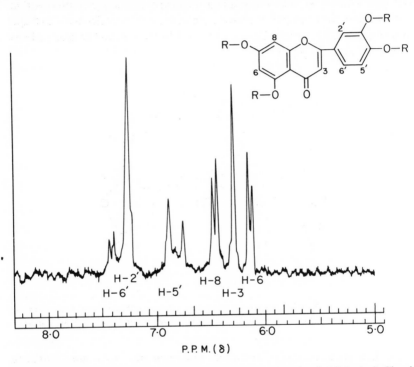

Fig. 1.7 Proton NMR spectrum of the flavone luteolin (as the trimethylsilyl ether). Chemical shifts are relative to TMS (Mabry, 1969). Note that the six protons in luteolin give well separated shifts. If luteolin is further substituted, the position of substitution is clear from the disappearance of one or other of the six signals. Tambuletin (see text) is 8-substituted and its NMR spectrum is similar in this region to luteolin except that the doublet signal at 6·5 δ has disappeared.

The detection of signals from carbon atoms in the NMR apparatus is possible, due to the presence of small amounts (*ca.* 1·1%) of carbon-13 along with carbon-12 in natural plant substances. The smaller magnetic moment generated by ^{13}C compared to that of a proton means that the signal is much weaker. Only since the technical advances of pulsed NMR and Fourier Transform analysis has ^{13}C-NMR spectroscopy become generally available and, even with these advances, more instrument time may be required than with proton NMR. This procedure is fairly widely used in structural analysis, although the need to have a sample weighing at least 10 mg is still a limitation.

Similar solvents are used, as in proton NMR, but the range of ^{13}C resonances is much greater, namely 0–200 p.p.m. downfield from TMS compared with a range from 0–10 p.p.m. for proton resonances. Thus, ^{13}C-NMR spectra are more highly resolved and, in most cases, each carbon within the molecule (see Fig. 1.8) can be assigned to a separate signal. As with proton NMR (see Table 1.4), differently substituted carbon atoms give shifts within specific ranges; for example, aliphatic carbon atoms give shifts between 0 and 40 p.p.m., aromatic carbons between 110 and 150 p.p.m. and ketonic carbons between 160 and 230 p.p.m.

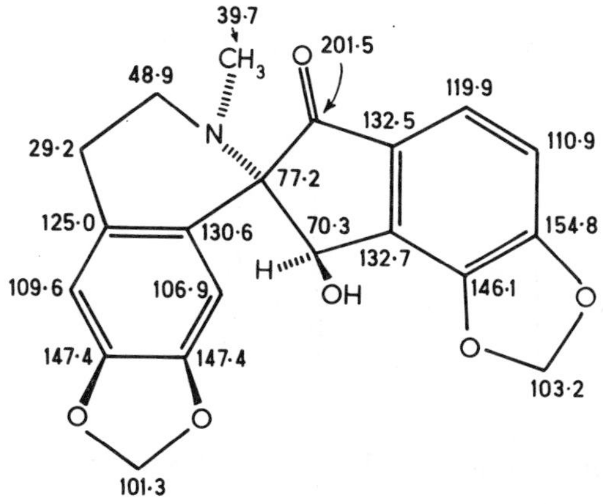

Fig. 1.8 Carbon-13 shifts relative to TMS (in p.p.m.) for the different carbon atoms in the molecule of the spirobenzylisoquinoline alkaloid sibiricine from *Corydalis sibirica* (Fumariaceae).

^{13}C-NMR spectroscopy is essentially complementary to proton NMR and the combination of the two techniques provides a very powerful means of structural elucidation for new terpenoids, alkaloids or flavonoids. It is useful in the analysis of glycosides, in indicating the linkage between sugar moieties and their configurations. Both proton and ^{13}C-NMR measurements have been successfully applied to structural and other analyses of proteins and other macromolecules (see Jones, 1980). There is no simple guide to NMR spectroscopy written for the plant scientist. There are, though, many works dealing with its application to organic chemistry (e.g. Jackman, 1959; Scheinmann, 1970) or to biochemistry (e.g. James, 1975).

1.4.6 Criteria for phytochemical identifications

As already mentioned above, a known compound discovered in a new plant source can be identified on the basis of chromatographic and spectral comparison with authentic material. Authentic samples may be obtained commercially from a firm of suppliers (see Appendix B), by re-isolation from a known source or, as a last resort, by request from the scientist who originally isolated and described it. The extent of the comparison to be made varies according to the class of compound being studied but as a general principle, it is desirable to use as many criteria as possible to be certain of the correctness of the identification (see Table 1.5).

Chromatographic comparison should be based on co-chromatography of the compound with authentic material, without separation, in at least four systems. If TLC is the main basis of comparison, then there are obvious advantages in employing different adsorbents (e.g. cellulose and silica gel) as well as different solvents on the same adsorbent (Table 1.5). Whenever possible, it is desirable to compare unknown and standard by three distinct chromatographic criteria, e.g. by retention time by GLC and HPLC and R_F on TLC or by R_F on PC and TLC and relative mobility on electrophoresis. Again, for spectral comparisons, two or more procedures should be adopted wherever possible. Ideally, UV, IR, MS and ^{1}H-NMR spectra should all be compared.

With new plant substances, it is usually possible to establish the structure on the basis of spectral and chromatographic measurements, especially in relationship to those made on known compounds in the same series. Confirmation of structure may be possible by chemical conversion to a known

Table 1.5 The type of criteria needed for the identification of known plant constituents. The identification of 6-hydroxyluteolin 7-methyl ether in leaves of *Crocus minimus*

Criterion	Property recorded
1. Physical properties	Yellow powder, m.p. 245–6°C
2. Molecular formula by MS	Molecular ion found 316.0574 $C_{16}H_{12}O_7$ requires 316.0582
3. MS fragmentation pattern	Fragment ion by demethylation at 301.0344 ($C_{15}H_9O_7$ requires 301.0345), etc.
4. UV spectral properties (and shifts with alkali, etc.)	Maxima at 254, 273, 346 nm, etc.
5. Colour on TLC plate	Yellow in daylight Dark brown in UV $\pm$ NH$_3$
6. TLC on cellulose	R_F 0·73 in n-BuOH—HOAc—H$_2$O (4:1:5) R_F 0·59 in 50% HOAc R_F 0·67 in CHCl$_3$—HOAc—H$_2$O (90:45:6)
7. TLC on polyamide	R_F 0·36 in C$_6$H$_6$—MeCOEt—MeOH (4:3:3)
8. Chemical conversion	Demethylation with pyridinium chloride to 6-hydroxyluteolin

compound. At one time, an essential step in structural identification was the determination of molecular formula by micro-analysis, with at least a carbon, hydrogen determination. Such micro-analyses are still very desirable, but when only micrograms of material are available, it is possible to use a precise mass measurement on the molecular ion, determined by mass spectrometry (see Section 1.4.4 and Table 1.5). Derivatization is also valuable with new compounds, e.g. the preparation of acetates, methyl ethers, etc., since their analysis provides useful confirmation of the molecular formula of the original substance.

1.5 ANALYSIS OF RESULTS

1.5.1 Qualitative analysis

Much plant analysis is devoted to the isolation and identification of secondary constituents in a particular species or group of species with the expectation that some of the constituents may be novel or of an unusual structure. In such cases, it is important to recognize that many of the more readily isolated components are either commonly present or universal in occurrence. Sucrose may crystallize out from an aqueous plant concentrate and sitosterol from a phytosterol fraction. The more interesting components are often those present in lower amounts.

When an apparently new structure has been obtained, it is necessary to check carefully that it has not been reported before. Reference may be made to the various literature compilations available (e.g. the recent encyclopedia of terpenoids, Glasby (1982)) but a thorough search in *Chemical Abstracts* is also desirable.

Another motive for phytochemical analysis is the characterization of an active principle responsible for some toxic or beneficial effect shown by a crude plant extract when tested against a living system. In such cases, it is essential to monitor the extraction and separation procedures at each stage in order to follow the active material as it is purified. The activity can sometimes vanish during fractionation, due to its lability and a pure crystalline compound may be eventually obtained which lacks the activity of the original extract. The possibility of damage to the active principle during isolation and characterization must always be borne in mind.

Similarly, it is essential to realize that the production of artifacts is a common feature of plant analysis. Many of the compounds that occur in plant tissues are quite labile and almost inevitably may undergo change during extraction. The plastid pigments, the chlorophylls and carotenoids, are susceptible to modification during chromatography (see Chapters 3 and 4). All plant glycosides are liable to some hydrolysis, either enzymic or non-enzymic, during isolation, while esters may undergo transesterification in the presence of alcoholic solvents. The volatile terpenes are susceptible to molecular re-arrangements during steam distillation (Chapter 3) and the racemization of optically active constituents may occur, unless special precautions are taken. Again, proteins may be subject to protease attack during isolation procedures (Chapter 7).

Additionally, artifacts may be introduced unwittingly from laboratory equipment during purification. The most common additive is butyl isoph-thalate, which is a plasticizer that almost always contaminates plant extracts. It has actually been reported as a plant constituent, in spite of its obvious origin from a plastic wash bottle used by the operator during the isolation. In avoiding artifacts, it is necessary to check the original crude plant extract to see if a compound isolated only after extensive purification is actually present there.

1.5.2 Quantitative analysis

Equally as important as qualitative measurements on a plant extract are determinations of the amounts of the components present. In the simplest approach, quantitative data can be obtained by weighing the amount of plant material initially used (assuming dried tissue) and the amount of the product obtained. Such a yield as a percentage of the whole will be a minimal figure, since inevitably some material will be lost during purification. Losses can be estimated by adding a known weight of pure substance to the crude extract, repeating the purification and determining the amount recovered. If fresh tissue is extracted, a conversion factor (most plant leaves are 90% water) will be needed to express the result as percentage dry weight.

Quantitative measurements can also be conducted on dried, powdered plant material to determine the total content of sugar, nitrogen, protein, phenol, tannin and so on. Some of the procedures that can be used will be mentioned in later chapters. Such procedures are liable to error due to interference from other components present. Whether such determinations have much value in terms of, for example, the amount of herbivory a particular plant organ suffers needs to be assessed.

Ideally, in quantitative measurements, the amounts of the individual components within a particular class of constituent need to be determined and this is now readily achieved by either GLC or HPLC. The amounts of fatty acids bound in neutral plant lipids can, for example, be determined in a thoroughly reproducible way after saponification, methyl ester formation and quantification of the methyl esters by GLC (Chapter 4). Similarly, HPLC measurements can be used to determine the amounts of flavonoid pigments in different varieties and genotypes of garden flowers (Chapter 2).

The importance of repeating measurements so that they can be seen to be statistically significant is obvious but not always appreciated. Variation in amounts due to environmental parameters needs to be eliminated and sampling has to be considered in relation to plant age and provenance. Guidance on these matters are available in most modern plant texts, but see also Paech and Tracey (1956–1964).

1.6 APPLICATIONS

1.6.1 General

While phytochemical procedures have today an established role in practically all branches of plant science, this has not always been so. Although these methods are obviously essential in all chemical and biochemical studies, their application in more strictly biological spheres has only come within the last two decades. Even in disciplines so remote from the chemical laboratory as systematics, phytogeography, ecology and palaeobotany, phytochemical methods have become important for solving certain types of problems and will undoubtedly be used here with increasing frequency in the future.

There is only room in this book to mention a few of the many applications of phytochemical methods. Some of the obvious applications in agriculture, in nutrition and the food industry and in pharmaceutical research must be taken for granted. The following applications are simply given to illustrate the value of phytochemical techniques in some of the major branches of plant science.

1.6.2 Plant physiology

The major contributions of phytochemical studies to plant physiology are undoubtedly in determining the chemical structures, biosynthetic origins and modes of action of natural growth hormones. As a result of continuing collaboration over the years between physiologists and phytochemists, five classes of growth regulators are now recognized: the auxins, cytokinins, abscisins, gibberellins and ethylene. Methods of detection, which vary from GLC through TLC to PC, are discussed in later pages as follows: auxins (p. 206), cytokinins (p. 212), abscisins (p. 113) and gibberellins (p. 118). One of the more remarkable aspects of the gibberellin group of hormones is the large number of known structures (over sixty), all apparently with a similar range of growth properties. The need for very precise methods of detection and distinction between gibberellins led to the development of combined GC–MS for their analysis. A general book on methods for the isolation of plant growth substances, edited by Hillman (1978), can be consulted for more details. The necessary requirements for accurate hormone analysis are critically considered by Reeve and Crozier (1980). An excellent review of the latest techniques, including the use of radio-immunoassay, is that of Horgan (1981).

1.6.3   Plant pathology

Phytochemical techniques are primarily important to the pathologist for the chemical characterization of phytotoxins (products of microbial synthesis produced in higher plants when invaded by bacteria or fungi) and of phytoalexins (products of higher plant metabolism formed in response to microbial attack). A range of different chemical structures are involved in both cases. The most familiar phytotoxins are lycomarasmin and fusaric acid, amino acid derivatives which are wilting agents in the tomato. Other toxins that have been isolated are glycopeptides, naphthaquinones or sesquiterpenoids (Durbin, 1981). Some phytotoxins are chemically labile so that special precautions have to be taken during their isolation and identification.

Phytoalexins also have different structures, according to the plant source (Bailey and Mansfield, 1982). They may be sesquiterpenoid (rishitin from *Solanum tuberosum*), isoflavanoid (pisatin from *Pisum sativum*), acetylenic (wyerone acid from *Vicia faba*) or 'phenolic' (orchinol from *Orchis militaris*). Identifications of isoflavanoids and acetylenes are described in Chapters 2 and 5 respectively and a procedure for phytoalexin induction is included as a practical experiment in Chapter 2.

'Pre-infective' substances (naturally occurring secondary constituents) are considered by some plant pathologists to be important in imparting disease resistance to plants. Phenolic compounds, such as phloridzin in apple, tannin in raspberry, are mainly implicated here. Methods of identification of such compounds are covered in detail in Chapter 2.

1.6.4   Plant ecology

Two research areas where secondary plant constituents are significant in plant ecology are in plant–animal and plant–plant interactions. The analytical problems in both cases are difficult because of the very limited amounts of biological material at the disposal of the phytochemist. For example, in following the fate of secondary compounds in insects feeding on plants, it is necessary to analyse different organs of the insect to see where the compounds are stored; such analyses are often complicated and time-consuming.

Compounds so far known to be involved in plant–animal interactions are primarily alkaloids and cardiac glycosides, mustard oil glycosides, cyanogens, steroids or volatile terpenes. The plant compounds may variously act as feeding attractants or repellents, have hormonal effects on the insects or provide the insects with a useful defence mechanism against predation (Harborne, 1982).

Plant–plant interactions involve so-called allelopathic substances which one plant exudes from its roots or leaves in order to prevent the growth of other plant species in its vicinity. The compounds are either volatile terpenes (e.g.

cineole) or simple phenolic acids, depending on whether the plant is growing in a semi-tropical or a temperate climate. The phytochemical study of allelopathy can be difficult since it requires determinations on whole leaf extracts, natural leaf leachates and on soil samples too. The possible rapid turnover of active substances in the soil also provides another analytical hazard in this field.

Applied aspects of research on plant–animal interactions include the control of insect predation on crop plants by natural or synthetic pesticides and phytochemical analysis may be required to follow the fate of these pesticides in the environment. Recent developments here are reviewed in Hutson and Roberts (1983).

1.6.5 Palaeobotany

Phytochemistry has only recently been applied to the study of fossil plants but there seems little doubt that it will play an increasing role, for example, in testing various hypotheses on the early origins of land plants. Some of the present achievements of phytochemical techniques include the identification of partially degraded chlorophyll pigments in lignite deposits 50 million years old, of carbohydrates in Palaeozoic plants 250–400 million years old and of hydrocarbons in Triassic *Equisetum* dating back 200 million years (Chaloner and Allen, 1970). Pollen wall material (pollenins) from fossil plants have also been analysed successfully (Shaw, 1970) to yield recognizable fatty acids and phenolic acids on degradation. Identification of terpenes in fossil resins and ambers has also yielded new data of considerable phylogenetic interest (Thomas, 1970). The application of phytochemical techniques to palaeobotany has recently been reviewed by Niklas (1980).

1.6.6 Plant genetics

In the past, phytochemistry has contributed to higher plant genetics in providing the means of identifying anthocyanin, flavone and carotenoid pigments occurring in different colour genotypes of garden plants. The results have shown that the biochemical effects of these genes have a simple basis and have indicated the probable pathway of pigment synthesis in these organisms (Alston, 1964). The inheritance in plants of other chemical attributes (alkaloids, terpenes, etc.) has also been successfully mapped out by phytochemical analysis.

A more recent contribution of phytochemistry to genetics is in the identification of hybrid plants and in the elucidation by chemical means of their parental origin. Phytochemistry has also won increasing recognition as a useful tool, together with cytology, to be used in the analysis of genetic variation within plant populations (cf. Harborne and Turner, 1984).

1.6.7   Plant systematics

One of the most rapidly developing fields in phytochemistry at the present time is the hybrid discipline between chemistry and taxonomy, known as biochemical systematics or chemotaxonomy. Basically, it is concerned with the chemical survey of restricted groups of plant, mainly for secondary constituents but also for macromolecules and the application of the data so obtained to plant classification (Harborne and Turner, 1984).

Perhaps the most useful class of compounds for such study are the flavonoids. Surveys of many other classes of compounds (notably of alkaloids, non-protein amino acids, terpenes and sulphur compounds) have also yielded potentially useful new information for taxonomic purposes. Accurate methods are essential, both in preliminary screening of plants and in the more detailed analysis of individual components.

Chemical analyses of the amino acid sequences of plant proteins have also been brought to bear on systematic problems, at the higher levels of classification. Results have been obtained with cytochrome C, plastocyanin and ferredoxin; the sequencing of plant nucleic acids has also yielded data of taxonomic interest (Jensen and Fairbrothers, 1983).

REFERENCES

General references

Bailey, J.A. and Mansfield, J.W. (eds) (1982) *Phytoalexins*, Blackie, Glasgow.

Bobbitt, J.M. (1963) *Thin Layer Chromatography*, Reinhold Pub. Co., New York.

Brand, J.C.D. and Eglinton, G. (1965) *Applications of Spectroscopy*, Oldbourne Press, London.

Burchfield, H.P. and Storrs, E.E. (1962) *Biochemical Application of Gas Chromatography*, Academic Press, New York.

Cross, A.D. and Jones, R.A. (1969) *An Introduction to Practical IR Spectroscopy*, 3rd edn, Butterworths, London.

Culberson, C.F. (1969) *Chemical and Botanical Guide to Lichen Products*, North Carolina Univ. Press, Chapel Hill, U.S.A.

Durbin, R.D. (ed.) (1981) *Toxins in Plant Disease*, Academic Press, New York.

Gillam, A.E. and Stern, E.S. (1957) *Electronic Absorption Spectroscopy*, 2nd edn, Edward Arnold, London.

Hamilton, R.J. and Sewell, P.A. (1982) *Introduction to High Performance Liquid Chromatography*, 2nd edn, Chapman and Hall, London.

Harborne, J.B. (1967) *Comparative Biochemistry of the Flavonoids*, Academic Press, London.

Harborne, J.B. (1982) *Introduction to Ecological Biochemistry*, 2nd edn, Academic Press, London.

Harborne, J.B. and Turner, B.L. (1984) *Plant Chemosystematics*, Academic Press, London.

Heftmann, E. (1983) *Chromatography: Fundamentals and Applications of Chromatographic and Electrophoretic Techniques*, Elsevier, Amsterdam.

Hillman, J.R. (ed.) (1978) *Isolation of Plant Growth Substances*, Cambridge University Press, Cambridge.

Horgan, R. (1981) Modern methods for plant hormone analysis. *Prog. Phytochem.*, **7**, 137–70.

Jackman, L.M. (1959) *Applications of NMR Spectroscopy in Organic Chemistry*, Pergamon Press, Oxford.

James, T.L. (1975) *NMR in Biochemistry: Principles and Applications*, Academic Press, New York.

Kirchner, J.G. (1978) *Thin Layer Chromatography*, 2nd edn, John Wiley, New York.

Lederer, E. and Lederer, M. (1957) *Paper Chromatography*, 2nd edn, Elsevier, Amsterdam.

Linskens, H.F. (1959) *Papier Chromatographie in der Botanik*, Springer Verlag, Berlin.

Morton, R.A. (1975) *Biochemical Spectroscopy*, Adam Hilger, London.

Paech, K. and Tracey, M.V. (eds) (1956–1964) *Moderne Methoden der Pflanzenanalyse*, Springer Verlag, Berlin.

Reeve, D.R. and Crozier, A. (1980) Quantitative analysis of plant hormones. *Encycl. Pl. Physiol. New Series* **9**, 203–80.

Sargent, J.R. (1969) *Methods in Zone Electrophoresis*, 2nd edn, BDH Chemicals Ltd., Poole, England.

Scheinmann, F. (ed.) (1970) *An Introduction to Spectroscopic Methods for the Identification of Organic Compounds* (Vol. I), Pergamon Press, Oxford.

Scheinmann, F. (ed.) (1974) *An Introduction to Spectroscopic Methods for the Identification of Organic Compounds* (Vol. II), Pergamon Press, Oxford.

Scott, A.I. (1964) *Interpretation of the Ultraviolet Spectra of Natural Products*, Pergamon Press, Oxford.

Sherma, J. and Zweig, G. (1971) *Paper Chromatography*, Academic Press, New York.

Simpson, C. (1970) *Gas Chromatography*, Kogan Page, London.

Stahl, E. (ed.) (1969) *Thin Layer Chromatography*, 2nd edn, George Allen and Unwin, London.

Touchstone, J.C. and Dobbins, M.F. (1978) *Practice of Thin Layer Chromatography*, John Wiley, Chichester.

Truter, E.V. (1963) *Thin Film Chromatography*, Cleaver Hume Press, London.

Turner, W.B. (1971) *Fungal Metabolites*, Academic Press, London.

Turner, W.B. and Aldridge, D.C. (1983) *Fungal Metabolites II*, Academic Press, London.

Waller, G.R. (ed.) (1972) *Biochemical Applications of Mass Spectrometry*, Wiley Interscience, New York.

Waller, G.R. and Dermer, O.C. (1980) *Biochemical Applications of Mass Spectrometry, First Supplement*, John Wiley, Chichester.

Williams, D.H. and Fleming, I. (1966) *Spectroscopic Methods in Organic Chemistry*, McGraw-Hill, London.

Supplementary references

Alston, R.E. (1964) in *Biochemistry of Phenolic Compounds* (ed. J.B. Harborne), Academic Press, London, pp. 171–204.

Bate-Smith, E.C. and Westall, R.G. (1956) *Biochim. Biophys. Acta.* **4,** 427.

Chaloner, W.G. and Allen, K. (1970) in *Phytochemical Phylogeny* (ed. J.B. Harborne), Academic Press, London, pp. 21–30.

Eglinton, G. (1970) in *Introduction to Spectroscopic Methods for the Identification of Organic Compounds* (Vol. I) (ed. F. Scheinmann), Pergamon Press, Oxford, pp. 123–144.

Glasby, J.S. (1982) *Encyclopedia of the Terpenoids*, John Wiley, Chichester.

Harborne, J.B. (1969) *Phytochemistry*, **8,** 419.

Harborne, J.B., Lebreton, P., Combier, H., Mabry, T.J. and Hammam, Z. (1971) *Phytochemistry*, **10,** 883.

Harley, R.M. and Bell, M.G. (1967) *Nature (Lond.)*, **213,** 1241.

Hershenson, H.M. (1956) *UV and Visible Absorption Spectra* (Vol. I, 1930–54), Academic Press, New York.

Hershenson, H.M. (1959) *Infra Red Absorption Spectra* (Vol. I, 1945–57), Academic Press, New York.

Hershenson, H.M. (1961) *UV and Visible Absorption Spectra* (Vol. II, 1955–9), Academic Press, New York.

Hershenson, H.M. (1964) *Infra Red Absorption Spectra* (Vol. II, 1958–62), Academic Press, New York.

Hershenson, H.M. (1966) *UV and Visible Absorption Spectra* (Vol. III, 1960–3), Academic Press, New York.

Hostettmann, K. (1981) in *Natural Products as Medicinal Agents* (eds J.L. Beal and E. Reinhard), Hippokrates Verlag, Stuttgart, pp. 79–92.

Hutson, D.H. and Roberts, T.R. (eds) (1983) *Progress in Pesticide Biochemistry and Toxicology* (Vol. 3), John Wiley, Chichester.

Jensen, U. and Fairbrothers, D.E. (eds) (1983) *Proteins and Nucleic Acids in Plant Systematics*, Springer-Verlag, Berlin.

Jones, R. (1980) in *An Introduction to Spectroscopy for Biochemists* (ed. S.B. Brown), Academic Press, London.

Knights, B.A. (1965) *Phytochemistry*, **4,** 857.

Lang, L. (1959) *Absorption Spectra*, Academy of Science Press, Budapest.

Mabry, T.J. (1969) in *Perspectives in Phytochemistry* (eds J.B. Harborne and T. Swain), Academic Press, London, pp. 1–46.

Niklas, K.J. (1980) *Progr. Phytochem.*, **6,** 143–81.

Paris, R., Durand, M. and Bounet, J.L. (1960) *Ann. Pharm. Franc.*, **18,** 769.

Peereboom, J.W.C. (1971) in *Comprehensive Analytical Chemistry* (ed. C.L. Wilson and D.W. Wilson) (Vol. IIC), Elsevier, Amsterdam, pp. 1–129.

Phillipson, D.J.D. (1982) *Phytochemistry*, **21,** 2441–56.

Poindexter, E.H. and Carpenter, R.D. (1962) *Phytochemistry*, **1,** 215.

Sanford, K.J. and Heinz, D.E. (1971) *Phytochemistry*, **10,** 1245.

Shannon, J.S. and Letham, D.S. (1966) *New Zealand J. Sci.*, **9,** 833.

Shaw, G. (1970) in *Phytochemical Phylogeny* (ed. J.B. Harborne), Academic Press, London, pp. 31–58.

Thomas, B.R. (1970) in *Phytochemical Phylogeny* (ed. J.B. Harborne), Academic Press, London, pp. 59–80.

Van Sumere, C.F. (1969) *Revue des Fermentations et des Industries Alimentaires (Brussels)*, **24,** 91–139.

CHAPTER TWO

Phenolic Compounds

2.1 INTRODUCTION

The term phenolic compound embraces a wide range of plant substances which possess in common an aromatic ring bearing one or more hydroxyl substituents. Phenolic substances tend to be water-soluble, since they most frequently occur combined with sugar as glycosides and they are usually located in the cell vacuole. Among the natural phenolic compounds, of which several thousand structures are known, the flavonoids form the largest group but simple monocyclic phenols, phenylpropanoids and phenolic quinones all exist in considerable numbers. Several important groups of polymeric materials in plants – the lignins, melanins and tannins – are polyphenolic (see Fig. 2.1) and occasional phenolic units are encountered in proteins, alkaloids and among the terpenoids.

While the function of some classes of phenolic compound are well established (e.g. the lignins as structural material of the cell wall; the anthocyanins as flower pigments), the purpose of other classes is still a matter of speculation. Flavonols, for example, appear to be important in regulating control of growth in the pea plant (Galston, 1969) and their adverse effects on insect feeding (Isman and Duffey, 1981) have indicated that they may be natural resistance factors; neither of these functions, however, has yet been firmly established.

To the plant biochemist, plant phenols can be a considerable nuisance, because of their ability to complex with protein by hydrogen bonding. When plant cell constituents come together and the membranes are destroyed during isolation procedures, the phenols rapidly complex with protein and as a result, there is often inhibition of enzyme activity in crude plant extracts. On the other hand, phenols are themselves very susceptible to enzymic oxidation and phenolic material may be lost during isolation procedures, due to the action of specific 'phenolase' enzymes present in all plants. Extraction of the phenols from plants with boiling alcohol normally prevents enzymic oxidation occurring and this procedure should be adopted routinely.

The classic procedure for detecting simple phenols is by means of the intense green, purple, blue or black colours many of them give in solution when 1% aqueous or alcoholic ferric chloride is added. This procedure, modified by using a fresh aqueous mixture of 1% ferric chloride and 1% potassium

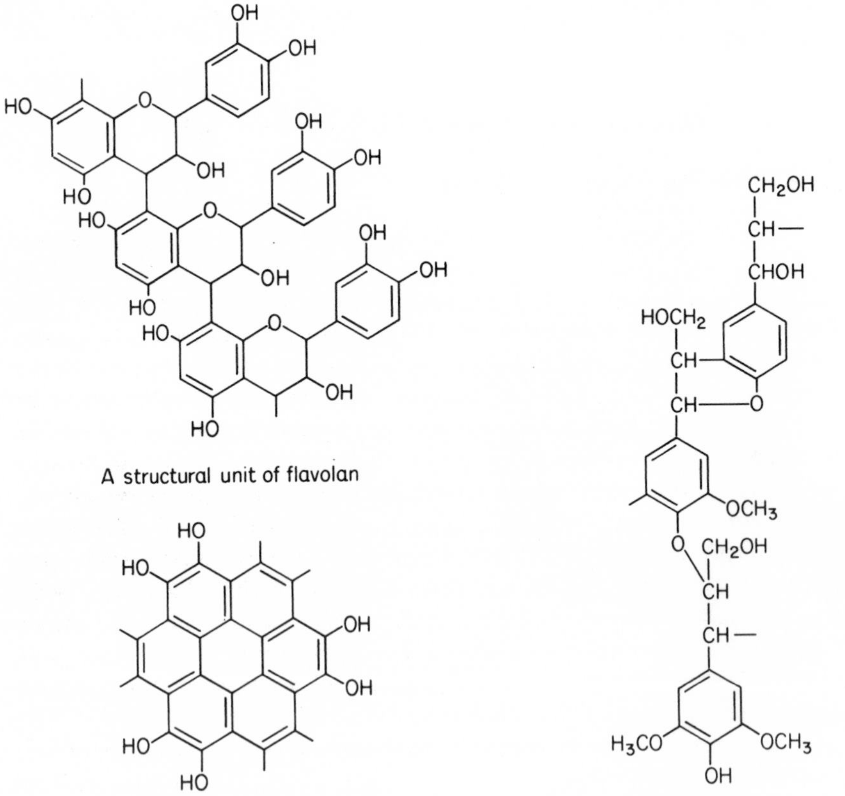

A structural unit of flavolan

A structural unit of plant melanin

A structural unit of lignin

Fig. 2.1 Partial structures for the phenolic polymers of plants.

ferricyanide, is still used as a general means of detecting phenolic compounds on paper chromatograms. However, the majority of phenolic compounds (and especially the flavonoids) can be detected on chromatograms by their colours or fluorescences in UV light, the colours being intensified or changed by fuming the papers with ammonia vapour. The phenolic pigments are visibly coloured and they are thus particularly easily monitored during their isolation and purification.

Phenolic compounds are all aromatic, so that they all show intense absorption in the UV region of the spectrum. In addition, phenolic compounds characteristically exhibit bathochromic shifts in their spectra in the presence of alkali. Spectral methods are, therefore, especially important for the identification and quantitative analysis of phenols.

Of the numerous texts devoted to the plant phenols, the reader is especially referred to the simple student treatise *Plant Phenolics* (Ribereau-Gayon, 1972) and to the more advanced reference volumes of Harborne (1964) and Swain *et al.* (1979). A critical review of the chromatography of phenolic compounds is that of Harborne (1983).

2.2 PHENOLS AND PHENOLIC ACIDS

2.2.1 Chemistry and distribution

The free phenols and phenolic acids (for formulae, see Fig. 2.2) are best considered together, since they are usually identified together during plant analysis. Acid hydrolysis of plant tissues releases a number of ether-soluble phenolic acids, some of which are universal in their distribution. These acids are either associated with lignin combined as ester groups or present in the alcohol-insoluble fraction of the leaf; alternatively they may be present in the alcohol-soluble fraction bound as simple glycosides. Universal among the angiosperms are *p*-hydroxybenzoic acid, protocatechuic acid, vanillic acid and syringic acid. Gentisic acid is also widespread; less common are salicylic and *o*-protocatechuic, two acids found characteristically in the Ericaceae. Finally, there is gallic acid, recently reported as a natural inhibitor of flowering, in leaves of *Kalanchoe* (Pryce, 1972a). Gallic acid is found in many woody plants, bound as gallotannin, but it is a very reactive substance. More common in acid-hydrolysed plant extracts is its dimeric condensation product, ellagic acid, formed from ellagitannins (Seikel and Hillis, 1970) present in direct extracts.

By contrast with the above acids, free phenols are relatively rare in plants. Hydroquinone is probably the most widely distributed; others, such as catechol, orcinol, phloroglucinol and pyrogallol, have been reported from only a few sources. The simple phenols are included here, because their identification is important in relation to determining the structure of flavonoids.

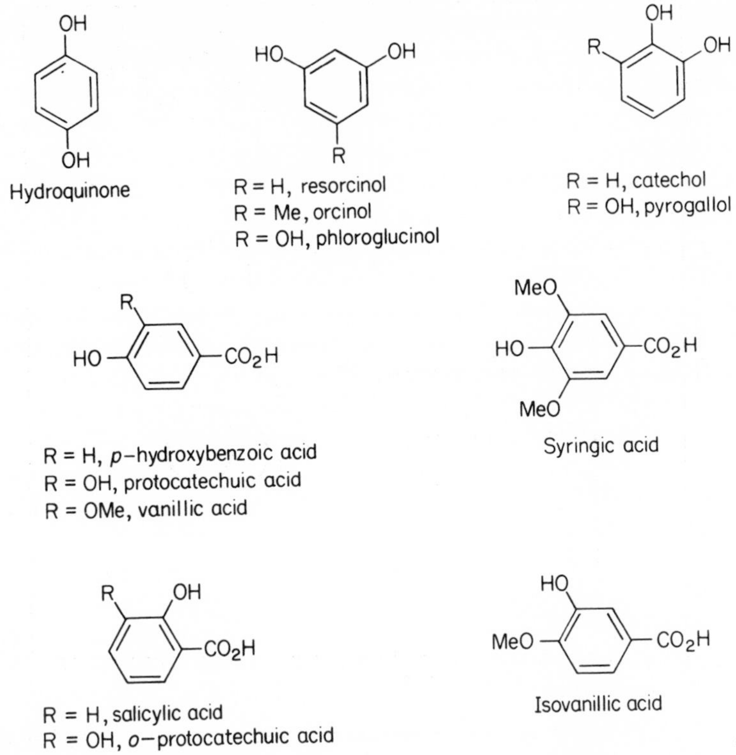

Hydroquinone

R = H, resorcinol
R = Me, orcinol
R = OH, phloroglucinol

R = H, catechol
R = OH, pyrogallol

R = H, *p*-hydroxybenzoic acid
R = OH, protocatechuic acid
R = OMe, vanillic acid

Syringic acid

R = H, salicylic acid
R = OH, *o*-protocatechuic acid

Isovanillic acid

Fig. 2.2 Structures of phenols and phenolic acids.

Alkaline degradation or reductive cleavage of flavonoids and other complex phenols yields, depending on the particular substitution pattern present, one or more of the simple phenols and phenolic acids shown in Fig. 2.2.

One group of natural phenolic polymers, the plant melanins (see Fig. 2.1), also yield a simple phenol, catechol, on alkaline degradation. Unlike animal melanins which are based on 3,4-dihydroxyphenylalanine, most plant melanins such as the black pigments of seed coats or in fungal rust spores are nitrogen-free polymers. They are characterized by heating them at 200–300°C in nitrogen when they give catechol and by alkali fusion, when they yield catechol, protocatechuic and salicylic acids (Nicolaus and Piattelli, 1965).

2.2.2 Recommended techniques

The best method of separating and identifying simple 'phenolics' is by TLC. They are normally detected after acid or alkaline hydrolysis of plant tissue (fresh or dried) or of a concentrated aqueous-alcoholic plant extract. Acid

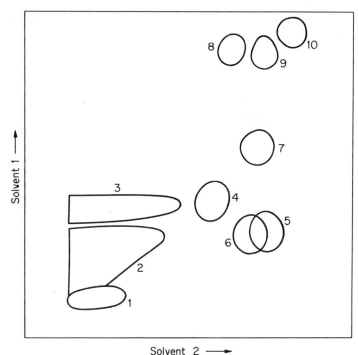

Fig. 2.3 Thin layer chromatography separation on silica gel of simple plant phenols. Key: solvent 1, 10% acetic acid in chloroform; solvent 2, 45% ethyl acetate in benzene; 1, gallic acid; 2, 3,4-dihydroxybenzoic acid; 3, 2,5-dihydroxybenzoic acid; 4, rhododendrol; 5, hydroquinone; 6, orcinol; 7, *p*-hydroxybenzoic acid; 8, syringic acid; 9, vanillic acid; 10, salicylic acid. Compounds 1, 2 and 5 give a blue colour with Folin reagent. Compounds 3, 4, 6, 7, 8, 9 and 10 give a blue colour with Folin reagent after fuming with ammonia. Compound 6 gives a pink colour with vanillin–HCl.

hydrolysis is carried out with 2 M HCl for 0·5 h, the resultant solution being cooled and filtered before extraction. Alkaline hydrolysis is carried out with 2 M NaOH at room temperature under nitrogen for 4 h, acidification being necessary before extraction. In either case, the phenols are taken into ether and the ether extract washed, dried and evaporated to dryness. The residue, dissolved in ether, is then chromatographed two-dimensionally on silica gel in acetic acid–chloroform and ethyl acetate–benzene and separately on cellulose MN 300 in benzene–methanol–acetic acid and aqueous acetic acid (see Fig. 2.3). Typical R_F values for the common phenols and phenolic acids in these four solvents are shown in Table 2.1. The value of the method is indicated by the separation achieved for the three isomeric *C*-methyl-resorcinols, orcinol (6-methylresorcinol) and the 4- and 2-methyl isomers (Harborne and Williams, 1969).

Table 2.1 R_F and spectral properties of simple phenolics

Phenolic	TLC R_F ($\times 100$) in solvent*				Colour	EtOH λ_{max}	EtOH–NaOH λ_{max}
	1	2	3	4			
Simple phenols					*Vanillin–HCl*		
Orcinol	19	62	46	67	bluish-pink	276,282	294
4-Methylresorcinol	25	63	59	65	brick red	282	291
2-Methylresorcinol	40	64	58	73	bluish-pink	275,280	288
Resorcinol	17	59	48	74	red	276,283	293
Catechol	35	66	58	72	} none	279	decomposition
Hydroquinone	18	58	34	69		295	decomposition
Pyrogallol	08	15	19	72		266	decomposition
Phloroglucinol	05	47	09	62		269,273	350
Phenolic acids					*Folin*		
Gallic	05	40	05	40	blue	272	decomposition
Protocatechuic	19	44	19	52	blue	260,295	240,283
Gentisic	33	44	41	61	} blue after ammonia fuming	237,335	308
p-Hydroxybenzoic	55	80	60	62		265	278
Syringic	79	58	74	52		271	298
Vanillic	82	73	70	57		260,290	285,297
Salicylic	91	82	86	66		235,305	225,297

*Solvents 1 and 2 are on silica gel: HOAc–CHCl$_3$ (1:9) and EtOAc–benzene (9:11); solvents 3 and 4 on cellulose MN 300: benzene–MeOH–HOAc (45:8:4) and 6% aq–HOAc.

Phenols absorb in the short UV, and can be detected in light of wavelength 253 nm as dark absorbing spots on plates spread with silica gel containing fluorescent indicator. It is, however, usually preferable to detect them by means of a more specific reagent, the best of which is that of Folin-Ciocalteu. With this reagent (which is available commercially, as a spray solution), phenols with catechol or hydroquinone nuclei appear as blue spots immediately after spraying; other phenols show up as blue to grey spots when the plate is fumed with ammonia vapour. Two other reagents are also useful. Vanillin–HCl (1 g vanillin in 10 ml conc. HCl) and vanillin–H$_2$SO$_4$ (10% vanillin in EtOH–conc. H$_2$SO$_4$, 2:1) give a range of pink colours with resorcinol and phloroglucinol derivatives. Gibbs reagent (2% 2,6-dichloroquinone chloroimide in chloroform) followed by fuming the plate with 2 M NH$_4$OH gives a variety of colours with different phenols. It can be used to distinguish vanillic (pink colour) from isovanillic acid (blue colour), isomers which have rather similar R_F values.

Confirmation of a phenolic identification is readily achieved by UV spectral comparisons (Table 2.1) in alcoholic and alkaline solutions. If there is sufficient material, further confirmation can be done by GLC and IR measurements.

Authentic markers are readily obtainable commercially.

2.2.3 Alternative techniques

(a) *Paper chromatography*

This has been much employed for phenol separations, but has the disadvantage that in many of the best solvents (e.g. butanol–acetic acid–water) simple phenols tend to cluster together near the front. A successful two-dimensional system on paper is benzene–acetic acid–water (6:7:3) and sodium formate–formic acid–water (10:1:200) (Ibrahim and Towers, 1960). The benzene solvent must be used fresh, since there is streaking if the solvent mixture is allowed to equilibrate. This solvent is useful for one-dimensional separation of isomeric phenolic acids. For example, 2,3-dihydroxy, 2,4-dihydroxy, 2,5-dihydroxy (gentisic) and 2,6-dihydroxybenzoic acids have the following R_Fs respectively, 48, 33, 25 and 17; protocatechuic (3,4-dihydroxy) R_F 22, separates from 3,5-dihydroxy, R_F 29; and vanillic, R_F 68, separates from isovanillic, R_F 37. Two other solvents for one-dimensional separation of phenols are water and butanol–2 M NH_4OH (1:1, top layer). Reio (1964) has devised a system for the identification of simple 'phenolics' of fungal origin based on PC in six solvents.

(b) *Gas liquid chromatography*

GLC has not been widely used for phenol separations, partly because most phenols have to be converted to suitable derivatives (trimethylsilyl ethers or acetates) to make them sufficiently volatile. However, it is an essential technique when complex mixtures of simple phenols occur in any one plant

Table 2.2 Relative gas liquid chromatography retention times of some phenols in tobacco leaf

	GLC *on column*		
Phenolic	DEGS	PPE	TXP*
Phenol	13	17	16
o-Cresol	16	23	21
m-Cresol	19	28	24
p-Cresol	19	29	26
Guaiacol	46	55	54
Catechol	100	100	100
Hydroquinone	125	129	121
4-Methylcatechol	125	139	135
4-Methyl,2,6-dimethoxy-phenol	147	186	170

*For key, see text. Retention time of catechol is taken as standard.

tissue. Cured tobacco leaf, for example, contains thirty-eight phenols ranging from phenol itself, *p*-cresol and catechol to 4-methyl-2.6-dimethoxyphenol (Irvine and Saxby, 1969). Separations of the acetates were obtained using three columns: polyphenyl ether OS124 (PPE), trixylenyl phosphate (TXP) and diethylene glycol succinate (DEGS); all programmed at 130–60°C or 150–80°C at 1·5° min^{-1} (Table 2.2). Complex mixtures of phenols also occur in willow barks (Salicaceae) and GLC on columns of OV-1, OV-17 and OV-25 (2–4%) has been used for such separations (Steele and Bolan, 1972).

GLC is a technique to use when its greater sensitivity, as compared to TLC, is important. It has been employed, for example, by Robertson *et al.* (1968) to detect the minute amounts of phenolic acids formed in tissue cultures of potato infected with *Phytophthora infestans*. A detailed guide to the GLC of phenols has been published by Preston and Pankratz (1978).

(c) *Other procedures*

Reverse-phase HPLC, using columns of μBondapak C_{18} and Spherisorb C_{18} and the solvents water–methanol–acetic acid (12:6:1) and water–acetic acid–*n*-butanol (342:1:14) respectively, has successfully resolved mixtures of phenolic acids and phenolic aldehydes (*cf.* Harborne, 1983). HPLC is recommended for the separation of acylphloroglucinols, which occur in *Dryopteris* ferns and which are rather labile when chromatographed by TLC, and one system that works is a μBondapak C_{18} column with the solvent mixture tetrahydrofuran–H_3PO_4–water (650:1:350) (Widen *et al.*, 1980).

Direct determination of simple *o*- or *p*-dihydroxybenzene derivatives in crude plant extracts is possible by transforming them to the semiquinone state by aerial oxidation in alkaline solution and then identifying them by their electron-spin resonance spectra (Pedersen, 1978). The method is semi-quantitative and quite sensitive: it can be used to detect hydroquinone and catechol as well as dihydroxycinnamic acids such as caffeic acid (see next section).

2.2.4 Practical experiment

(a) *Detection of phenolic acids and lignins in plant tissues*

Lignins are phenolic polymers present in the cell walls of plants, which are responsible, with cellulose, for the stiffness and rigidity of plant stems. Lignins are especially associated with woody plants, since up to 30% of the organic matter of trees consists of lignin. On oxidation with nitrobenzene, lignins yield three simple phenolic aldehydes related to the common phenolic acids of plants. These are *p*-hydroxybenzaldehyde, vanillaldehyde and syringaldehyde. There is, however, a significant difference between angiosperm and gymnosperm lignin in that, with very few exceptions, the lignins of the latter group of plants specifically lack the syringyl groups which are present in angiosperm lignins. This difference in the structural components in lignins is reflected in a difference in the phenolic acids produced on acid treatment of the leaves of gymnosperms or angiosperms. The former contain only *p*-hydroxybenzoic and vanillic acids, whereas the latter usually have syringic acid in addition. It is a simpler procedure to detect the presence/absence of syringic acid in leaf hydrolysates than analyse the respective lignins, so this provides a way of demonstrating the differences in lignin composition.

The close association between the lignins of trees and the occurrence of phenolic acids in the leaves of the same trees is also reflected in the fact that, among other plant groups, the absence of lignin is correlated with the absence of phenolic acids. Ferns contain lignin and hence *p*-hydroxybenzoic and vanillic acids. On the other hand, all bacteria, fungi and algae and most mosses lack lignification in their tissues and also fail to yield phenolic acids on acid treatment. The test for phenolic acids thus provides an alternative means of testing for lignin.

(b) *Procedure*

(1) Collect fresh leaf samples (approximately 20 g) of two angiosperms and two gymnosperms. Also gather a fern (e.g. bracken), a moss and the fruiting body of a fungus.
(2) In each case, immerse the sliced tissue in 2 M HCl in a test tube and heat for 30 min in a boiling water bath.
(3) Cool, extract into ether. Pipette off the ether extract and concentrate it to dryness. Dissolve each residue in 1–2 drops 95% ethanol.
(4) Chromatograph each hydrolysed extract on a silica gel plate in two dimensions, using acetic acid–chloroform (1:9) in the first direction and ethyl acetate–benzene (9:11) in the second. Chromatograph marker solutions of *p*-hydroxybenzoic, vanillic and syringic acids along the side of the plate, in each direction.
(5) Spray the plate with Gibbs reagent and then spray with dilute ammonia. The acids appear as blue or mauve spots on a white background.
(6) Record the presence of syringic acid in the angiosperm tissues only and note the absence of acids from lower plant tissues.

(c) *Additional experiments*

(1) There is a simple colour test, known as the Maüle test, for detecting the presence of syringyl groups in angiosperm woods. Wood shavings are treated successively at room temperature with solutions of 1% aqueous $KMnO_4$, 2 M HCl and 2 M NH_4OH. If syringyl groups are present, a

rose-red colour develops. When gymnosperm woods are similarly treated, a brown colour develops.

(2) The extraction and degradation of lignins from plants is a more complicated procedure, but has been adapted so that it can be carried out on relatively small plant samples. Extraction can be carried out, for example, with boiling 95% ethanol containing 3% HCl. The purified lignin is then oxidized with nitrobenzene and the aldehydes formed (*p*-hydroxybenzaldehyde, vanillaldehyde and syringaldehyde) can be separated and detected by PC, TLC or HPLC. One TLC system on silica gel is benzene–HOAc (9:1) with detection as the 2,4-dinitrophenylhydrazones. Other solvents for extracting lignins from woods are dioxan–H_2O (9:1) or thioglycolic acid (for references, see Swain, 1979).

2.3   PHENYLPROPANOIDS

2.3.1   Chemistry and distribution

Phenylpropanoids are naturally occurring phenolic compounds which have an aromatic ring to which a three-carbon side-chain is attached. They are derived biosynthetically from the aromatic protein amino acid phenylalanine and they may contain one or more C_6–C_3 residues. The most widespread are the hydroxycinnamic acids, which are important not only as providing the building blocks of lignin but also in relation to growth regulation and to disease resistance. Among the phenylpropanoids are included hydroxycoumarins, phenylpropenes and lignans. Structures of some typical plant phenylpropanoids are illustrated in Fig. 2.4.

Four hydroxycinnamic acids are common, in fact almost ubiquitous, in plants: ferulic, sinapic, caffeic and *p*-coumaric acids. Their separation is simple and they are readily detected on paper chromatograms because they fluoresce (different shades of blue and green), in UV light. At least six other cinnamic acids are known, but these are relatively rare in their occurrence; examples are isoferulic (3-hydroxy-4-methoxycinnamic), *o*-coumaric and *p*-methoxycinnamic acids. Hydroxycinnamic acids usually occur in plants in combined form as esters; and they are obtained in best yield by mild alkaline hydrolysis, since with hot acid hydrolysis material is lost due to the fact that they undergo decarboxylation to the corresponding hydroxystyrenes. Caffeic acid occurs regularly as the quinic acid ester, chlorogenic acid, but isomers are known (e.g. isochlorogenic acid) and derivatives with sugars (e.g. caffeoylglucose) and with organic acids (e.g. rosmarinic acid, caffeoyltartaric acid) have also been described. The different combined forms of caffeic acid are of particular interest from the chemotaxonomic point of view (Harborne, 1966).

The most widespread plant coumarin is the parent compound, coumarin itself, which occurs in over twenty-seven plant families. It is common in many

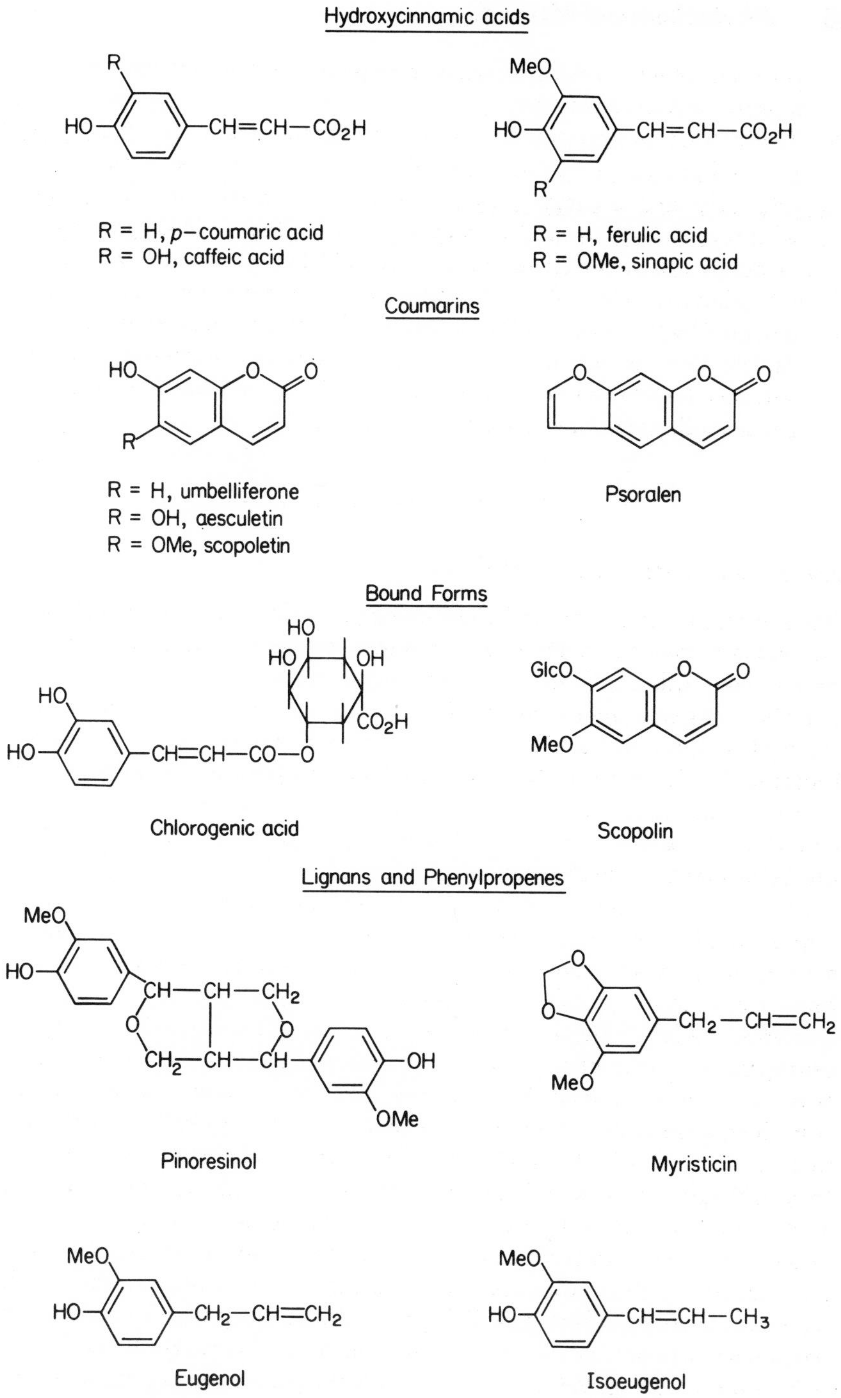

Fig. 2.4 Structures of phenylpropanoids.

grasses and fodder crops and is familiar as the sweet-smelling volatile material released from new mown hay. Hydroxycoumarins are also found in many different plant families; the common ones are based on umbelliferone (7-hydroxy), aesculetin (6,7-dihydroxy) or scopoletin (6-methoxy-7-hydroxy-coumarin). Rarer hydroxycoumarins are daphnetin (7,8-dihydroxy) from *Daphne* and fraxetin (6-methoxy-7,8-dihydroxy) from the ash tree, *Fraxinus*. More complex coumarins occur in plants, for example, the furanocoumarins typified by psoralen (see Fig. 2.4), but these are generally restricted to a few families, such as the Rutaceae and the Umbelliferae (Murray *et al.*, 1982). Lignans, dimeric C_6–C_3 compounds such as pinoresinol, which is illustrated in Fig. 2.4, are mainly found in heartwoods. Some two hundred structures have been reported. Their identification is only briefly touched upon here.

One other group of phenylpropanoids must be mentioned – the phenyl-propenes – because of their important contribution to the volatile flavours and odours of plants. The phenylpropenes are usually isolated in the 'essential oil' fraction of plant tissues, together with the volatile terpenes. They are lipid-soluble, as distinct from most other phenolic compounds. Some structures are widespread, such as eugenol, the major principle of oil of cloves. Others are restricted to a few families. Anethole occurs in anise and fennel (both Umbelliferae) and myristicin, first described as a principle of nutmeg, *Myristica fragrans*, Myristicaceae, is also found in a number of umbellifers. Pairs of allyl and propenyl isomers (e.g. eugenol and isoeugenol, see Fig. 2.4) are known, sometimes occurring together in the same plant. Isomerization of the allyl to the propenyl form can be achieved in the laboratory, but only under drastic conditions (e.g. with strong alkali). Such isomerization is very unlikely to occur during normal conditions of isolation (i.e. extraction with ether, etc.)

2.3.2 Recommended techniques

(a) *Hydroxycinnamic acids and hydroxycoumarins*

These are usually detected together, as with the simple 'phenolics', after acid or alkaline hydrolysis of a plant extract. They can be extracted into ether or ethyl acetate and the extract is then washed, dried and taken to dryness. Chromatography of the residue either one-dimensionally on paper or two-dimensionally on plates of micro-crystalline cellulose is equally satisfactory for detection. A typical TLC separation of cinnamic acids is illustrated in Fig. 2.5. Paper chromatographic data are given in Table 2.3 for the cinnamic acids and Table 2.4 for the coumarins.

These compounds are very easily detected, since they give characteristic fluorescent colours in UV light, which are intensified by further treatment with ammonia vapour. An advantage of PC is that these colour changes are more easily observed than on TLC plates. Cinnamic acids can be dif-

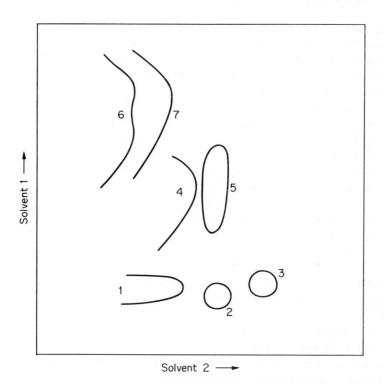

Fig. 2.5 Thin layer chromatography separation on microcrystalline cellulose of cinnamic acids. Solvent 1 = benzene-acetic acid–water (6:7:3). Solvent 2 = 15% acetic acid in water. Key: 1, caffeic acid; 2, gallic acid; 3, cinnamic acid; 4, *p*-coumaric acid; 5, *o*-coumaric acid; 6, sinapic acid; 7, ferulic acid. Compounds 1, 4, 5, 6 and 7 can be detected as different shades of blue under UV light and fuming with ammonia. Compounds 2 and 3 are dark absorbing spots in UV light. Compound 2 gives a blue colour with Folin reagent on fuming with ammonia.

Table 2.3 R_F colour and spectral data for Hydroxycinnamic acids

Cinnamic acid	R_F (×100) in*				Colour		EtOH λ_{max}	EtOH– NaOH λ_{max}
	BAW	BN	BEW	Water	UV	UV+*ammonia*		
p-Coumaric	92	16	88	42,85	none	mauve	227,310	335
Caffeic	79	04	79	26,69	blue	light blue	243,326	decomposition
Ferulic	88	12	82	33,75	blue	bright blue	235,324	344
Sinapic	84	04	88	62	blue	blue-green	239,325	350
o-Coumaric	93	21	85	82	yellow	yellow-green	227,275,325	390
p-Methoxycinnamic	95	17	87	23	dark absorbing		274,310	298
Isoferulic	89	12	67	37	mauve	yellow	295,323	345
3,4,5-Trimethoxycinnamic	95	18	87	75	mauve	dark	232,303	293

*Key: BAW = *n*-BuOH–HOAc–H$_2$O (4:1:5, top layer); BN = *n*-BuOH–2 MNH$_4$OH (1:1, top layer). BEW = *n*-BuOH–ethanol–water (4:1:2·2).

Table 2.4 Colour and spectral data of hydroxycoumarins

Coumarin	R_F ($\times 100$) in			Fluorescence in UV *light*	EtOH λ_{max}
	BAW	Water	10% HOAc		
Aglycones					
Coumarin*	92	67	76	none	212, 274, 282,† 312
Umbelliferone	89	57	60	bright blue	210, 240,† 325
Aesculetin	79	28	45	blue	230, 260, 303, 351
Scopoletin	83	29	51	blue-violet	229, 253, 300, 346
Daphnetin	81	61	54	pale yellow	215, 263, 328
Glycosides	BAW	Water	BN		
Aesculin	53	56	13	clear blue	224, 252, 293, 338
Cichoriin	53	61	10	pale pink	228, 255, 290, 345
Scopolin	53	64	44	mauve	227, 250, 288, 339

For key to solvents, see Table 2.3.
*Detected by spraying paper with 5% aqueous NaOH; intense yellow-green fluorescence develops when dried paper is placed under UV light for 5–10 min.
†Inflection.

ferentiated from coumarins by their less intense fluorescence and by the fact that they invariably give two spots when chromatographed in aqueous solvents, due to the separation of the *cis*- and *trans*-isomers. Isomerization of cinnamic acids (and their derivatives) occurs during isolation on exposure of extracts to light (and particularly to UV light) so that even if the natural material in the plant is all *trans*, an equilibrium mixture of *cis*- and *trans*- is normally obtained.

Identification can be confirmed by spectral measurements (Tables 2.3 and 2.4). For example, caffeic acid and its derivatives have characteristic absorption bands at 243 and 326 nm, with a distinctive shoulder at 300 nm to the long wave band (Fig. 2.6). Hydroxycoumarins absorb at longer wavelengths than cinnamic acids; aesculetin, the coumarin related to caffeic acid, has absorption bands at 230, 260, 303 and 351 nm. Measurements of the magnitudes of bathochromic shifts in spectra in the presence of alkali are also useful for distinguishing the different cinnamic acids and coumarins (see Fig. 2.6).

Separation of bound forms of cinnamic acids can be achieved using a combination of PC and TLC on silica gel. Such separations are illustrated here by reference to the bound forms of caffeic acid (Table 2.5). Separation of the glycosides of hydroxycoumarins can be carried out using similar procedures as for the free coumarins (Table 2.4).

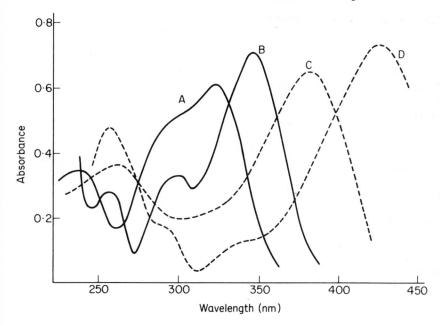

Fig. 2.6 Ultraviolet spectra of two phenylpropanoids. Key: A, Caffeoylquinic acid (chlorogenic acid) in 95% EtOH; B, aesculetin in 95% EtOH; C, caffeoylquinic acid in EtOH–NaOH; D, aesculetin in EtOH–NaOH.

Table 2.5 Separation of caffeic acid derivatives

		R_F (×100) *in*			
Caffeic derivative	*Structure*	BAW	BEW	Water	TLC*
Chlorogenic acid	caffeoylquinic acid	63	44	67,84	16
Isochlorogenic acid	dicaffeoylquinic acids	79	67	10	30
Rosmarinic acid	ester of caffeic and 3,4-dihydroxy-phenyllactic acid	83	62	27	65
l-Caffeoylglucose		50	63	61,72	14

Solvent key: BAW = *n*-BuOH–HOAc–H$_2$O (4:1:5); BEW = *n*-BuOH–EtOH–H$_2$O (4:1:2·2).
Colours in UV light, bright blue; UV+NH$_3$, green.
*TLC on silica gel in toluene–ethyl formate–formic acid (2:1:1).

(b) *Furanocoumarins*

By contrast with the simple hydroxycoumarins, furanocoumarins are generally lipid-soluble and can be isolated during extraction of dried plant material with ether or light petroleum. They occasionally occur in bound form as glycosides and have then to be released by prior acid hydrolysis. TLC on silica gel is most commonly used for their separation. Suitable solvents include pure chloroform, chloroform containing 1·5% ethanol, ether–benzene (1:1), and ether–benzene–10% acetic acid (1:1:1); times of development vary between 1 and 2 h. Typical separations (R_Fs relative to bergapten) that can be obtained for five common furanocoumarins in ether–benzene (1:1) and chloroform, respectively, are as follows: bergapten (100, 100), isobergapten (112, 167), pimpinellin (108, 86), isopimpinellin (97, 43) and sphondin (92, 90).

Furanocoumarins are detected by their blue, violet, brown, green or yellow colours in UV light. The colour may be intensified by spraying plates with 10% KOH in methanol, or 20% antimony chloride in chloroform. Furano-coumarins can be further identified by their UV absorption; unlike hydroxy-coumarins, they do not, as a rule, exhibit bathochromic spectral shifts in alkaline solution. Other techniques can be applied in furanocoumarin separation. For example, Reyes and Gonzalez (1970), in isolating twelve coumarins from the roots of *Ruta pinnata*, used PC in water and GLC on QF-1 at 174°C for distinguishing the different compounds. Procedures for surveying plant material for furanocoumarins are given by Crowden *et al.* (1969).

(c) *Phenylpropenes*

These compounds are detected, together with essential oils, in ether extracts of plant tissues. Particularly rich sources are fruits of the Umbelliferae and other plants used as spices or for flavouring. They are easily separated on silica gel plates in benzene, mixtures of benzene with chloroform (10%) or light petroleum (20%) or in *n*-hexane–chloroform (3:2). They give coloured spots

Table 2.6 R_F values and colours of phenylpropenes

	R_F ($\times 100$) *in*		*Colour with*	
*Phenylpropene**	*Benzene*	*n-Hexane–*CHCl$_3$(3:2)	*vanillin–*H$_2$SO$_4$	*Gibbs reagent*
Safrole	74	86	—	grey
Estragole	72	—	pink	—
Anethole	69	—	rose	—
Myristicin	50	71	brown	brown
Apiole	39	41	brown	brown
Thymol	38	—	rose	—
Eugenol	20	31	brown	brown
Isoeugenol	29	27	red	yellow
Methyleugenol	—	42	—	red-brown
Methylisoeugenol	—	42	—	purple
Elemicin	—	27	—	yellow

*These compounds can be further identified by the spectral measurements; myristicin, for example, has λ_{max} 278 and 285, whereas apiole has a single broad band at 280 nm.

when sprayed with vanillin–1 M H_2SO_4 (2 g vanillin, 1 g H_2SO_4 diluted to 100 ml with 96% ethanol) or with Gibbs reagent (Forrest *et al.*, 1972). Typical R_Fs and colours for most of the common phenylpropenes are given in Table 2.6. It will be noted that pairs of isomeric compounds (e.g. anethole and estragole, eugenol and isoeugenol, etc.) have similar R_F values but the shift in the position of the double bond in the side-chain considerably alters the colour reactions. For example, with vanillin–1 M H_2SO_4, the compounds which have the double bond adjacent to the benzene ring give reddish colours, whereas the isomers in which the double bond is farthest from the benzene ring (see Fig. 2.4) give brownish colours. Procedures for surveying plant tissues for myristicin and other phenylpropanoids in fruits of the Umbelliferae are described by Harborne *et al.* (1969).

(d) *Lignans*

Simple mixtures of these substances may be separated by PC in butanol–acetic acid–water and 15% acetic acid. More complex mixtures can be separated using butanol-acetic acid-aqueous molybdic acid on chromatography paper previously impregnated with dilute molybdic acid (Pridham, 1959). Complex mixtures may also be resolved by TLC on silica gel using such solvents as ethyl acetate–methanol (19:1) or benzene–ethanol (9:1).

Unfortunately, there does not seem to be a specific spray reagent, which distinguishes lignans from other simple phenols. Lignans can be seen as dark absorbing spots on paper in short UV light or can be revealed by spraying with 10% antimony chloride in chloroform. On TLC plates, they are detected by spraying with conc. H_2SO_4. They can be further identified by spectral means; they show absorption at 280–284 nm, this band being shifted to about 298 nm in the presence of alkali. A survey procedure for detecting the lignan glycoside arctiin in fruits of certain members of the family Compositae is described by Hänsel *et al.* (1964).

(e) *Authentic markers*

Most of the common phenylpropanoids are available commercially; e.g. the hydroxycinnamic acids, coumarin, umbelliferone, aesculin and chlorogenic acid. Others can be obtained from readily available plant materials, e.g. myristicin from nutmeg, apiole from parsley seed.

2.3.3 Alternative techniques

GLC is occasionally used for detecting phenylpropanoids, because of its greater sensitivity and value in quantitative analyses. The detection of the coumarin scopolin by GLC using electron capture has been described by

Andersen and Vaughn (1972) and the same authors (Vaughn and Andersen, 1971) have used this technique for measuring cinnamic acid, formed as a product of the enzymic action of phenylalanine ammonia lyase on phenyl-alanine. The method is ten times more sensitive than spectral measurements. Phenylpropenes can be separated by GLC on columns of 20% Reoplex 400 on Gas Chrom Q at 80–200°C/4°C min (Wagner and Holzl, 1968).

HPLC, with its high resolving power, is useful when complex mixtures of phenolics are encountered. Retention times of many phenylpropanoids on a Lichrosorb RP-18 column eluted with varying proportions of water–formic acid (19:1) and methanol have been recorded by van der Casteele *et al.* (1983). The application of HPLC to the determination of the many phenolics present in maize extracts is outlined by Andersen and Pedersen (1983).

2.3.4   Practical experiment

(a) *Detection of scopolin in blight-infected potato tuber*

One of the responses of plant tissue to infection by micro-organisms is a large increase in the synthesis of particular phenolic compounds. It is presumably a protective response to invasion, although it does not necessarily prevent the organism establishing itself in the tissue. As an example of this, the coumarin scopolin is formed in high concentrations in infected plants of the Solanaceae and is particularly easily observed in blighted potato tubers (Hughes and Swain, 1960).

Scopolin synthesis can be observed in partly blighted potatoes collected in the autumn, soon after harvest. On cutting such tubers in half, an intense blue UV fluorescent zone can be seen in the healthy tissue adjacent to the infected zone. Scopolin synthesis can be induced artificially by inoculating tubers of the variety Majestic with *Phytophthora infestans*, race 4, and incubating the tubers for 14 days at 18°C. The blue fluorescence is due to the local increase by 10–20-fold in scopolin synthesis. It is a simple procedure to identify scopolin in these tissues and measure its concentration.

(b) *Procedure*

(1) Macerate separately equal amounts of healthy tuber tissue and of tissue with the blue fluorescence in 95% ethanol and heat the macerates for 10 min on a boiling water bath. Filter, concentrate and refilter.

(2) Chromatograph aliquots of the control and diseased extracts two-dimensionally on paper in BAW and water. Note scopolin (R_F 53, 64) as a deep mauve spot in UV light, unaffected in colour by fuming with ammonia; the other major fluorescent material is chlorogenic acid (R_F 63, 67 and 84, blue changing to green after fuming with ammonia).

(3) Streak the extract of diseased tissue on Whatman 3MM paper and develop in BAW. Cut out the scopolin band, elute with 70% ethanol. Concentrate the eluate and then hydrolyse for 0·5 h in 2 M HCl. Cool, extract into ether and chromatograph the aglycone scopoletin against coumarin markers (Table 2.4). It should be possible to identify scopoletin by this means.

(c) *Additional experiments*

(1) The above experiment can be repeated on a quantitative basis and the amounts of scopolin in healthy and infected tissue measured. This can be done by cutting out the spots from two-dimensional chromatograms run on thick 3MM paper and measuring the absorption at the scopolin maximum (339 nm). Some interference by chlorogenic acid (λ_{max} 324 nm) is possible. This can be avoided by using a spectrofluorimeter (if available), since scopoletin fluoresces 30 000 times as strongly as chlorogenic acid at pH 5·8 (tris-maleate buffer).

(2) Similar comparisons can be made of 'phenolic' synthesis in leaves of healthy and infected tobacco or tomato plants; other diseases beside blight also have an effect on scopolin production. Environmental stress and mineral shortage (e.g. boron deficiency) may also cause similar changes in 'phenolic' metabolism.

2.4 FLAVONOID PIGMENTS

2.4.1 Chemistry and distribution

The flavonoids are all structurally derived from the parent substance flavone, which occurs as a white mealy farina on *Primula* plants, and all share a number of properties in common. Some ten classes of flavonoid are recognized (Table 2.7) and these different classes will be considered in more detail in subsequent sections. In this section, general procedures of identification are discussed and methods for distinguishing the different classes are outlined.

Flavonoids are mainly water-soluble compounds. They can be extracted with 70% ethanol and remain in the aqueous layer, following partition of this extract with petroleum ether. Flavonoids are phenolic and hence change in colour when treated with base or with ammonia; thus they are easily detected on chromatograms or in solution (Table 2.8). Flavonoids contain conjugated aromatic systems and thus show intense absorption bands in the UV and visible regions of the spectrum (Table 2.9). Finally, flavonoids are generally present in plants bound to sugar as glycosides and any one flavonoid aglycone may occur in a single plant in several glycosidic combinations. For this reason, when analysing flavonoids, it is usually better to examine the aglycones

Table 2.7 Properties of the different flavonoid classes

Flavonoid class	Distribution	Characteristic properties
Anthocyanins	scarlet, red, mauve and blue flower pigments; also in leaf and other tissues	water-soluble, visible max. 515–545 nm, mobile in BAW on paper*
Proanthocyanidins	mainly colourless, in heartwoods and in leaves of woody plants	yield anthocyanidins (colour extractable into amyl alcohol) when tissue is heated for 0·5 h in 2 M HCl
Flavonols	mainly colourless co-pigments in both cyanic and acyanic flowers; widespread in leaves	after acid hydrolysis, bright yellow spots in UV light on Forestal chromatograms; spectral max. 350–386 nm
Flavones	as flavonols	after acid hydrolysis, dull absorbing brown spots on Forestal chromatograms; spectral max. 330–350 nm
Glycoflavones	as flavonols	contain C–C linked sugar; mobile in water unlike normal flavones
Biflavonyls	colourless; almost entirely confined to the gymnosperms	on BAW chromatograms dull absorbing spots of very high R_F†
Chalcones and aurones	yellow flower pigments; occasionally present in other tissues	give red colours with ammonia (colour change can be observed *in situ*), visible max. 370–410 nm
Flavanones	colourless; in leaf and fruit (especially in *Citrus*)	give intense red colours with Mg/HCl; occasionally an intense bitter taste
Isoflavones	colourless; often in root; only common in one family, the Leguminosae	mobile on paper in water; no specific colour tests available

*Betalains replace anthocyanins as purple pigments in one order of plants, the Centrospermae; they can be distinguished by the very low mobility in BAW and spectral range is 530–554 nm (see also p. 68).
†Some highly methylated flavones behave similarly.

present in hydrolysed plant extracts before considering the complexity of glycosides that may be present in the original extract.

Flavonoids are present in all vascular plants but some classes are more widely distributed than others; while flavones and flavonols are universal, isoflavones and biflavonyls are found in only a few plant families (Table 2.7). Detailed references to the natural distribution of the flavonoids may be found in Harborne (1967), Harborne *et al.* (1975) and Harborne and Mabry (1982). The identification of flavonoids is the subject of a book by Mabry *et al.* (1970), but a most valuable guide for the beginner is a recent volume in the Biological Techniques series by Markham (1982).

Table 2.8 Colour properties of flavonoids in visible and ultraviolet light

Visible colour	Colour in UV light alone	Colour in UV light with ammonia	Indication
Orange Red Mauve	dull orange, red or mauve	blue	anthocyanidin 3-glycosides
	fluorescent yellow cerise or pink	blue	most anthocyanidin 3,5-diglycosides
Bright yellow	dark brown or black	dark brown or black	6-hydroxylated flavonols and flavones; some chalcone glycosides
		dark red or bright orange	most chalcones
	bright yellow or yellow-green	bright orange or red	aurones
Very pale yellow	dark brown	bright yellow or yellow brown	most flavonol glycosides
		vivid yellow-green	most flavone glycosides
		dark brown	biflavonyls and unusually substituted flavones
None	dark mauve	faint brown	most isoflavones and flavanonols
	faint blue	intense blue	5-desoxyisoflavones and 7,8-dihydroxyflavanones
	dark mauve	pale yellow or yellow-green	flavanones and flavanonol 7-glycosides

General chromatographic sprays: 5% alc. $AlCl_3$–response: yellow-green fluorescence in UV; aqueous $FeCl_3$–$K_3Fe(CN)_6$ (1:1)–response: blue on yellow background; Folin reagent–response (+NH_3): blue on white background.

Table 2.9 Spectral characteristics of main flavonoid classes

Principal maxima (nm)	Subsidiary maxima (nm) (with relative intensities)	Indication
475–560	ca. 275 (55%)	anthocyanins
390–430	240–270 (32%)	aurones
365–390	240–260 (30%)	chalcones
350–390 ⎱ 250–270 ⎰	ca. 300 (40%)	flavonols
330–350 ⎱ 250–270 ⎰	absent	flavones and biflavonyls
275–290 ⎱ ca. 225 ⎰	310–330 (30%)	flavanones and flavanonols
255–265	310–330 (25%)	isoflavones

2.4.2 Preliminary classification

Flavonoids are present in plants as mixtures and it is very rare to find only a single flavonoid component in a plant tissue. In addition, there are often mixtures of different flavonoid classes. The coloured anthocyanins in flower petals are almost invariably accompanied by colourless flavones or flavonols and recent research has established that the flavones are important co-pigments, being essential for the full expression of anthocyanin colour in floral tissues. Mixtures of anthocyanins are also the rule, particularly in the flowers of ornamental plants, and any one flower tissue may contain up to ten different pigments.

Classification of flavonoid type in a plant tissue is based initially on a study of solubility properties and colour reactions. This is followed by a one-dimensional chromatographic examination of a hydrolysed plant extract and two-dimensional chromatography of a direct alcoholic extract. Finally, the flavonoids can be separated by chromatographic procedures and the individual components identified by chromatographic and spectral comparison with known markers. Novel compounds discovered during surveys require more detailed chemical and spectral examination (Harborne, 1967a; Mabry *et al.*, 1970; Markham, 1982).

2.4.3 Practical experiment

(a) *Examination of flavonoid aglycones in hydrolysed plant extracts*

A general procedure for surveying plant tissue for flavonoids has been evolved in the author's laboratory and the present system is a modification of that developed by Bate-Smith (1962) who has surveyed over 1000 angiosperm species for their flavonoids.

(b) *Procedure*

(1) A small amount of plant tissue (usually leaf or flower) is immersed in 2 M HCl and heated in a test tube for 30–40 min at 100°C.
(2) The cooled extract is then filtered if necessary and extracted with ethyl acetate.
(3) If the solution is coloured (either because the original tissue was coloured with anthocyanin or because colour has formed from anthocyanidin during acid treatment), then the aqueous extract is further heated to remove the last traces of ethyl acetate and re-extracted with a small volume of amyl alcohol.
(4) The ethyl acetate is concentrated to dryness, taken up in 1–2 drops ethanol and aliquots chromatographed one-dimensionally, alongside

authentic markers, in five solvents: Forestal (acetic acid–conc. HCl–water; 30:3:10), 50% HOAc (50% aqueous acetic acid), BAW (*n*-butanol–acetic acid–water, 4:1:5, top layer), PhOH (phenol saturated with water) and water.

(5) The amyl alcohol extract, which should be coloured, is concentrated to dryness, taken up in a few drops of 1% methanolic HCl and aliquots chromatographed in Forestal and in Formic (formic acid–conc. HCl–water, 5:2:3).

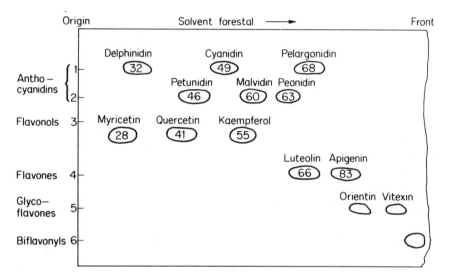

Fig. 2.7 Forestal chromatogram of the common flavonoid aglycones of plants. Colour Key: 1, mauve, red and orange respectively in visible light; 2, mauve, mauve and red in visible light; 3, bright yellow in UV light; 4, 5, dull brown in UV light changing to bright yellow-green with NH$_4$OH; 6, dull brown unchanged by NH$_4$OH.

A composite Forestal chromatogram, illustrating the chromatographic behaviour of most types of flavonoid aglycone, is illustrated in Fig. 2.7. The results of the Forestal chromatogram are the most important, data from the other chromatographic runs being used to confirm or supplement these results. For example, the results in 50% HOAc are similar to those in Forestal, except that there is a clearer separation of the aglycones from traces of unhydrolysed glycoside which may still be present in the 'hydrolysed' extract. On the BAW chromatogram also, there is clear separation of unhydrolysed *O*-glycoside and flavone *C*-glycosides (medium to low R_F) from the aglycones (high R_F). The phenol chromatogram is useful, since methylated flavones have very high R_F values and can thus be distinguished. Finally, the water chromatogram is useful for distinguishing glycoflavones, isoflavones and flavanones, which are mobile, from other classes of flavonoid which remain at the origin. On the basis of these results and from preliminary examination (Table 2.2), it is usually possible to determine the presence/absence of the common flavonoids such as the flavonols quercetin and kaempferol in plant tissues. A chromatographic study of the direct alcoholic extract is also useful, as indicated below.

2.4.4   Examination of flavonoids in alcoholic plant extracts

A popular procedure for routinely screening plant tissues for their flavonoid patterns is by two-dimensional PC of concentrated alcoholic extracts using the solvents BAW and 5% acetic acid. A standard marker for use on such chromatograms is the flavonol glycoside rutin; it is useful since it occupies a position approximately in the middle of the chromatogram and also is, itself, very common in plants and thus one of the most likely compounds to be found during survey work.

Many other variations of the above standard procedure have been suggested, but in our experience this remains the most generally applicable. Almost exactly similar results can be obtained by replacing the paper support by thin layers of cellulose MN 300, with the advantage that separations can be achieved in a shorter time. For extraction, powdered dried plant tissue (e.g. from herbarium sheets if necessary) can be extracted with small volumes of 70% ethanol at room temperature for 8–24 h and this extract can usually be applied directly to the chromatogram. Fresh tissue is extracted for 5–10 min with boiling 95% ethanol and the extract subsequently concentrated on a watchglass in an air draught. During evaporation, chlorophyll and other impurities deposit on the watchglass and the aqueous concentrate can be collected in a capillary and applied directly to the chromatogram.

A typical separation of flavonoids on a two-dimensional paper chromatogram is illustrated in Fig. 2.8. Some spots can be seen in visible light, but the majority only appear during examination in UV light, with the additional fuming of the chromatogram with ammonia when reversible colour changes can be observed. Many of the colours are quite characteristic for particular flavonoids; for example, it is possible with practice to distinguish between the 3-glycosides of kaempferol, quercetin and myricetin by subtle differences in their colour responses. The relative positions the different spots take up on the chromatogram give a good indication (see legend to Fig. 2.8) of the nature of the flavonoid present.

For comparative purposes, two-dimensional chromatograms of a range of plant samples must be developed under identical conditions, using freshly prepared BAW solvent for an overnight run and maintaining some control of temperature in the cabinet or room used for chromatography.

The flavonoid data obtainable from two-dimensional chromatograms have

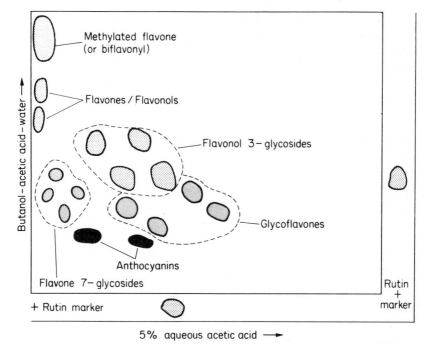

Fig. 2.8 Typical two-dimensional chromatogram of flavonoids in a plant extract. The above chromatogram indicates approximate position common flavonoids take up on a chromatogram. During extraction, some hydrolysis of glycosides may occur, so that small amounts of flavone and flavonol aglycones are usually found. Colour key: ☐ Dull brown, bright yellows and yellow-green. ☐ Dull brown, intense yellow or green. ■ Visible colour; for other colours, see key to Fig. 2.7.

been used directly by some taxonomists for comparing the chemistry of different species within genera or for solving some particular problem of classification. However, it is generally advisable to carry out more detailed flavonoid analysis, as indicated in the following three sections, before applying the data to systematic problems.

2.5 ANTHOCYANINS

2.5.1 Chemistry and distribution

The anthocyanins are the most important and widespread group of colouring matters in plants. These intensely coloured water-soluble pigments are responsible for nearly all the pink, scarlet, red, mauve, violet and blue colours in the petals, leaves and fruits of higher plants. The anthocyanins are all based chemically on a single aromatic structure, that of cyanidin, and all are derived from this pigment by addition or subtraction of hydroxyl groups or by methyl-ation or by glycosylation (see Fig. 2.9).

There are six common anthocyanidins (anthocyanin aglycones formed when anthocyanins are hydrolysed with acid), the magenta-coloured cyanidin being by far the most common. Orange-red colours are due to pelargonidin with one *less* hydroxyl group than cyanidin, while mauve, purple and blue

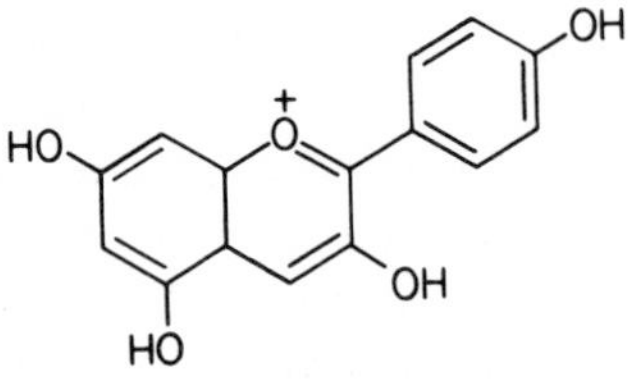

Pelargonidin

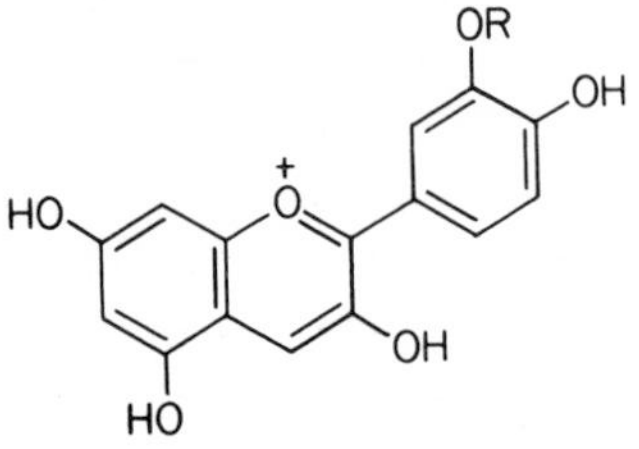

Cyanidin, R = H
Peonidin, R = Me

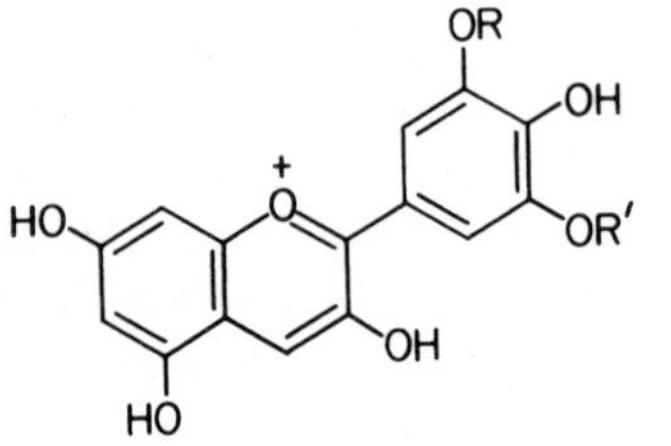

Delphinidin, R = R' = H
Petunidin, R = Me, R' = H
Malvidin, R = R' = Me

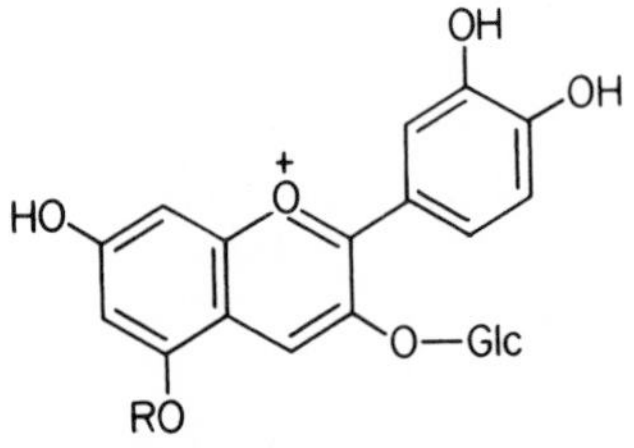

Cyanidin 3-glucoside, R = H
Cyanidin 3,5-diglucoside,
R = glucose

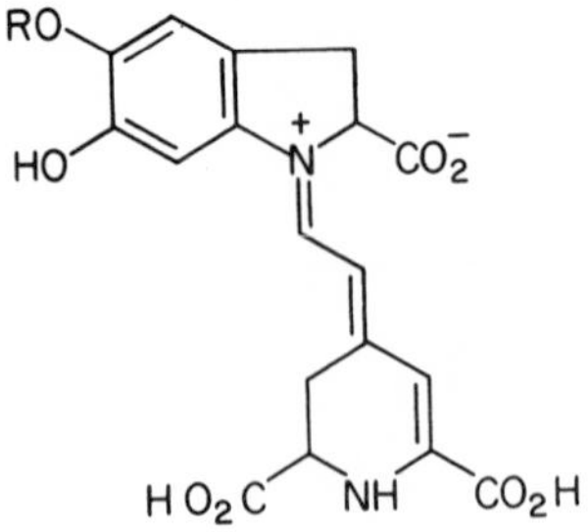

Betanidin, R = H
Betanin, R = glucose

Fig. 2.9 Structures of anthocyanins and betacyanins.

colours are generally due to delphinidin, which has one *more* hydroxyl group than cyanidin. Three anthocyanidin methyl ethers are also quite common: peonidin derived from cyanidin; and petunidin and malvidin, based on delphinidin. Each of these six anthocyanidins occurs with various sugars attached as a range of glycosides (i.e. as anthocyanins). The main variation is in the nature of the sugar (often glucose, but may also be galactose, rhamnose, xylose or arabinose), the number of sugar units (mono, di- or tri-glycosides) and the position of attachment of sugar (usually to the 3-hydroxyl, or to the 3- and 5-hydroxyls, see Fig. 2.9).

Delphinidin, cyanidin and, to a much less extent, pelargonidin may also be formed in acid-hydrolysed plant tissue from colourless polymeric tannins present, from compounds originally known as leucoanthocyanidins and now as proanthocyanidins or flavolans. Production of anthocyanidin in this way constitutes the main method for detecting these colourless substances in plants. They are regularly present in wood and leaf of woody plants, but very often appear in the flowers of these plants. In such cases, if the tissue is already coloured, it is not always clear in preliminary analysis whether the anthocyanidin detected is derived from anthocyanidins or from anthocyanin pigment already present.

While anthocyanins are nearly universal in vascular plants (they have been detected in a few mosses, in young fern fronds as well as in gymnosperms and angiosperms), they are replaced by a superficially similar group of pigments, the betacyanins, in one order of higher plants, the Centrospermae. The best known betacyanin is betanin (aglycone: betanidin), the major pigment of beetroot, *Beta vulgaris*. Betacyanins are chemically distinct from anthocyanins (see Fig. 2.9), being derived biosynthetically from the aromatic amino acid, 3,4-dihydroxyphenylalanine. The restriction of betacyanins to the Centrospermae (except one family, the Caryophyllaceae) is of considerable taxonomic interest and methods for distinguishing between anthocyanins and betacyanins are discussed in a later section.

A more detailed account of the plant anthocyanins may be found in Harborne (1967a).

2.5.2 Recommended techniques

(a) *General precaution*

Anthocyanins are unstable in neutral or alkaline solution and, even in acid solution, the colour may fade slowly due to exposure to light. Anthocyanins must therefore be extracted from plants with solvents containing acetic or hydrochloric acids (e.g. methanol containing 1% conc. HCl) and solutions should be stored in the dark and preferably refrigerated.

(b) *Anthocyanidins*

Fresh petal (or other cyanic tissue) is heated in 2 M HCl in a test tube for 40 min at 100°C. The coloured extract is cooled and decanted from the plant tissue. The cooled extract is twice washed with ethyl acetate to remove flavones, the ethyl acetate layers being discarded and the remaining aqueous layer being heated at 80°C for 3 min to remove the last traces of ethyl acetate. The pigment is then extracted into a small volume of amyl alcohol, which can be pipetted off and concentrated to dryness by heating on a watchglass on a boiling waterbath. (*N.B.* overheating destroys the pigment). The anthocyanidin in the residue is dissolved in 2–4 drops of methanolic HCl and chromatographed one-dimensionally on paper in Forestal and formic acid solvents. The six common anthocyanidins can be distinguished on the basis of differences in R_F and colour (Table 2.10 and see Fig. 2.7). The two pigments cyanidin and petunidin run fairly close together and, if a mixture of these two pigments is suspected, a longer time should be allowed for chromatographic separation (24–30 h, instead of the usual 16–18 h).

Anthocyanidins can be further identified by measuring their visible spectra and by noting whether the spectrum undergoes a bathochromic shift in the presence of aluminium chloride, a specific test for the catechol nucleus (Table 2.10). Reductive cleavage or alkaline degradation can also be carried out, to yield different phenolic acids according to the nature of the hydroxylation pattern in the original pigment. Identification procedures for simple phenols and phenolic acids are given in Section 2.2.

Table 2.10 Properties of the common anthocyanidins

Pigment*	R_F (× 100) in			Visible colour	Visible max. (nm) in MeOH– HCl	Colour shift with AlCl₃†
	Forestal	Formic	BAW			
Pelargonidin	68	33	80	red	520	−
Cyanidin	49	22	68	magenta	535	+
Peonidin	63	30	71		532	−
Delphinidin	32	13	42		546	+
Petunidin	46	20	52	purple	543	+
Malvidin	60	27	58		542	−

Solvent key: Forestal = conc.HCl–HOAc–H_2O (3:30:10): Formic = conc.HCl–HCO_2H–H_2O (2:5:3); BAW = *n*-BuOH–HOAc–H_2O (4:1:5).
*There are about six other known anthocyanidins, but these are all rare in occurrence. For the properties, see Harborne (1967a).
†This bathochromic shift to blue colours with $AlCl_3$ may be observed in the visible spectrum by measuring the spectrum in methanol–0·01% HCl and then adding 5% alc. $AlCl_3$; it may also be detected by spraying or dipping chromatograms (which have been thoroughly dried to remove acid vapour) in 5% alc. $AlCl_3$.

(c) *Anthocyanins*

Extraction of small amounts of fresh coloured petals by crushing in a narrow specimen tube with a minimum amount of methanol containing 1% conc. HCl provides (within 10–15 min) an extract sufficiently concentrated for direct PC. Alternatively, larger amounts of tissue may be macerated for 5 min in methanol–HCl, the macerate filtered or centrifuged and the extract concentrated *in vacuo* at 35–40°C, until the volume is reduced to about one-fifth of the original extract. The anthocyanins are chromatographed one dimensionally on paper in BAW, BuHCl and 1% HCl, against one or more marker solutions. Colours, R_Fs and sources of some of the common anthocyanins are shown in Table 2.11. One of the main factors determining R_F is the number of sugars attached to the anthocyanin, and increasing glycosylation reduces the R_F in BAW and BuHCl but leads to a higher R_F in 1% HCl (compare the R_Fs of the cyanidin mono-, di- and triglycosides in Table 2.11). Glycosides of the different anthocyanidins can be recognized by their visual colours on

Table 2.11 R_F values and sources of some common anthocyanins*

| Anthocyanin | R_F ($\times 100$) *in* | | | |
	BAW	BuHCl	1% HCl	*Petal source*
Monoglycosides				
Pelargonidin 3-glucoside	44	38	14	Chinese Aster
Cyanidin 3-glucoside	38	25	07	Chrysanthemum
Malvidin 3-glucoside	38	15	06	*Primula polyanthus*
Diglycosides				
Pelargonidin 3,5-diglucoside	31	14	23	*Pelargonium*
Cyanidin 3-rhamnosylglucoside	37	25	19	Snapdragon
Peonidin 3,5-diglucoside	31	10	17	Peony
Delphinidin 3,5-diglucoside	15	03	08	*Verbena*
Triglycosides				
Cyanidin 3-rhamnosylglucoside-5-glucoside	25	08	36	Cape Primrose
Cyanidin 3-(2^G-glucosylrhamnosylglucoside)	26	11	61	*Begonia coccineus*
Acylated Diglucoside				
Pelargonidin 3-(*p*-coumarylglucoside)-5-glucoside	40	46	19	*Monarda didyma*

Solvent key: BAW = *n*-BuOH–HOAc–H$_2$O (4:1:5); BuHCl = *n*-BuOH–2 M HCl (1:1, top layer); 1% HCl = conc.HCl–H$_2$O (3:97).
*Visible colour depends on nature of the aglycone: pelargonidin glycosides are orange-red, cyanidin glycosides deep red, etc. The 3,5-diglycosides of pelargonidin, peonidin and malvidin can be further distinguished by the fluorescent colours they give in UV light, other glycosides give dull to bright colours.

chromatograms, since pelargonidin glycosides are orange, cyanidin and peonidin glycosides magenta and delphinidin glycosides mauve.

Some anthocyanins occur in more complex combined form, acylated with a simple organic acid or with a cinnamic acid. They can be identified by the effect that acylation has on R_F value (Table 2.11) and by the fact that, on re-chromatography, they give rise to two spots due to the lability of the acyl group. A test for acylation in anthocyanins is alkaline hydrolysis at room temperature (under nitrogen), which removes the acyl group and causes a change in R_F.

(d) *Authentic markers*

Anthocyanidins are not readily available commercially, but marker solutions can be obtained from readily available plant sources as follows: pelargonidin from pelargonium flowers or red radish; cyanidin from copper beech leaf or blackberry; peonidin from peony flower; delphinidin from blue delphinium flower or skin of eggplant (aubergine); petunidin from black grapes or red wine; malvidin from mallow flowers. Sources of some anthocyanins are given in Table 2.11. For further details of pigment sources, see Harborne (1967a).

2.5.3 Alternative techniques

(a) *Thin layer chromatography*

Anthocyanidins will separate on silica gel in ethyl acetate–formic acid–2 M HCl (85:6:9), but the colour of the spots fade fairly rapidly after separation. It is important to use standard-grade silica gel, since the traces of metal present aid the separation of peonidin and malvidin from cyanidin and delphinidin. Two-dimensional TLC of anthocyanidins on microcrystalline cellulose has been carried out in formic acid–conc. HCl–water (10:1:3), and amyl alcohol–acetic acid–water (2:1:1) (Nybom, 1964).

Anthocyanins can be separated by TLC on cellulose or on mixtures of cellulose and silica gel (Asen, 1965) using similar solvents to those on paper.

(b) *High performance liquid chromatography*

This method using reverse-phase columns such as μBondapak C_{18} or LiChrosorb RP-18 is ideal for the quantitative analysis of anthocyanins in floral tissues (Strack *et al.*, 1980). It has also been applied successfully to the sepal extracts of Poinsettia, where cultivars can be distinguished by quantitative differences in the five pigments present (Stewart *et al.*, 1980). Acidic solvents are necessary, such as water–acetic acid–methanol (71:10:19) or varying proportions of 1·5% H_3PO_4, 20% acetic acid and 25% acetonitrile in water.

2.5.4 Practical experiment

(a) *Distinction between anthocyanins and betacyanins*

As mentioned above, while anthocyanins are almost universal in their occurrence in plants, they are replaced by a different type of water-soluble purple pigment in the Centrospermae. Plants in all but one of the families of this order have nitrogen-containing betacyanins. The most common betacyanin is betanin, the pigment of the beetroot. Other members of the Centrospermae from which betacyanins have been isolated are various cacti, bougainvillea, pokeweed, *Phytolacca* and *Amaranthus*.

Betacyanins differ from anthocyanins in being much more unstable to acid hydrolysis, in their colours at different pHs and in their chromatographic and electrophoretic properties, so that it is easy to distinguish them by simple colour tests. Since the two classes of pigment are mutually exclusive in their occurrence (i.e. they never occur together in the same plant), these tests can be carried out on crude pigment extracts. Betacyanins also differ from anthocyanins in that, when ingested in food (e.g. as beetroot) they are not metabolized by some 14% of the human population and their colour is excreted unchanged in the urine; hence the condition known as 'beeturia'.

(b) *Procedure*

(1) Students can prepare methanolic HCl extracts from any *one* of the many anthocyanin sources (e.g. elderberry, raspberry, blackberry, copper beech leaves, almost any red to purple flowering plant), solution (A) and from beetroot, solution (B) as being the most accessible betacyanin source. After maceration, the extracts should be filtered (with celite) and concentrated *in vacuo* to about one fifth of the original volume. Students may then be provided with four 'unknown' purple-red pigment extracts, solutions (C–F) and told to identify the nature of the pigment present.

(2) Tests, as indicated in Table 2.12, are carried out on all six solutions (A–F). The results obtained with A and B are then used to decide what pigments are present in (C–F). The first four tests shown in Table 2.12 are sufficient for most purposes; the latter two tests can be added if the necessary time and equipment are available.

(c) *Additional experiments*

(1) Anthocyanins in coloured petals in any series of ornamental plants available may be separated by PC in BAW and 1% HCl of methanolic HCl extracts of fresh tissues. Marker solutions of pelargonidin 3,5-diglucoside (from pelargonium flowers), cyanidin 3,5-diglucoside (most

red roses) and delphinidin 3,5-diglucoside (*Verbena*, *Delphinium*) should be included on all chromatograms. The nature of the aglycone and the approximate number of sugar residues can be determined from R_Fs and visible colours for each of the flower anthocyanins which separate.

(2) Anthocyanins in different colour types of the same plant may be compared by PC in BAW and the biochemical effects of genes controlling anthocyanin synthesis be noted. Plants with a wide range of colour types suitable for such study include snapdragon, sweet pea, *Streptocarpus* and anemone. In the sweet pea, *Lathyrus odoratus*, for example it will be found that purple, violet and lilac varieties contain glycosides of malvidin, petunidin and delphinidin, carmine varieties glycosides of cyanidin and peonidin and scarlet shades glycosides of pelargonidin.

Table 2.12 Test for distinguishing anthocyanin and betacyanin*

Test	Anthocyanin response	Betacyanin response
1. Heat with 2 M HCl for 5 min at 100°C	colour stable (can be extracted into amyl alcohol)	colour vanishes
2. Add 2 M NaOH dropwise	changes to blue-green and slowly fades	changes to yellow
3. Chromatography in 1% aqueous HCl	low to intermediate R_F	high R_F
4. Chromatography in BAW	moderate R_F (10–40)	very low R_F (00–10)
5. Visible spectrum in Methanol–HCl	maximum in the range 505–535 nm	maximum in the range 532–554 nm
6. Paper electrophoresis at pH 2–4	moves towards cathode	moves towards anode

*These colour tests can be carried out on crude pigment extracts; for anthocyanin, any deep red-coloured flowers may be extracted or certain fruits (e.g. blackberry, raspberry, strawberry); for betacyanin, beetroot extract is convenient. The responses are those shown for such crude extracts; purified pigments may behave slightly differently.

2.6 FLAVONOLS AND FLAVONES

2.6.1 Chemistry and distribution

Flavonols are very widely distributed in plants, both as co-pigments to anthocyanins in petals and also in leaves of higher plants. Like the anthocyanins, they occur most frequently in glycosidic combination. Although two or three hundred flavonol aglycones are known, only three are at all common: kaempferol (corresponding in hydroxylation pattern to the anthocyanidin pelargonidin); quercetin (cf. cyanidin) and myricetin (cf. delphinidin). Like the corresponding anthocyanidins these three flavonols can be separated clearly by simple PC. The other known flavonols are mostly simple structural variants on the common flavonols and are of limited natural occurrence. In the case of quercetin, a number of O-methylated derivatives are known, the 3'-methyl ether (isorhamnetin) and the 5-methyl ether (azaleatin) being but two examples. Addition of a hydroxyl in the 8-position to the structure of quercetin gives gossypetin, one of the few flavonols which are pigments in their own right, providing yellow flower colour in the primrose, and in the cotton plant.

There is a considerable range of flavonol glycosides present in plants. More than a hundred different glycosides of quercetin alone have been described. By far the commonest is quercetin 3-rutinoside, known as rutin, which is of pharmaceutical interest in relation to the treatment of capillary fragility in man. Some of the best known glycosides of quercetin are listed in Table 2.14 (see p. 73).

Flavones only differ from flavonols in lacking a 3-hydroxyl substitution; this affects their UV absorption, chromatographic mobility and colour reactions and simple flavones can be distinguished from flavonols on these bases (Tables 2.8, 2.9). There are only two common flavones apigenin and luteolin, corresponding in hydroxylation pattern to kaempferol and quercetin. The flavone tricetin, corresponding to myricetin, is known but it is of very rare occurrence. More common are two methyl ethers: chrysoeriol, the 3'-methyl ether of luteolin, and tricin, the 3', 5'-dimethyl ether of tricetin.

Flavones occur as glycosides but the range of different glycosides is less than in the case of the flavonols. A common type is the 7-glucoside, exemplified by luteolin 7-glucoside. Flavones, unlike flavonols, also occur, remarkably, with sugar bound by a carbon-carbon bond. A series of such glycosylflavones have been described, one example being orientin, the 8-C-glucoside of luteolin. The carbon-carbon bond is very resistant to acid hydrolysis, so that it is relatively easy to distinguish these C-glycosides from O-glycosides, which are more readily hydrolysed.

One other structural variant in the flavone series must be mentioned, the biflavonyl. These dimeric compounds are formed by carbon-carbon or

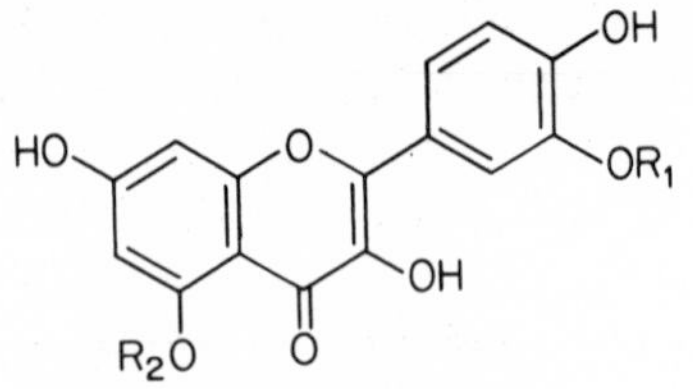

Quercetin, R₁ = R₂ = H
Isorhamnetin, R₁ = Me, R₂ = H
Azaleatin, R₁ = H, R₂ = Me

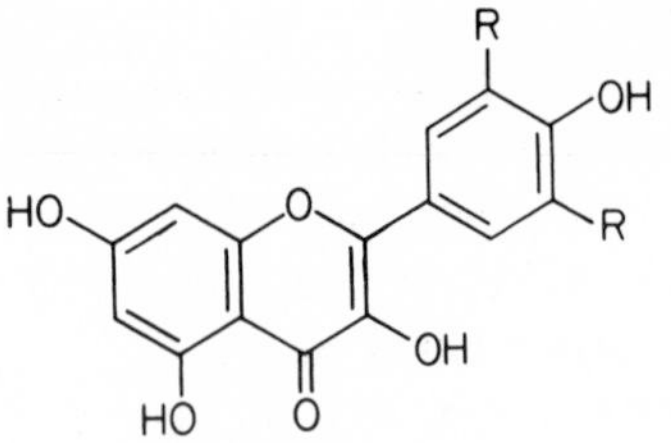

Kaempferol, R = H
Myricetin, R = OH

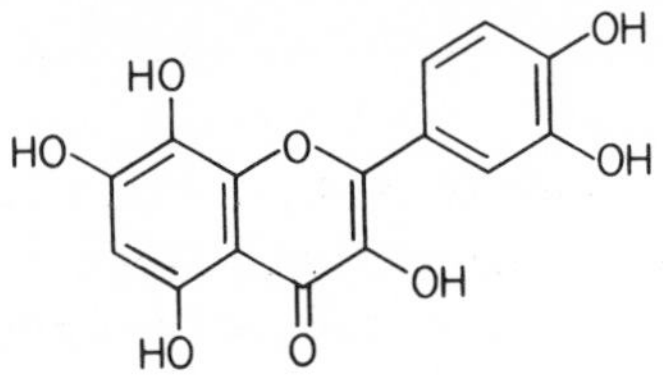

Gossypetin

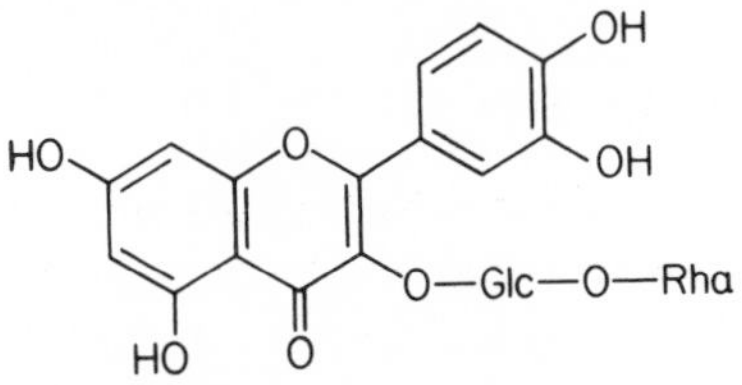

Quercetin 3-rutinoside

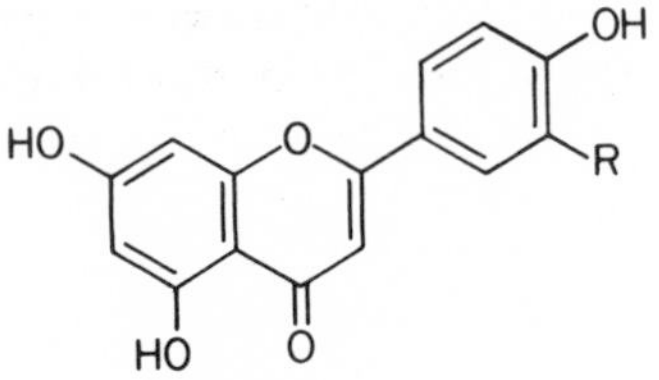

Apigenin, R = H
Luteolin, R = OH
Chrysoeriol, R = OMe

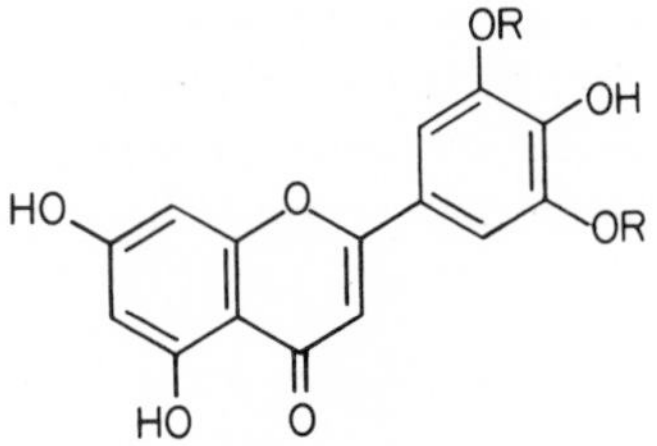

Tricetin, R = H
Tricin, R = Me

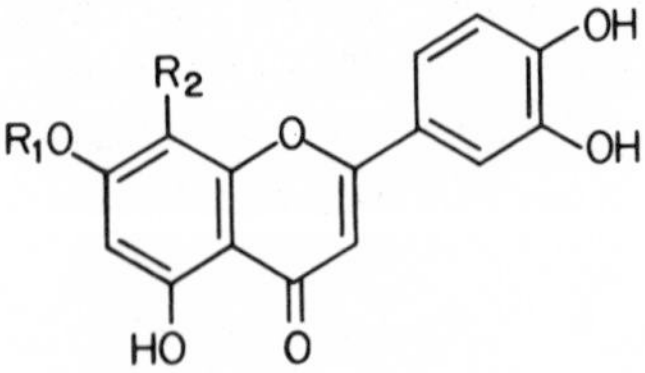

Luteolin 7-glucoside,
R₁ = Glc, R₂ = H
Orientin, R₁ = H, R₂ = Glc

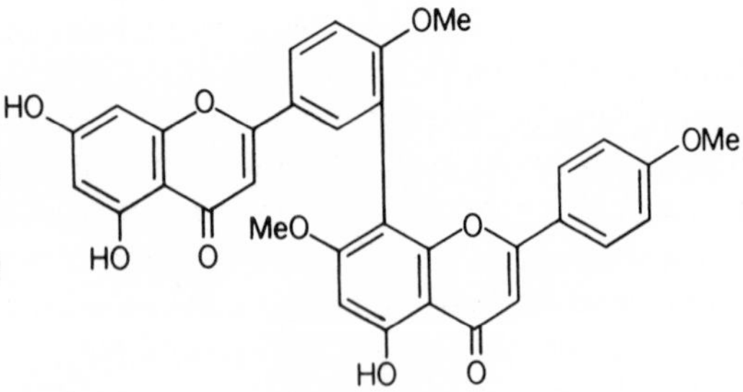

Kayaflavone

Fig. 2.10 Structures of flavonols and flavones.

carbon-oxygen coupling between two flavone (usually apigenin) units. Most also carry *O*-methyl substituents, a typical example being kayaflavone (Fig. 2.10). Biflavonyls occur almost exclusively in the gymnosperms, but they have been found occasionally in angiosperms and methods for their identification are of general importance.

The structure of some representative flavonols and flavones are shown in Fig. 2.10. A more detailed account of the phytochemistry of these flavonoids can be found in Harborne *et al.* (1975).

2.6.2 Recommended techniques

(a) *Acid hydrolysis*

Plant tissue is hydrolysed with 2 M HCl for 30–40 min at 100°C. The cooled solution is extracted twice with ethyl acetate and the combined extracts taken to dryness and the residue taken up in a small volume of ethanol, for chromatography. Hydrolysis can be carried out on fresh or dried tissue and, in the case of leaves, it is possible to use material from herbarium sheets: successful results have been obtained on 100-year-old specimens. It must be remembered that acid hydrolysis of the glycosides may not be complete under the above standard conditions, so that it is not unusual to find some glycoside material as well as the aglycones on chromatograms.

(b) *Flavonols and flavones*

The five common substances – apigenin, luteolin, kaempferol, quercetin and myricetin – are readily separated and identified by PC in Forestal and other standard phenolic solvents, as indicated in Table 2.13 (see also Fig. 2.7). Identification can be confirmed by spectral measurements.

Other flavonols and flavones mostly differ in either R_F, colour or spectrum from the common structures and can usually be distinguished by these simple procedures. Isorhamnetin (quercetin 3'-methyl ether) is not readily identified if present as a mixture with kaempferol and quercetin. In this case, it is best to run a chromatogram in chloroform–acetic acid–water (90:45:6) when isorhamnetin moves ahead of both kaempferol and quercetin. More highly methylated flavonols than isorhamnetin are probably best separated by TLC on polyamide or silica gel (see below).

The yellow flower pigment gossypetin is readily distinguished during flavonoid surveys by its relatively low R_F and dull black colour in UV light (Table 2.13). The isomeric 6-hydroxyquercetin, called quercetagetin, has almost the same R_F in a range of solvents, but can be distinguished by dipping the chromatogram in saturated alcoholic sodium acetate and drying at room temperature; within 40 min, gossypetin gives a blue-grey colour, while quercetagetin remains yellow (Harborne, 1969). The two isomers also have different spectral properties (gossypetin, λ_{max} 262, 278, 341 and 386 nm; quercetagetin λ_{max} 259, 272 and 365 nm).

Table 2.13 Properties of common flavonols and flavones

Flavonoid	R_F ($\times$ 100) in			Colour in UV and UV plus ammonia	Spectral max. in EtOH (nm)	Shift with sodium borate
	Forestal	BAW	PhOH			
Flavonols						
Kaempferol	55	83	58 ⎫	bright yellow	268,368	0
Quercetin	41	64	29 ⎬		255,374	+
Myricetin	28	43	13 ⎭		256,378	+
Isorhamnetin	53	74	66	bright yellow	254,369	0
Azaleatin	49	48	50	fluorescent yellow	254,369	+
Gossypetin	26	31	12	dull black	262,278, 341,386	+
Flavones						
Apigenin	83	89	88 ⎫	dull ochre	269,336	0
Luteolin	66	78	66 ⎬	→ bright	255,268,350	+
Chrysoeriol	77	82	90 ⎬	yellow or	252,269,350	0
Tricin	72	73	87 ⎭	yellow-green	248,269,355	0
Glycosylflavones	*Water*					
Vitexin	06	41	63 ⎫	dull ochre	as apigenin	
Isovitexin	16	56	79 ⎭	→ bright		
Orientin	02	31	43 ⎫	yellow or	as luteolin	
Iso-orientin	09	41	51 ⎭	yellow-green		
Biflavonyl						
Kayaflavone	00	98	99	dull brown	as apigenin	

Solvent key: Forestal = conc.HCl–HOAc–H_2O (3:30:10); BAW = *n*-BuOH–HOAc–H_2O (4:1:5); PhOH = PhOH–H_2O (3:1).

(c) *Glycosylflavones*

These *C*-glycosides are not as soluble in ethyl acetate as the flavone aglycones and they may remain in the aqueous layer after hydrolytic treatment. It is advisable, if their presence is suspected, to either salt them out into ethyl acetate by adding ammonium sulphate or to further extract the aqueous layer with amyl alcohol, into which layer they are transferred. Orientin and the apigenin analogue vitexin separate from the parent flavones luteolin and apigenin in most solvents (Table 2.13), although they have the same colours. Another feature of glycosylflavones is that they undergo isomerization during acid hydrolysis and they give two spots on chromatograms. Under the influence of acid, the pyrone ring of the flavone is opened and when it recloses a mixture of both 8-*C*-glucoside and 6-*C*-glucoside is present. R_F values of the isomers formed from orientin and vitexin are shown in the table.

(d) *Biflavonyls*

These compounds migrate to the front in most solvents on paper. However, they can be separated satisfactorily on paper in butanol–2 M NH₄OH (1:1), a solvent in which most of the common flavones are relatively immobile. TLC on silica gel in toluene–ethyl formate–formic acid (5:4:1) is another useful procedure for biflavonyls. More obstinate biflavonyl mixtures can be separated on silica gel using benzene–pyridine–formic acid (36:9:5) (Khan *et al.*, 1971). For separating and purifying biflavonyls from gymnosperm leaf extracts, a recommended procedure used in the author's laboratory is: PC in butanol–acetic acid–water, TLC on silica gel consecutively in the two solvents mentioned above, and a final purification by PC in butanol–NH₄OH.

(e) *O-Glycosides*

These are conveniently separated by PC in solvents such as BAW, water and 15% acetic acid (Table 2.14). Two-dimensional separations are carried out in BAW and 5% or 15% acetic acid (see Fig. 2.8).

Table 2.14 R_F values of some quercetin glycosides

Quercetin glycoside*	R_F (× 100) *in*			Colour in UV → UV *plus ammonia*
	BAW	Water	15% HOAc	
3-Arabinoside	70	07	31	
3-Xyloside	65	06	32	
3-Glucoside	58	08	37	dull brown
3-Galactoside	55	09	35	→ bright
3-Rhamnoside	72	19	49	yellow
3-Glucuronide	40	69	38	
3-Rutinoside	45	23	51	
7-Glucoside	32	00	10	bright yellow
4'-Glucoside	48	01	12	dull yellow

*Many have trivial names, e.g. 3-rutinoside is rutin, 3-glucoside is isoquercitrin, 3-rhamnoside is quercitrin, etc.

(f) *Authentic markers*

Most of the common flavonol and flavone aglycones are available from commercial suppliers. Alternatively, since they occur so widely in plants, it is not difficult to find convenient sources. Luteolin, for example, is the major flavone of carrot leaves and apigenin of parsley seed. Rutin and certain other flavonol glycosides can also be obtained commercially.

2.6.3   Alternative techniques

If TLC is to be used instead of PC, it is as well to employ microcrystalline cellulose as adsorbent in the first instance and the same range of solvents as on paper. For more difficult separations, another adsorbent to try is polyamide, with a solvent mixture such as chloroform–methanol–butanone (9:4:2). Silica gel plates are only of value for separating those flavones in which only one or two hydroxyl groups are free [e.g. the simple mono- or dihydroxy-flavones of *Primula* farinas (Harborne, 1968) or kaempferol and quercetin mono- and di-methyl ethers (Wollenweber and Egger, 1971)]. Solvents that can be used include 10% acetic acid in chloroform and 45% ethyl acetate in benzene. Separations on polyamide plates may be conducted with benzene–petroleum (b.p. 100–140°C)–butanone–methanol (60:27:7:7).

HPLC is a useful complementary technique to PC or TLC and will some-times separate flavonoid mixtures not resolvable by other methods. It is also valuable for quantitative analyses; a typical separation is illustrated in Fig. 1.3.

Paper electrophoresis is required for screening plants for the presence of flavonoid potassium sulphate salts, which have been found in over twenty plant families (Harborne, 1977). Since they carry a negative charge, the sulphates move towards the anode, when electrophoresis is carried out at 10 V cm^{-1} for 2 h in a formic acid–acetic acid buffer, pH 2·2; ordinary flavones remain at the origin. Electrophoresis can also be employed for distinguishing flavones with glucuronic acid or malonic acid substituents; these compounds are immobile at pH 2·2 but move towards the anode when a pH 4 buffer is used (Harborne, 1983).

The joint application of UV and mass spectral techniques are often sufficient for the complete identification of flavonols and flavones on a micro scale (Markham, 1982). Proton and ^{13}C-NMR spectral measurements are also widely used for purposes of characterization when larger amounts of material are available (Mabry *et al.*, 1970; Markham and Chari, 1982).

2.6.4   Practical experiment

(a) *Flavonol detection: identification of rutin*

Almost every higher plant contains a characteristic pattern of flavone and flavonol glycosides in leaf or flower and thus these substances are ideal taxonomic markers for use in studying problems of plant classification, hybridization or phytogeography. While much can be learnt by comparing the patterns of flavonoid spots on two-dimensional chromatograms of extracts of different plant species or plant populations, it is still very desirable to identify the major components present. This can be done by using a simple series of phytochemical techniques and, for learning these procedures, it is suggested that they should be applied in the first instance to a well known flavonoid, the quercetin glycoside, rutin.

Rutin or quercetin 3-rutinoside was first isolated from buckwheat, *Fagopyrum esculentum* and this plant is still used as a commercial source. Rutin is undoubtedly the most widespread of all quercetin glycosides and probably occurs in up to 25% of any given local flora. Easily available sources include: flowers of magnolia, pansy, *Viola,* and horse-chestnut, leaf of tobacco, plane, rhubarb, tea and French bean, *Phaseolus vulgaris.* Material of any of these plants should be collected and extracted with either hot 95% EtOH (for fresh tissue) or 70% ethanol (for dried tissue) for 30 min and the extract concentrated to a small volume.

(b) *Procedure*

(1) Produce a two-dimensional chromatogram (run in BAW and 5% HOAc) of the direct extract, using a solution of authentic rutin as a marker (see Fig. 2.8). Note which spot on the chromatogram is rutin by comparing R_Fs with the marker spot on the side of the chromatogram.

(2) Using this knowledge, isolate this 'spot', by chromatographing a large aliquot of the original extract as a streak on 2 or 3 sheets of Whatman no. 3 paper in the solvent BAW. After drying the papers and examining them in UV light, cut out the flavonol glycoside band corresponding to rutin. Elute the material from the paper with 70% ethanol, concentrate the eluate and rerun on one sheet of thick paper in water. Dry the paper, elute the rutin 'band' and concentrate the eluate.

(3) Co-chromatograph a small sample of this solution with an authentic marker solution of rutin in at least four solvents and note absence of any separation.

(4) Measure the UV spectrum of the final eluate, diluting the sample with 95% ethanol as much as is necessary. Add 2–4 drops of 2 M NaOH to the cell solution and measure the spectrum in the presence of alkali, noting the magnitude of the bathochromic shift. Compare these results with spectral measurements on a 95% ethanol solution of authentic rutin. (Note that impurities from the paper used in chromatography may interfere with the UV absorption in the 260–270 nm region.) Other spectral shifts may be noted by adding successively powdered sodium acetate and boric acid and by adding two drops 5% alcoholic aluminium chloride (Table 2.15; for further details, see Mabry *et al.,* 1970).

(5) Hydrolyse a sample of the final eluate with 2 M HCl for 30 min. Extract

with ethyl acetate and identify quercetin in this layer by chromatographic comparison with an authentic sample (see p. 72, Table 2.13). Identify the sugars in the aqueous residue, after removal of acid and concentration, by chromatographic comparison with authentic sugar markers (see Table 6.1).

(6) Finally, oxidize another sample with hydrogen peroxide and identify rutinose by the procedure already outlined elsewhere (see Chapter 6, p. 235).

Table 2.15 UV and visible spectral shifts for rutin

Ethanol solution	Spectral maxima (nm)			Spectral effect	Structural diagnosis
	Band I	Band II	Band III		
Alone	259	266sh	363	12 nm hypsochromic shift, compared to quercetin (band III 375 nm)	3-OH substituted
Plus 2 drops 2M NaOH	272	327	415	52 nm bathochromic shift (band III)	4'-OH free
Plus 2 drops 5% AlCl₃	275	303sh	433	70 nm bathochromic shift (band III)	5-OH free
Plus powdered NaOAc	271	325	393	12 nm shift in band I	7-OH free
Plus NaOAc and H₃BO₃	262	298	387	20 nm bathochromic shift (band III)	3',4'-diOH free

2.7 MINOR FLAVONOIDS, XANTHONES AND STILBENES

2.7.1 Chemistry and distribution

The chalcones, aurones, flavanones, dihydrochalcones and isoflavones are designated 'minor flavonoids' because each of these classes is of limited natural distribution. Occurrence is either sporadic (e.g. the flavanones) or else limited to a very few plant groups (e.g. isoflavones to the Leguminosae and

Iridaceae). However, they may well be more widely distributed than is at present thought and it is important to know how to detect them during phytochemical surveys of plant phenols.

Chalcones and aurones together comprise the 'anthochlors', yellow pigments which can be detected by the fact that a change to orange or red colour is observed when a yellow petal is fumed with the alkaline vapour of a cigar or with a vial of ammonia. These compounds occur characteristically in the Compositae (especially in *Coreopsis*), but they have also been recorded in over ten other families. A typical chalcone is butein and a common aurone is aureusidin; both occur naturally as glycosides. Dihydrochalcones have a different distribution pattern from chalcones, being mainly confined to the Rosaceae and Ericaceae. The dihydrochalcone phloridzin, isolated from the apple, is associated with disease resistance in this plant.

Flavanones are isomeric with chalcones and the two classes are interconvertible *in vitro*. Chalcones are frequently found in nature together with the flavanone analogues, but the converse is not always true. Flavanones, for example, accumulate in quantity in *Citrus* fruits without being accompanied by chalcones. Some flavanones have important taste properties; for example, naringin of the seville orange is very bitter (see p. 235).

Isoflavones, of which over two hundred are known, are isomeric with the flavones but are much rarer in their occurrence. They occur almost entirely in one sub-family (the Papilionoideae) of the Leguminosae. Isoflavonoids can be divided into three classes depending on their physiological properties. Compounds such as 7,4'-dihydroxyisoflavone (daidzein) and 5, 7, 4'-trihydroxyisoflavone (genistein) are weak natural oestrogens, present in clover. Complex isoflavans such as rotenone are powerful natural insecticides, while related coumestans such as pisatin are phytoalexins, protective substances formed in plants in response to disease attack (Bailey and Mansfield, 1982).

Xanthones are yellow phenolic pigments which are similar in colour reactions and chromatographic mobility to flavonoids. For this reason, their detection and analysis is included in this section. Chemically, they are different (see structure of mangiferin in Fig. 2.11) and they are readily distinguished from flavonoids by their characteristic spectral properties. Almost all the known xanthones are confined to four families: the Guttiferae, Gentianaceae, Moraceae and Polygalaceae (Carpenter *et al.*, 1969). One xanthone, however, namely mangiferin, which is *C*-glycosylated, is very widespread, occurring in ferns as well as in higher plants (Hostettmann and Wagner, 1977).

Hydroxystilbenes are biogenetically related to the chalcones, but have one less carbon atom in the basic skeleton, which is $C_6-C_2-C_6$. As heartwood constituents in relatively few plants, little attention has been paid to them. Recently, however, it has been discovered that a dihydrostilbene carboxylic acid is an important growth inhibitor in liverworts and algae. This phenolic substance, lunularic acid (see Fig. 2.11) replaces in these plants the sesquiter-

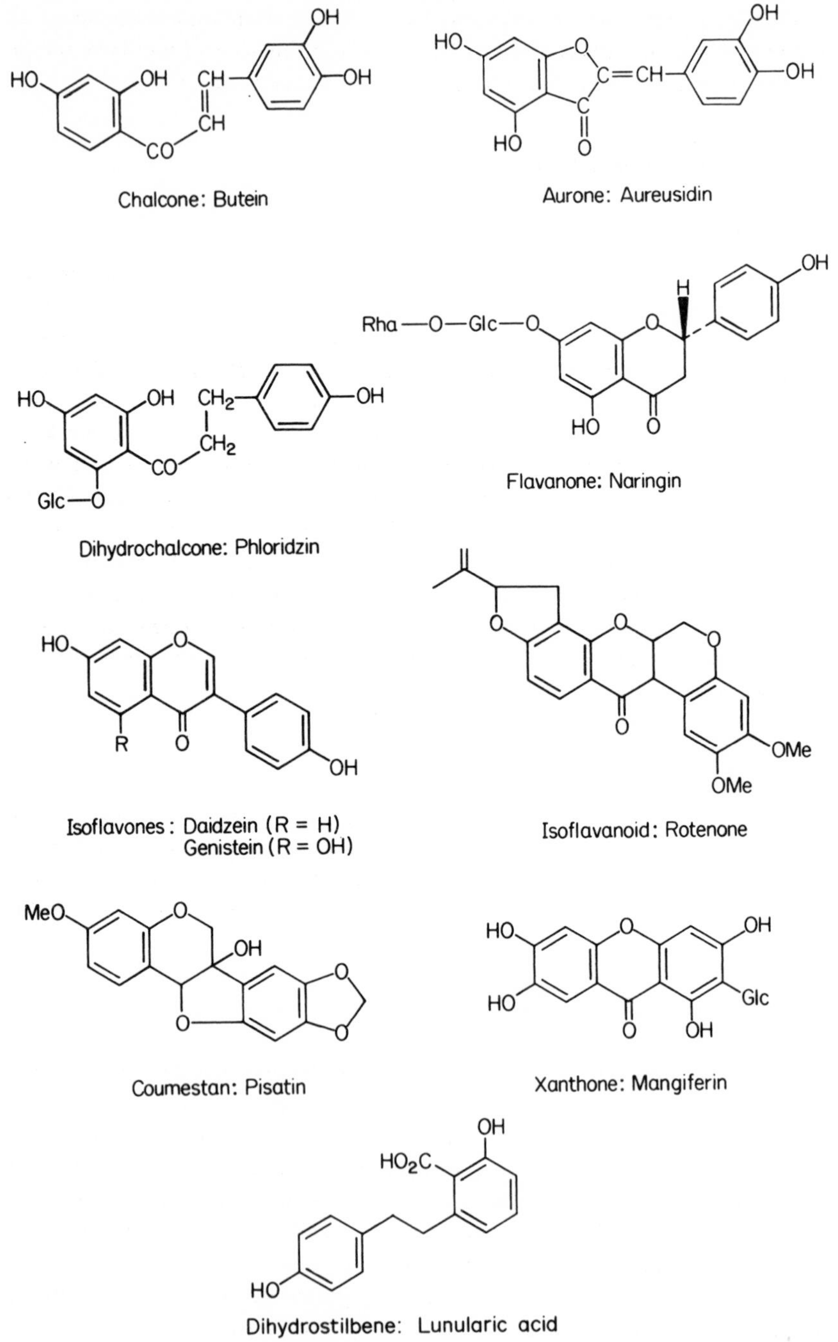

Chalcone: Butein

Aurone: Aureusidin

Rha—O—Glc—O

Flavanone: Naringin

Dihydrochalcone: Phloridzin

Isoflavones : Daidzein (R = H)
Genistein (R = OH)

Isoflavanoid: Rotenone

Coumestan: Pisatin

Xanthone: Mangiferin

Dihydrostilbene: Lunularic acid

Fig. 2.11 Structures of minor flavonoids.

pene abscisic acid, which functions as the major dormancy hormone in all other plant groups (Pryce, 1971, 1972). The identification of lunularic acid in plant materials is thus of considerable current physiological interest.

2.7.2 Recommended techniques

(a) *Chalcones*

These are yellow phenolic pigments, which give intense deep brown UV colours when chromatographed on paper; on fuming the paper with ammonia, the colour may change to a rich deep red, although a few chalcones fail to respond in this way. Chalcones are easily separated by PC in the usual solvents (Table 2.16). TLC may be carried out on silica gel, buffered with sodium acetate, using benzene–ethyl acetate–formic acid (9:7:4). Chalcone agly-cones can be distinguished from their glycosides, since only the latter pigments are mobile on paper in water. Chalcones exhibit a broad peak between 365 and 390 nm in the visible spectrum, which distinguishes them from aurones (visible λ_{max} 390–430 nm). These spectra undergo large bathochromic shifts in the presence of alkali and metal salts. In handling chalcones during isolation, it must be remembered that they are isomerized by acid to flavanones; solutions of some are also slowly oxidized in the air to the cor-responding aurones.

(b) *Aurones*

These, like chalcones, appear on paper chromatograms as yellow spots in daylight. In the UV, they are very different; the colour of aurones is an intense bright yellow, changing with ammonia to a bright orange-red. Chromat-ographic and spectral properties of some common aurones are shown in Table 2.16.

ʹ(c) *Flavanones*

These are colourless substances, which cannot be detected during chromato-graphic surveys unless chromogenic sprays are employed. It is true that some flavanones give bright yellow-green or light blue colours on paper, when viewed in UV light with the help of ammonia vapour, but this is not reliable enough to be used as a diagnostic test. An important colour test in alcoholic solution is reduction with magnesium powder and conc. HCl; only flavanones among the flavonoids give *intense* cherry-red colours. This procedure can be applied to paper chromatograms or TLC plates by spraying first with an alcoholic solution of sodium borohydride (*ca.* 1%) and then later spraying with ethanolic aluminium chloride (Koeppen, 1965). A suitable procedure for detecting 3-hydroxyflavanones (or dihydroflavonols) on paper is by use of zinc dust spread on the paper, followed by spraying with 2 M HCl. On chromatograms originally run in water, mauve spots will appear at R_F 30 (for dihydrokaempferol) 28 (for dihydroquercetin) and 24 (for dihydromyricetin).

Table 2.16 Properties of minor flavonoids

Compound and class	R_F ($\times$ 100) in			Colours in UV and UV plus ammonia	Spectral max. in EtOH (nm)	Alkaline $\Delta\lambda$
	BAW	Water	PhOH			
Chalcones						
Isoliquiritigenin	89	00	90	dark	235,372	+70
Butein	78	00	66	brown	263,382	unstable
Isoliquiritigenin 4'-glucoside	61	06	80	→ deep	240,372	+53
Isosalipurposide	67	04	36	red	240,369	+72
Aurones	BAW	30% HOAc	PhOH			
Sulphuretin	80	19	70	bright	257,279,399	+65
Aureusidin	57	10	29	yellow	254,269,399	unstable
Aureusidin 4-glucoside	49	25	45	→ orange	255,267,405	+45
Aureusidin 6-glucoside	28	16	35	red	272,322,405	+85
Flavanones	BAW	Water	30% HOAc			
Naringenin	89	16	66	faint	224,290,325	+37
Hesperitin	89	11	67	purple	226,288,300	+40
Hesperidin	48	50	85	→ light	225,286,330	+77
Naringin	59	62	87	yellow-green	226,284,330	+143
Dihydroquercetin	78	28	67	faint purple	228,289,325	+25
Dihydrochalcone						
Phloridzin	75	30	73	faint yellow	224,285,315†	+41
Isoflavones	BAW	Water	TLC*			
Daidzein	92	08	36	see text	250,260,302	+30
Genistein	94	04	41		263,325†	+10
Xanthone						
Mangiferin	45	12	—	apricot → yellow-green	242,258, 316,364	—

*TLC on silica gel in 11% methanol in chloroform.
†Inflection.

Flavanones have different spectral properties from other flavonoids, with one intense peak at about 225 nm, another at either 278 or 288 nm and a weak peak or inflection above 300 nm. In alkaline solution, some flavanones undergo ring opening to the corresponding chalcone, which will be observed by an unusually large (up to 110 nm) alkaline shift in the spectrum.

(d) *Dihydrochalcones*

These are separated by PC in the usual solvents. They are detected by spraying the papers with diazotized *p*-nitroaniline and with alcoholic $AlCl_3$. Phloridzin gives an orange-red colour with the first reagent and an intense greenish fluorescence with the second reagent.

(e) *Isoflavones*

These are difficult to characterize, since they do not respond specifically to any one colour reaction. Some isoflavones (e.g. daidzein) give a brilliant light blue colour in UV light in the presence of ammonia, but most others (e.g. genistein) appear as dull purple absorbing spots, changing to dull brown with ammonia. PC is of limited value for separating isoflavones, although their significant mobilities in water distinguish them from most other flavonoid aglycones. For plant surveys, TLC on silica gel in 11% methanol in chloroform is recommended, detecting the spots with Folin Ciocalteu reagent. Isoflavone identifications can be confirmed by spectral measurements (Table 2.16). A method for their quantitative determination in plant extracts, using Sephadex G-25 columns, is described by Glencross *et al.* (1972). Quantitation by HPLC on a Lichrosorb RP-8 column is also possible (Koster *et al.*, 1983).

(f) *Xanthones*

These are commonly separated by TLC on silica gel, using chloroform–acetic acid (4:1), chloroform–benzene (7:3) or chloroform–ethyl acetate (varying proportions). They can be detected by their colours in UV light with and without ammonia or by using a general 'phenolic' spray. Mangiferin differs from practically all other xanthones in being water-soluble and it separates well on paper (Table 2.16). Xanthones have distinct spectral properties, with maxima at 230–245, 250–265, 305–330 and 340–400 nm. Like those of flavonoids, the spectra undergo characteristic bathochromic shifts with alkali, $AlCl_3$ and sodium acetate–boric acid, which vary according to the number and position of the hydroxyl substituents.

(g) *Stilbenes*

Stilbenes show intense purple fluorescence in UV light changing to blue with NH_3, have spectral maxima at about 300 nm and can be separated on paper or by TLC. As an example, resveratrol (4,3′,5′-trihydroxystilbene) has a λ_{max} at

305 with an inflection at 320 nm, shifting to 318 and 330 nm in the presence of alkali, and R_Fs of 78 in BAW and 67 in 15% HOAc.

The growth inhibitor lunularic acid, a dihydrostilbene, has R_F 40 in EtOAc–CHCl$_3$–HOAc (15:5:1), 32 in benzene–MeOH–HOAc (20:4:1) and 45 in di-isopropyl ether-HOAc (19:1) on silica gel plates. It can also be identified by GLC of its methyl ester dimethyl ether. Suitable columns are 1·5% E-60 or 1% OV-17 absorbed on Gas Chrom Q (80–100 mesh) and conditions should include a nitrogen flow rate of 60 ml min^{-1} and the injection heater at 250°C (Pryce, 1971). For quantitative analysis, it is better to directly derivatize for GLC (using silylation with BSA), than to purify plant extracts by TLC, since up to 50% of material can be lost on silica gel plates (Gorham, 1977). Lunularic acid can also be analysed directly on a nanogram scale by reverse-phase HPLC on a Partisil ODS column using 50% MeCN in H$_2$O containing 0·1% HOAc as solvent and monitoring at 285 nm (Abe and Ohta, 1983).

(h) *Authentic markers*

2′, 4′, 6′, 4-tetrahydroxychalcone 2′-glucoside (isosalipurposide) can be obtained from yellow carnations and from willow bark. The aurone aureusidin occurs as its 6-glucoside (aureusin) in yellow snapdragon flowers. Both aurones (sulphurein) and chalcones (butein 4′-glucoside, coreopsin) occur in yellow dahlias. The flavanones hesperidin and naringin are available commercially, as is dihydroquercetin. Simple isoflavone markers (e.g. daidzein, genistein) can be isolated from clover, *Trifolium repens*, from gorse leaf or from broom.

2.7.3   Practical experiment

(a) *Induction of the isoflavanoid phytoalexin pisatin in pea leaves*

Phytoalexins are low-molecular-weight antibiotics produced *de novo* in higher plants in response to infection by micro-organisms, particularly by fungi. They may also be formed under stress conditions and can be induced abiotically by treating tissues with inorganic salts. Phytoalexins produced by a given plant can prevent the growth and development of non-pathogenic fungi, whereas pathogenic organisms can withstand or overcome their toxic effects and can develop within the plant. The resistance of a plant to infection depends on the rate and degree of phytoalexin accumulation and on the varying sensitivity of micro-organisms to the phytoalexins.

The type of phytoalexin produced in a given plant varies according to the family to which that plant belongs. In the Leguminosae, isoflavonoids such as pisatin are commonly formed (Ingham, 1981). In this experiment, pisatin will be characterized. This was the first phytoalexin to be fully identified and is produced in large quantities in fungally infected leaves of pea, *Pisum sativum*. It is also produced by other *Pisum* species and by many species of *Lathyrus* (Robeson and Harborne, 1980). The phytoalexin can be isolated from pea leaves using the drop-diffusate technique, which takes advantage of the fact that phytoalexins can diffuse into water droplets resting on the surface of excised leaves. The droplets contain an organism or chemical capable of initiating phytoalexin production. After suitable incubation (up to 3 days) the droplets are collected and the phytoalexin extracted.

(b) *Procedure*

(1) Excised leaflets from greenhouse-grown pea plants are floated, adaxial side uppermost, on water in clear plastic sandwich boxes (*ca.* 15×25 cm). The leaves are then inoculated with a conidial suspension (*ca.* $2 \cdot 5 \times 10^5$ spores ml^{-1}) of *Helminthosporium carbonum* in 0·05% Tween 20. If a fungal spore suspension is not available, it can be replaced by a 0·2% aqueous solution of copper sulphate. Control leaves receive drops of distilled water containing 0·05% Tween 20.

(2) The boxes are covered with plastic sheets and incubated at room temperature under continuous diffuse lighting for up to 72 h.

(3) The droplets are then separately collected and extracted into ethyl acetate (×3), the extracts being taken to dryness *in vacuo* at 40°C.

(4) Dissolve the two extracts (control and infected) separately in 0·2–0·4 ml ethanol and apply each as a streak about 2 cm from the bottom of a silica gel TLC plate, with fluorescent indicator. Develop the plate by ascent in chloroform containing 2% methanol for 60–90 min.

(5) Remove the plate, mark the front, and leave to dry. On examining in UV light, the phytoalexin should be clearly visible as a dark absorbing band which can be marked with a pencil. For the control, mark a corresponding area of blank silica gel.

(6) Remove the phytoalexin band and the control. Elute with ethanol (2×3 ml aliquots), concentrating the eluates *in vacuo*. Dissolve the two residues in 2 ml ethanol and measure the UV spectrum, using the control eluate as blank. Add 5 drops of conc. HCl and record the spectral shift. Pisatin has maxima at 213, 280, 286 and 309 nm and in the presence of HCl characteristically undergoes dehydration to dehydropisatin, which has maxima at 248, 342 and 358 nm.

(7) From the spectral absorbance, it is possible to calculate the amount of phytoalexin formed as follows:

$$\mu\text{g pisatin ml}^{-1} \text{ diffusate} = \frac{D \times [A_{310} - A_{350}] \times B}{C}$$

where D = dilution of sample, A_{310} and A_{350} = absorbance at given wavelength, B = pisatin constant = 47, C = volume of diffusate (in ml).

2.8   TANNINS

2.8.1   Chemistry and distribution

Tannins occur widely in vascular plants, their occurrence in the angiosperms being particularly associated with woody tissues. By definition, they have the ability to react with protein, forming stable water-insoluble co-polymers. Industrially, tannins are substances of plant origin which because of their ability to cross-link with protein are capable of transforming raw animal skins into leather. In the plant cell, the tannins are located separately from the proteins and enzymes of the cytoplasm but when tissue is damaged, e.g. when animals feed, the tanning reaction may occur, making the protein less accessible to the digestive juices of the animal. Plant tissues high in tannin are, in fact, largely avoided by most feeders, because of the astringent taste they impart. One of the major functions of tannins in plants is thought to be as a barrier to herbivory.

Chemically, there are two main types of tannin (Table 2.17), which are distributed unevenly throughout the plant kingdom. The condensed tannins occur almost universally in ferns and gymnosperms and are widespread among the angiosperms, especially in woody species. By contrast, hydrolysable tannins are limited to dicotyledonous plants and here are only found in a relatively few families. Both types of tannin, however, can occur together in the same plant, as they do in oak bark and leaf.

Table 2.17 Classification of plant tannins

Nomenclature	Structure	Molecular weight range	Protein precipitation
Condensed tannins			
Proanthocyanidins* (or flavolans)	Oligomers of catechins and flavan-3,4-diols	1000–3000	++++
Hydrolysable tannins			
Gallotannins	Esters of gallic acid and glucose	1000–1500	+++++
Ellagitannins	Esters of hexahydroxydiphenic acid and glucose	1000–3000	++++
Prototannins			
Tannin precursors	Catechins (and gallocatechins) flavan-3,4-diols	200–600	±

*The term leucoanthocyanidin (or leucoanthocyanin) was widely used at one time for these tannins, but is now more strictly reserved for the monomeric flavan-3,4-diols, which lack effective tannin action.

Condensed tannins or flavolans (see Fig. 2.1) can be regarded as being formed biosynthetically by the condensation of single catechins (or gallo-catechins) to form dimers and then higher oligomers, with carbon–carbon bonds linking one flavan unit to the next by a 4–8 or 6–8 link. Most flavolans have between two and twenty flavan units. The name proanthocyanidin is used alternatively for condensed tannins because on treatment with hot acid, some of the carbon–carbon linking bonds are broken and anthocyanidin monomers are released. Most proanthocyanidins are procyanidins, which means that they yield cyanidin on acid treatment. Prodelphinidins and pro-pelargonidins are also known, as are mixed polymers which yield cyanidin *and* delphinidin on acid degradation (see Section 2.5 for structures). Characterization thus involves the identification of the anthocyanidin or anthocyanidins so released.

Hydrolysable tannins are mainly of two classes (Table 2.17), the simplest being the galloylglucose depside, in which a glucose core is surrounded by five or more galloyl ester groups. A second type occurs where the core molecule is a dimer of gallic acid, namely hexahydroxydiphenic acid, again with glucose attachment; on hydrolysis, these ellagitannins yield ellagic acid. Within these two classes it is possible to further subdivide the known compounds according to their biogenetic origin (Haddock *et al.*, 1982).

The chemical elucidation of tannins has been difficult to accomplish and it is only within the last decade that some of the structural complexities have been fully appreciated. There are, for example, significant stereochemical differences between the proanthocyanidins of monocotyledonous and dicotyledonous plants (Ellis *et al.*, 1983). Detailed accounts of plant tannins include those of Bate-Smith (1973, 1977), Haslam (1966, 1982) and Swain (1979).

2.8.2 Recommended techniques

(a) *Condensed tannins*

Proanthocyanidins can be detected directly in green plant tissues (i.e. in the absence of cyanic pigmentation) by immersion in boiling 2 M HCl for 0·5 h. The production of a red colour extractable into amyl or butyl alcohol is evidence of their presence. Chromatography of the concentrated amyl alcohol extract on TLC plates and the recognition of pelargonidin, cyanidin or delphinidin is confirmatory evidence (see Section 2.8.3).

For more detailed studies, fresh plant tissues are extracted with hot 50–80% aqueous methanol. Such extraction, even in favourable cases, will yield only a proportion of the total tannin, since some will be irreversibly bound to other polymers within the cell. If dried plant tissues are used, the yield of tannin may be less, due to increased 'fixing' of the tannin *in situ*. The mixture of tannins present in such crude extracts can be monitored by 2-D paper chromatography, using the top phase of butan-2-ol–acetic acid–water (14:1:5) followed by 6% acetic acid. The tannins can be detected as dark purple spots in short UV light, which react positively to any one of the standard phenolic reagents (see Section 2.2) (Haslam, 1966). Monomeric catechins and other simple phenolics may also be revealed in these procedures. Alternatively, the pattern of condensed tannin may be examined by HPLC, using for example a LiChrosorb RP-8 column eluted with water–methanol mixtures (Lea, 1980). These crude extracts should also be capable of precipitating bovine serum albumin when added to a 5% aqueous solution.

The individual proanthocyanidins can be separated on a preparative scale by chromatography on Sephadex. In a typical procedure, a preliminary purification is achieved on Sephadex G-50, the column being eluted with acetone-water (1:1) containing 0·1% ascorbate to protect the tannin from aerial oxidation. This eluate is then applied to a Sephadex LH-20 column in aqueous methanol (1:1), the proanthocyanidins being recovered with aqueous acetone (7:3). Further fractionation according to molecular weight is then possible on G-50 or G-100; the molecular weight can be determined at the same time if standards are available (Jones *et al.*, 1976).

The purified polymers are then characterized by chemical degradation and by optical rotation measurements. The molecular weight and composition of mixed polymers can be directly determined by ^{13}C-NMR spectral measurements (Porter *et al.*, 1979).

(b) *Hydrolysable tannins*

Preliminary detection in leaf and other tissue after acid hydrolysis is by recognition of gallic acid and/or ellagic acid in concentrated ether or ethyl acetate extracts. The detection of gallic acid is illustrated earlier in this chapter in Fig. 2.5. Ellagic acid appears on Forestal chromatograms between myricetin and quercetin (see Fig. 2.7) as a violet-coloured spot in UV light, darkening in the presence of NH_3 vapour.

Large-scale extraction is best carried out with aqueous acetone to avoid the hydrolysis of ester linkages in the native tannins. The crude extracts can be monitored for ester components by 2-D paper chromatography in the same solvent systems as for condensed tannins (see above). Hydrolysable are usually well differentiatd from condensed tannins by differences in mobility (compare procyanidin B1 with R_Fs 25/50 respectively and pentagalloyl-glucose with R_Fs 50/06); they are similarly detected by any one of the general phenolic sprays. In addition, galloyl esters usually appear as violet fluorescent spots, enhanced in colour by NH_3 treatment; furthermore, they more specifically give rose pink colours when treated on paper with a saturated potassium iodate solution (Haslam, 1965). By contrast, ellagitannins give a specific colour reaction with nitrous acid (ice-cool 10% $NaNO_2$ to which acetic acid is added); this is a distinctive carmine red which changes with time to an indigo blue (Bate-Smith, 1972). Separation of hydrolysable tannins by HPLC has also been successful. Gradient elution on a RP-18 LiChrosorb 10 μm column, using increasing acetonitrile concentrations in water containing 0·05% H_3PO_4, separates galloyl esters from each other according to the number of galloyl residues and their position of attachment (Haddock *et al.*, 1982).

(c) *Total tannin*

No quantitative estimation of tannins in a given plant tissue will be accurate unless it is realized that the presence of other phenols may interfere with non-specific chemical methods and that total extractability, particularly of the condensed tannins, can rarely be achieved in practice. Measurements made on so-called total tannin extracts need to be repeated on the plant residues left after extraction (Swain, 1979). For analysis, carefully dried tissue should first be finely powdered and a weighed amount extracted with 3 portions successively of aqueous methanol (1:1), using about 0·1 ml solvent mg^{-1}. One or more of the following procedures may be used, depending on what types of tannin are present.

(1) Tannic acid equivalent (TAE). A suitable volume of the extract is made up to 1 ml and mixed with 1 ml of a fresh sample of diluted human blood (1:50 with water). After centrifugation to remove the tannin–protein precipitate, the residual haemoglobin is determined by its absorbance at 578 nm. The TAE can then be calculated from standard measurements made on a known tannic acid sample. This procedure, giving relative astringency of the plant extract, is a direct measure of total soluble tannin (Bate-Smith, 1973).

(2) Proanthocyanidin content. A known volume of the extract is concentrated to about a third in volume and is heated with *n*-butanol containing 5% conc. HCl (0·5 ml extract with 4 ml reagent) for 2 h at 95°C. The absorbance is then measured in a spectrophotometer at 545 nm (for cyanidin) and at 560 nm (for delphinidin). Comparison with standard anthocyanidin solutions gives a measure of proanthocyanidin initially present. Allowance, however, has to be made for the facts that the yield of anthocyanidin is not quantitative and that the yield varies according to the structural type of the original proanthocyanidins (Swain, 1979).

(3) Ellagitannin content. 0·5 ml of the extract is treated with 2·0 ml 0·1 M HNO_2 (from $NaNO_2$ and acetic acid) at room temperature under N_2; the blue colour that develops is measured after 15 min at 600 nm (Bate-Smith, 1972).

(4) Gallotannin content. The extract sample (0·5 ml) is reacted with 1·5 ml

12% KIO_3 in 33% methanol at 15°C, the red-brown colour being measured immediately at 550 nm (Bate-Smith, 1977). This measures the galloyl groups present in any galloyl-based tannin.

Other chemical tests for functional groups have been used in the past for tannin assays, such as the Folin–Denis procedure for total phenols and the vanillin–HCl procedure for catechins (see Goldstein and Swain, 1965) but these are being replaced more and more by procedures based on tannin–protein interactions (as in procedure 1 above). For further details of assays based on tannin–protein precipitations, see Martin and Martin (1983).

2.8.3 Practical experiment

(a) *Detection of plant proanthocyanidins*

Proanthocyanidins (condensed tannins) are widely distributed, mainly in woody plants, and may be an important barrier to mammals or insects feeding because of their astringency and tanning properties. Evidence suggests that the concentration in leaf tissues must be above 2% dry weight before tannins become feeding deterrents. When the leaf is crushed during eating, the tannins complex with the leaf protein. After ingestion, the tanned protein may not be fully absorbed, because the proteolytic enzymes are unable to break it down fully to its amino acid components.

Proanthocyanidins are flavan polymers which can be recognized by the fact that when heated with acid, leaves containing them turn red. The colour, due to anthocyanidin, can be extracted into amyl alcohol and the pigments identified.

It is possible to determine presence/absence of proanthocyanidins in a number of species with green leaves. Positive results should be obtained with any fern, gymnosperm or woody angiosperm, while all herbs should respond negatively. With positive samples, the anthocyanidins produced can be identified by TLC and by spectroscopy.

(b) *Experimental procedure*

(1) Cut up leaf samples separately, transfer to test tubes and cover with 2 M HCl. Hydrolyse in boiling water bath for 30 min.

(2) Into a 500 ml beaker put 20 ml Forestal (HOAc–H_2O–HCl, 30:10:3) and cover with petri dish lid.

(3) Remove test tubes from the water bath, *allow to cool*. If red colour is not present, note absence and discard sample. If red colour *is* present, transfer liquid to a centrifuge tube with a pipette and extract with 1 ml ethyl acetate. Centrifuge; pipette off, and discard ethyl acetate layer.

(4) Add *5 drops* amyl alcohol and shake thoroughly. Take up a small amount of the alcohol layer in a capillary tube and spot (*ca.* 6 applications) on a small cellulose plate. Samples should be spotted about 1 cm apart. Marker solutions of pelargonidin and cyanidin/delphinidin should also be spotted on the plate.

(5) Develop by ascent in the beaker containing Forestal for *ca.* 1 hour.

(6) Add a few more drops of amyl alcohol to each positive sample, shake and remove alcohol layer. Dilute with 1% methanolic HCl and scan in visible region (400–700 nm) using methanolic HCl as a blank. Record position of visible maximum (550 nm). This confirms the anthocyanidin nature of the pigment. Add alkali dropwise to solution and note colour changes.

(7) Remove plate from beaker, dry. Examine in visible light. Record colour and positions of spots in each sample.

2.9 QUINONE PIGMENTS

2.9.1 Chemistry and distribution

The natural quinone pigments range in colour from pale yellow to almost black and there are over 450 known structures (Thomson, 1971). Although they are widely distributed and exhibit great structural variation, they make relatively little contribution to colour in higher plants. Thus, they are frequently present in bark, heartwood or root, or else they are in tissues (such as the leaves) where their colours are masked by other pigments. By contrast, in bacteria, fungi and lichens they do make some contribution to colour; for example, the fruiting bodies of many Basidiomycetes are pigmented by quinones. Their distribution in higher plants has been mainly studied because certain anthraquinones, those of senna pods and cascara bark, have a cathartic action and are used in the Pharmacopiea of many countries as purgatives.

Quinones are coloured and contain the same basic chromophore, that of benzoquinone itself, which consists of two carbonyl groups in conjugation with two carbon–carbon double bonds. For their identification, quinones can conveniently be divided into four groups: benzoquinones, naphthaquinones, anthraquinones and isoprenoid quinones. The first three groups are generally hydroxylated, with 'phenolic' properties, and may occur *in vivo* either in combined form with sugar as glycoside or in a colourless, sometimes dimeric, quinol form. In such cases, acid hydrolysis is necessary to release the free quinones. The isoprenoid quinones are involved in cellular respiration (ubiquinones) and photosynthesis (plastoquinones) and are thus universally distributed in plants. The structures of quinones typifying the four main groups are shown in Fig. 2.12. A very full account of the natural occurrence and properties of all plant quinones can be found in Thomson (1971).

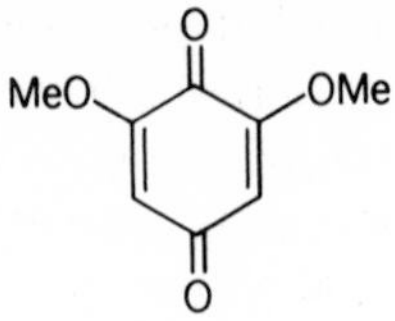

2,6—Dimethoxybenzoquinone

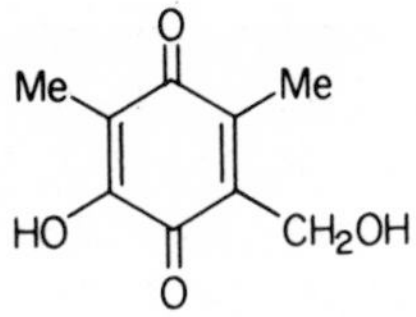

Shanorellin

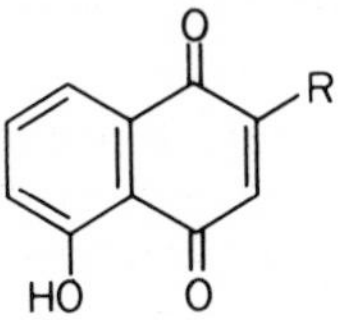

Juglone, R = H
Plumbagin, R = Me

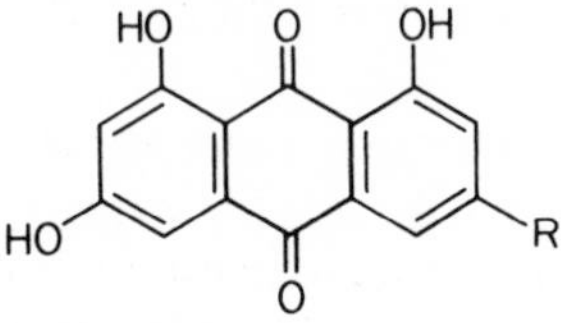

Emodin, R = Me
Emodic acid, R = CO_2H

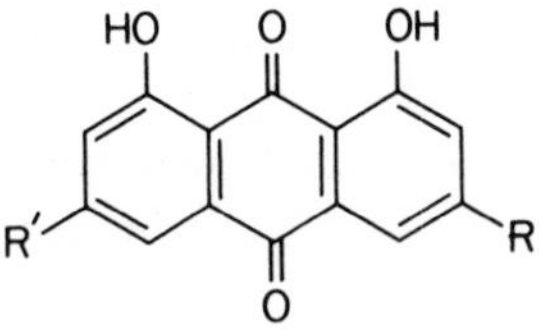

Aloe—Emodin, R = CH_2OH, R′=H
Physcion, R = Me, R′ = OMe

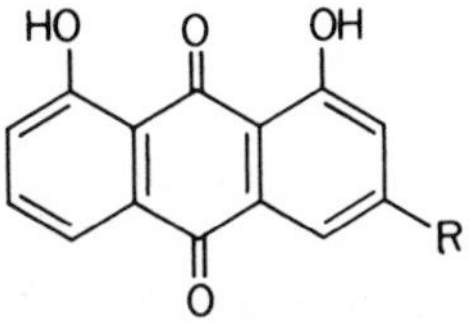

Chrysophanol (R = Me)
Rhein (R = CO_2H)

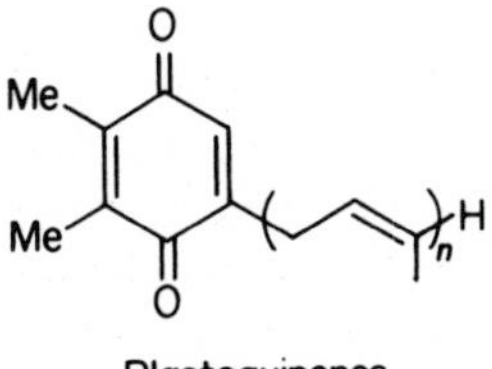

Plastoquinones
(n = 3 to 9)

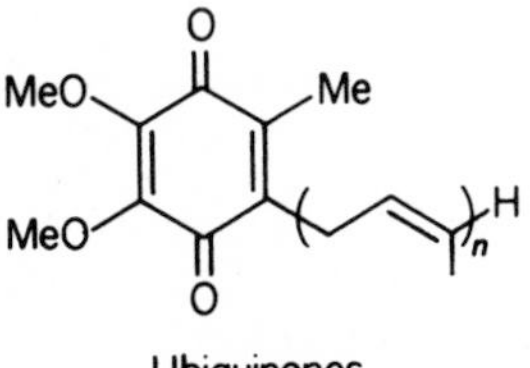

Ubiquinones
(n = 1 to 12)

Fig. 2.12 Structures of Quinones.

2.9.2 Recommended techniques

(a) *Preliminary detection*

Unlike flavonoid or carotenoid pigments, the hydroxyquinones (as distinct from isoprenoid quinones) have a sporadic distribution in plants. Some means of preliminary separation from other types of plant pigment is therefore called for. Hydroxyquinones may appear during paper chromatographic surveys as yellow or orange pigments which move to the front in butanolic solvents (e.g. BAW) and which are generally immobile in aqueous solvents. They have dull absorbing colours in UV light and may not necessarily respond to treatment with ammonia. Those present as glycosides may be slightly water-soluble but, by and large, quinones are more likely to be lipid-soluble and would be extracted from a crude plant extract with carotenoids and chlorophylls. They separate on TLC silica gel plates and are normally well clear of carotenoids and chlorophylls if they are chromatographed together in a crude petroleum extract.

For confirming that a pigment is a quinone, simple colour reactions are still very useful. Reversible reduction to a colourless form and restoration of the colour by aerial oxidation is characteristic. Reduction can be accomplished with sodium borohydride and re-oxidation can be effected simply by shaking the solution in air. This procedure can be applied to a chromatogram; when sprayed with leucomethylene blue, quinones give blue spots on a white background. For most quinones, reduction in slightly alkaline solution is more striking and the reoxidation occurs more quickly in air. Quinones give intense bathochromic shifts in alkali, but this is not characteristic of the class since other phenolic pigments do this as well. Colours produced range from orange and red to violet and blue; even green colours are formed in some cases (1,2,3-trihydroxyanthraquinones).

(b) *Separation and identification*

TLC on silica gel is a general procedure for separating quinones. However, there is so much structural variation that no one system, or even range of related systems, can be applied to quinones generally. Simple benzoquinones and naphthaquinones are very lipid-soluble and they may separate in pure benzene, pure chloroform, pure petroleum or simple mixtures of these solvents. On the other hand, highly hydroxylated anthraquinones are very polar and complex solvent mixtures are required to make them mobile. Since quinones are coloured, there is no difficulty in detecting them in visible light on TLC plates; examination in UV light, however, may be useful and provide a more sensitive means of detection.

HPLC can also be used for quinone analysis. Benzo- and naphthaquinones have been separated on Micropak Si-10 columns in 1% isopropanol in petroleum, while anthraquinones separate on Micropak CH-5 columns in methanol–water (1:1) acidified to pH 3 (Rittich and Krska, 1977).

For quinone identification, spectral measurements are essential (Thomson, 1971). The UV and visible spectrum indicates the class of quinone present, since the number and position of absorption bands increase with the complexity of the structure (see below). Quinones give characteristically strong carbonyl bands in the IR and, on MS, can be recognized by the ready loss of one and two molecules of carbon monoxide from the parent ion peak.

(c) *Benzoquinones*

These separate well by TLC on silica gel in lipid solvents. The choice of solvent depends on the particular pigment mixture that is being separated and can only be determined by trial and error. For example, benzoquinone itself can be separated from its methyl and ethyl derivatives in *n*-hexane–ethyl acetate (17:3) in which it has R_F 21. Again, shanorellin, the orange-yellow pigment of the ascomycete *Shanorella spirotricha*, can be separated from its methyl and acetyl derivatives in benzene–acetic acid (9:1) or chloroform–acetic acid (9:1) (Wat and Towers, 1971).

The spectra of benzoquinones characteristically have one strong band between 260 and 290 nm and one band of less intensity between 375 and 410 nm. 2,6-Dimethoxybenzoquinone, which occurs in wheatgerm, has, for example, maxima at 287 and 377 nm, whereas shanorellin has maxima at 272 and 406 nm.

(d) *Naphthaquinones*

These also separate well on silica gel. For surveying the distribution of plumbagin in roots of members of the Plumbaginaceae, the present author used petroleum (b.p. 60–80°C)–ethyl acetate (7:3), in which solvent plumbagin has R_F 76 (Harborne, 1967b). In surveying the Droseraceae for plumbagin and 7-methyljuglone, Zenk *et al.* (1969) extracted fresh leaves with ether and developed the concentrated extracts on silica gel plates in benzene–petroleum (b.p. 30–50°C) (2:1); recorded R_Fs were 40 and 30 respectively. For separating eleven naphthaquinones produced by the fungus *Fusarium solani*, parasitic on citrus trees, Tatum and Baker (1983) used silica gel and solvents such as hexane–acetone–acetic acid (15:5:0·3), benzene–nitromethane–acetic acid (75:25:2) and chloroform–ethyl acetate–hexane–acetic acid (10:5:5:0·3).

Naphthaquinones all have three or four spectral maxima: one or two below 300 nm, one at about 330 to 340 nm and one above 400 nm. Juglone (5-hydroxynaphthaquinone), first isolated from walnut, *Juglans regia*, has λ_{max} in ethanol at 249, 345 and 422 nm. By comparison, the closely related plumbagin has λ_{max} 220, 266 and 418 nm; in alkali, the visible maximum of plumbagin is at 526 nm and in the presence of alcoholic $AlCl_3$, the visible maximum is at 440 nm.

(e) *Anthraquinones*

In identifying these pigments from new plant sources, it is to be remembered that only a few anthraquinones occur at all regularly in plants. Emodin is the most frequent, occurring as it does in at least six higher plant families and also in a number of fungi. The chromatographic separation and spectral properties of six of the more common anthraquinones are shown in Table 2.18. These pigments have all been identified for example, in various *Rumex* species (Fairbairn and El-Muhtadi, 1972).

Table 2.18 R_Fs and spectral properties of anthraquinones

Anthraquinone	R_F ($\times 100$) *in*		*Spectral maxima in* EtOH (nm)	*Carbonyl bands in* IR (cm^{-1})
	System 1	*System 2**		
Emodin	52	18	223,254,267,290,440	
Chrysophanol	76	53	225,258,279,288,432	1630 (chelated)
Physcion	75	42	226,255,267,288,440	and
Aloe-Emodin	36	53	225,258,279,287,430	1670 (free)
Rhein	24	03	230,260 — — 432	
Emodic Acid	18	00	227,252,274,290,444	

*Key: System 1, silica gel plates made up with 0·01 M NaOH (50 ml per 25 g silica gel), solvent benzene−ethyl acetate−acetic acid (75:24:1); system 2, polyamide plates, solvent methanol−benzene (4:1).

Separation of anthraquinone mixtures is difficult and, often, special techniques have been developed for those mixtures present in a particular plant source. Aloes, for example, contain both *O*- and *C*-glycosides (e.g. barbaloin, the *C*-glycoside of aloe-emodin), which makes separation difficult. Hörhammer and Wagner (1965) solved the problem by using ethyl acetate−methanol−water (100:16·5:13·5) on silica gel plates (development time 75−90 min). The glycosides are mobile in this system, while they remain at the origin in most other solvents (e.g. those given in Table 2.18). For separating mixtures of anthraquinones and their reduced forms, the anthrones and dianthrones which occur in *Rhamnus* and *Cassia*, a useful support is silica gel G mixed with Kieselguhr G (1:6) (Labadie, 1969). Solvents for such separations are petroleum (40−60°C)−ethyl formate−formic acid (90:4:1) and petroleum (40−60°C)−ethyl formate−conc. HCl (85:15:05). Finally, mention should be made of the excellent anthraquinone separations that can be achieved on tartaric acid-treated silica gel plates in chloroform−methanol (99:1). The plates are made up using 3·75% aqueous tartaric acid for the slurry; five plates require 28 g silica gel GF and 72 ml of the tartaric acid solution.

Anthraquinones are detected on chromatographic plates by their visible and UV colours. By spraying plates with 10% methanolic KOH, the original yellow and yellow-brown colours change to red, violet, green or purple. Spectrally, anthraquinones can be distinguished from other classes of quinone by the fact they have four or five absorption bands in the UV and visible regions. At least three of these lie between 215 and 300 nm and another one is above 430 nm. Typical spectral values are shown in Table 2.18.

(f) *Lichen substances*

Because of the very small amounts of plant materials available, special methods have had to be devised for the detection of quinones and other pigments in lichens. Anthraquinones such as emodin occur widely in some lichen groups and Santesson (1970) has developed a special 'lichen MS' method for detecting them. Lichen fragments are introduced directly into a LKB 9000 GC-MS apparatus at 120–160°C and anthraquinone mixtures can be detected, as long as the molecular weights are different. Confirmation is by micro-scale TLC on silica gel, using toluene–hexane (8:1) or benzene–chloroform (3:1).

Other lichen compounds such as the phenolic depsidones and depsides (Culberson, 1969) are similarly separated by TLC, although PC and microchemical colour tests were once widely used. TLC is on silica gel in solvents such as n-hexane–ether–formic acid (5:4:1), the compounds being revealed with a 10% H_2SO_4 spray and heating at 70°C. TLC data for 220 lichen substances have been collected by Culberson (1972).

(g) *Isoprenoid quinones*

These usually occur in plant tissues in very small amounts and special procedures are needed to separate them from other lipid materials. The type of procedure involved includes extraction of the plant with ethanol–ether (1:1), evaporation and extraction into petroleum, evaporation and extraction into heptane, followed by chromatography on a sodium aluminium silicate (Decalso) or alumina column and eluting with 5% ether in heptane. The quinone separations are monitored spectrophotometrically by measuring the difference in extinction at 255 nm, before and after adding sodium borohydride.

Isoprenoid quinones are generally separated and detected by one- or two-dimensional TLC on silica gel. The naturally occurring ubiquinones, for example, will separate on paraffin-coated Kieselguhr G plates in acetone–water (9:1) (Wagner *et al.*, 1962); R_F values of ubiquinones range from 31 (for U_{50}) to 71 (for U_{30}). Detection is by means of spraying with a saturated solution of antimony chloride in chloroform and heating for 10 min at 110°C; grey-violet spots are obtained on a white background. More detailed procedures are outlined by Threlfall and Whistance (1970), who analysed nine higher plant tissues for these ubiquinones.

Plastoquinones are separated by similar procedures as for ubiquinones. For example, the separation of a new derivative, plastoquinone-8, from horse chestnut leaves required preliminary purification on alumina columns, followed by TLC on silica gel G in benzene–petroleum (40–60°C) (2:3) and then on paraffin-coated plates of silica gel G in acetone–water (19:1) (Whistance and Threlfall, 1970).

Spectral procedures are essential for confirming the identity of these quinones. Ubiquinones have λ_{max} in petroleum at 270 nm, whereas plasto-quinones absorb at 254 and 261 nm. Full details of spectral properties are given by Morton (1965).

2.9.3 Practical experiment

(a) *Detection and bioassay of allelopathic quinones*

The term allelopathy is used whenever a higher plant species exudes chemicals into the environment which may inhibit the growth of a second competitor species. It is most marked when trees or shrubs prevent the growth of annual herbs in their vicinity. Walnut trees do this, for example, by synthesizing the naphthaquinone juglone in their tissues. If the inhibitor or toxin is a phenolic compound, it is usually stored in non-toxic combined form (e.g. as a glucoside) and conversion to toxicity occurs during its release into the soil. This is true of juglone in walnut and also of hydroquinone (the quinol of benzoquinone) which occurs in leaves of *Arctostaphylos* and *Arbutus* species as the glucoside arbutin.

In this experiment, juglone and hydroquinone are to be released from their bound forms, identified by TLC and their allelopathic effects compared by bioassay against lettuce seed germination. Comparison can be made with arbutin and with carvone, a terpenoid also implicated as an allelopathic agent.

(b) *Detection of toxins in walnut and Arctostaphylos*

(1) Immerse small portions of walnut green shoot tissue and *Arctostaphylos* leaves separately in 2 N aqueous HCl in two test tubes. Heat for 0·5 h at 100°C.

(2) Cool to room temperature, extract with ether, pipette off ether layer into small beaker, and evaporate to dryness. Take up residue in 1–2 drops ethanol.

(3) Spot up on TLC plate (silica gel), also running marker solutions of juglone and hydroquinone. Develop plates for 40 min in petroleum (b.p. 40–60°C)–ethyl acetate (7:3). Take out and dry. Examine plate in visible

light (juglone is yellow) and spray with Folin reagent (hydroquinone gives a dark grey spot).

(c) *Bioassay with lettuce seeds*

(1) Set up five experiments in Petri dishes lined with filter paper circles, with 50 lettuce seeds per dish. The filter paper in the control is soaked with water–ethanol (99:1). The other four are soaked respectively with solutions of juglone, hydroquinone, arbutin and carvone containing 4 mg ml^{-1} water–ethanol (99:1).
(2) Incubate at 26°C for 24–48 hours in the light (time depending on type of lettuce seed used).
(3) Remove dishes and count number of seeds which have germinated. From these results and those of the control, it is possible to place the four compounds in order according to their effectiveness as inhibitors.

(d) *Additional experiments*

Other plants containing bound quinones can be similarly screened for the release of free pigments. The roots of *Plumbago capensis*, a popular ornamental shrub in the tropics, yield the naphthaquinone plumbagin on acid treatment.

The effects of quinones on seed germination can be investigated with seeds other than lettuce. In addition, other effects on growth can be observed; shoot and root growth are also altered by the presence of quinones.

REFERENCES

General references

Harborne, J.B. (ed.) (1964) *Biochemistry of Phenolic Compounds*, Academic Press, London.
Harborne, J.B. (1967a) *Comparative Biochemistry of the Flavonoids*, Academic Press, London.
Harborne, J.B. (1983) Phenolic Compounds. In *Chromatography, Fundamentals and Applications, Part B* (ed. E. Heftmann), Elsevier, Amsterdam, pp. 407–34.
Harborne, J.B. and Mabry, T.J. (eds) (1982) *The Flavonoids: Advances in Research*, Chapman and Hall, London.
Harborne, J.B., Mabry, T.J. and Mabry, H. (eds) (1975) *The Flavonoids*, Chapman and Hall, London.
Haslam, E. (1966) *Chemistry of Vegetable Tannins*, Academic Press, London.
Haslam, E. (1982) Proanthocyanidins. in *Flavonoids: Advances in Research* (ed. J.B. Harborne and T.J. Mabry), Chapman and Hall, London, pp. 417–48.
Mabry, T.J., Markham, K.R. and Thomas, M.B. (1970) *The Systematic Identification of Flavonoids*, Springer-Verlag, Berlin.
Markham, K.R. (1982) *Techniques of Flavonoid Identification*, Academic Press, London.

Morton, R.A. (ed.) (1965) *Biochemistry of Quinones.* Academic Press, London.

Preston, S.T. and Pancratz, R. (1978) *A Guide to the Analysis of Phenols by Gas Chromatography,* 2nd edn, Polyscience Corp., Evanston, Ill.

Ribereau-Gayon, P. (1972) *Plant Phenolics,* Oliver and Boyd, Edinburgh.

Swain, T. (1979) Tannins and Lignins. in *Herbivores: Their Interaction with Secondary Plant Metabolites* (ed. G.A. Rosenthal and D.H. Janzen), Academic Press, New York, pp. 657–82.

Swain, T., Harborne, J.B. and Van Sumere, C.F. (eds) (1979) *Biochemistry of Plant Phenolics,* Plenum Press, New York.

Thomson, R.H. (1971) *Naturally Occurring Quinones,* Academic Press, London.

Supplementary references

Abe, S. and Ohta, Y. (1983) *Phytochemistry,* **22,** 1917.

Andersen, J.M. and Pedersen, W.B. (1983) *J. Chromatog.,* **259,** 131.

Andersen, R.A. and Vaughn, T.H. (1972) *Phytochemistry,* **11,** 2593.

Asen, S. (1965) *J. Chromatog.,* **18,** 602.

Bailey, J.A. and Mansfield, J.W. (eds) (1982) *Phytoalexins,* Blackie, Glasgow.

Bate-Smith, E.C. (1962) *J. Linn. Soc. Bot.,* **58,** 39.

Bate-Smith, E.C. (1972) *Phytochemistry,* **11,** 1755.

Bate-Smith, E.C. (1973) *Phytochemistry,* **12,** 907.

Bate-Smith, E.C. (1977) *Phytochemistry,* **16,** 1421.

Carpenter, I., Locksley, H.D. and Scheinmann, F. (1969) *Phytochemistry,* **8,** 2013.

Crowden, R.K., Harborne, J.B. and Heywood, V.H. (1969) *Phytochemistry,* **8,** 1963.

Culberson, C.F. (1969) *Chemical and Botanical Guide to Lichen Products,* Univ. North Carolina Press, Chapel Hill.

Culberson, C.F. (1972) *J. Chromatog.,* **72,** 113.

Ellis, C.J., Foo, L.Y. and Porter, L.J. (1983) *Phytochemistry,* **22,** 483.

Fairbairn, J.W. and El-Muhtadi, F.J. (1972) *Phytochemistry,* **11,** 263.

Forrest, J.E., Richard, R. and Heacock, R.A. (1972) *J. Chromatog.,* **63,** 439.

Galston, A.W. (1969) in *Perspectives in Phytochemistry* (eds J.B. Harborne and T.S. Swain), Academic Press, London, pp. 193–204.

Glencross, R.G., Festenstein, G.N. and King, H.G.C. (1972) *J. Sci. Food Agric.,* **23,** 371.

Goldstein, J. and Swain, T. (1965) *Phytochemistry,* **4,** 185.

Gorham, J. (1977) *Phytochemistry,* **16,** 249.

Haddock, E.A., Gupta, R.K., Al-Shafi, S.M.K., Layden, K., Haslam, E. and Magnolato, D. (1982) *Phytochemistry,* **21,** 1049.

Hänsel, R., Schulz, H. and Lenckest, C. (1964) *Z. Naturforsch.,* **19b,** 727.

Harborne, J.B. (1966) *Z. Naturforsch.,* **21b,** 604.

Harborne, J.B. (1967b) *Phytochemistry,* **6,** 1415.

Harborne, J.B. (1968) *Phytochemistry,* **7,** 1215.

Harborne, J.B. (1969) *Phytochemistry,* **8,** 177.

Harborne, J.B. (1977) *Progr. Phytochem.,* **4,** 189.

Harborne, J.B., Heywood, V.H. and Williams, C.A. (1969) *Phytochemistry,* **8,** 1729.

Harborne, J.B. and Williams, C.A. (1969) *Phytochemistry,* **8,** 2223.

Haslam, E. (1965) *Phytochemistry,* **4,** 495.

Hörhammer, L. and Wagner, H. (1965) *Dt. Apoth. Ztg.,* **105,** 827.

Hostettmann, K. and Wagner, H. (1977) *Phytochemistry*, **16**, 821.

Hughes, J.C. and Swain, T. (1960) *Phytopathology*, **50**, 398.

Ibrahim, R.K. and Towers, G.H.N. (1960) *Arch. Biochem. Biophys.*, **87**, 125.

Ingham, J.L. (1981) in *Advances in Legume Systematics* (eds R.M. Polhill and P.H. Raven). HMSO, London, pp. 599–626.

Irvine, W.J. and Saxby, M.J. (1969) *Phytochemistry*, **8**, 2067.

Isman, M.B. and Duffey, S.S. (1981) *Ent. exp. & appl.*, **31**, 370.

Jones, W.T., Broadhurst, R.B. and Lyttleton, J.W. (1976) *Phytochemistry*, **15**, 1407.

Khan, N.U., Ansari, W.H., Usmani, J.N., Ilyas, M. and Rahman, W. (1971) *Phytochemistry*, **10**, 2129.

Koeppen, B.H. (1965) *J. Chromatog.*, **18**, 604.

Koster, J., Strack, D. and Barz, W. (1983) *Planta med.*, **48**, 131.

Labadie, R.P. (1969) *Pharm. Weekblad.*, **104**, 257.

Lea, A.G.H. (1980) *J. Chromatog.*, **194**, 62.

Markham, K.R. and Chari, V.M. (1982) in *The Flavonoids: Advances in Research* (ed. J.B. Harborne and T.J. Mabry), Chapman and Hall, London, pp. 19–134.

Martin, J.S. and Martin, M.M. (1983) *J. Chem. Ecol.*, **9**, 285.

Murray, R.D.H., Mendez, J. and Brown, S.A. (1982) *The Natural Coumarins*, J. Wiley, Chichester.

Nicolaus, R.A. and Piattelli, M. (1965) *Rend. dell' Acad. Sci. Fis. Mat. Napoli*, **32**, 3.

Nybom, N. (1964) *Physiol. Plantarum*, **17**, 157.

Pedersen, J.A. (1978) *Phytochemistry*, **17**, 775.

Porter, L.J., Czochanska, Z., Foo, L.Y., Newman, R.H., Thomas, W.A. and Jones, W.T. (1979) *J. Chem. Soc., Chem. Comm.*, 375.

Pridham, J.B. (1959) *J. Chromatog.*, **2**, 605.

Pryce, R.J. (1971) *Planta*, **97**, 354.

Pryce, R.J. (1972a) *Phytochemistry*, **11**, 1911.

Pryce, R.J. (1972b) *Phytochemistry*, **11**, 1759.

Reio, L. (1964) *Svensk Kemisk Tidskrift*, **76**, 5.

Reyes, R.E. and Gonzalez, A.G. (1970) *Phytochemistry*, **9**, 833.

Rittich, B. and Krska, M. (1977) *J. Chromatog.*, **130**, 189.

Robertson, N.F., Friend, J., Aveyard, M.A., Brown, J., Huffee, M. and Homans, A.L. (1968) *J. Gen. Microbiol.*, **54**, 216.

Robeson, D.J. and Harborne, J.B. (1980) *Phytochemistry*, **19**, 2359.

Santesson, J. (1970) *Phytochemistry*, **9**, 2149.

Seikel, M.K. and Hillis, W.E. (1970) *Phytochemistry*, **9**, 1115.

Steele, J.W. and Bolan, M. (1972) *J. Chromatog.*, **71**, 427.

Stewart, R.N., Asen, S., Massie, D.R. and Norris, K.H. (1979) *Biochem. System. Ecol.*, **7**, 281.

Strack, D., Akavia, N. and Reznik, H. (1980) *Z. Naturforsch.*, **35c**, 533.

Tatum, J.H. and Baker, R.A. (1983) *Phytochemistry*, **22**, 543.

Threlfall, D.R. and Whistance, G.R. (1970) *Phytochemistry*, **9**, 355.

van der Casteele, K., Geiger, H. and Van Sumere, C.F. (1983) *J. Chromatog.*, **258**, 111.

Vaughn, T.H. and Andersen, R.A. (1971) *Biochim. Biophys. Acta*, **244**, 437.

Wagner, H. and Holzl, J. (1968) *Dtsch Apoth. Ztg.*, **42**, 1620.

Wagner, H., Horhammer, L. and Dengler, B. (1962) *J. Chromatog.*, **7**, 211.

Wat, C.K. and Towers, G.H.N. (1971) *Phytochemistry*, **10**, 1355.

Whistance, G.R. and Threlfall, D.R. (1970) *Phytochemistry*, **9,** 737.
Widen, C.J., Physalo, H. and Salovaara, P. (1980) *J. Chromatog.*, **188,** 213.
Wollenweber, E. and Egger, K. (1971) *Phytochemistry*, **10,** 225.
Zenk, M.H., Furbringer, M. and Steglich, W. (1969) *Phytochemistry*, **8,** 2199.

The Terpenoids

3.1 INTRODUCTION

An enormous range of plant substances are covered by the word 'terpenoid', a term which is used to indicate that all such substances have a common biosynthetic origin. Thus, terpenoids are all based on the isoprene molecule $CH_2{=}C(CH_3){-}CH{=}CH_2$ and their carbon skeletons are built up from the union of two or more of these C_5 units. They are then classified according to whether they contain two (C_{10}), three (C_{15}), four (C_{20}), six (C_{30}) or eight (C_{40}) such units. They range from the essential oil components, the volatile mono- and sesquiterpenes (C_{10} and C_{15}), through the less volatile diterpenes (C_{20}) to the involatile triterpenoids and sterols (C_{30}) and carotenoid pigments (C_{40}). Each of these various classes of terpenoid (Table 3.1) are of significance in either plant growth, metabolism or ecology.

Although terpenoids are derived biogenetically from the molecule of isoprene, which does occur as a natural product, this substance is not the *in vivo* precursor. Instead, the compound actually involved is isopentenyl pyro- phosphate, $CH_2{=}C(CH_3)CH_2CH_2OPP$, which is formed itself from acetate *via* mevalonic acid, $CH_2OH{-}CH_2C(OH,CH_3)CH_2CO_2H$. Isopentenyl pyrophosphate exists in living cells in equilibrium with the isomeric dimethyl- allyl pyrophosphate, $(CH_3)_2C{=}CHCH_2OPP$. In biosynthesis, a molecule of isopentenyl pyrophosphate and one of dimethylallyl pyrophosphate are linked together to give geranyl pyrophosphate (C_{10}), the key intermediate in mono- terpene formation; geranyl pyrophosphate and isopentenyl pyrophosphate are, in turn, linked to give farnesyl pyrophosphate (C_{15}), the key intermediate of sesquiterpene synthesis. Different combinations of these C_5, C_{10} and C_{15} units are then involved in the synthesis of the higher terpenoids, triterpenoids being formed from two farnesyl units and carotenoids from the condensation of two geranylgeranyl units (see Fig. 3.1). Most natural 'terpenoids' have cyclic structures with one or more functional groups (hydroxyl, carbonyl, etc.) so that the final steps in synthesis involve cyclization and oxidation or other structural modification.

Table 3.1 The main classes of plant terpenoids

Number of isoprene units	Carbon number	Name or class	Main types and occurrence
1	C_5	isoprene	detected in *Hamamelis japonica* leaf
2	C_{10}	monoterpenoids	monoterpenes in plant essential oils (e.g. menthol from mint)
			monoterpene lactones (e.g. nepetalactone)
			tropolones (in gymnosperm woods)
3	C_{15}	sesquiterpenoids	sesquiterpenes in essential oils
			sesquiterpene lactones (especially common in Compositae)
			abscisins (e.g. abscisic acid)
4	C_{20}	diterpenoids	diterpene acids in plant resins
			gibberellins (e.g. gibberellic acid)
6	C_{30}	triterpenoids	sterols (e.g. sitosterol)
			triterpenes (e.g. β-amyrin)
			saponins (e.g. yamogenin)
			cardiac glycosides
8	C_{40}	tetraterpenoids	carotenoids* (e.g. β-carotene)
n	C_n	polyisoprene	rubber, e.g. in *Hevea brasiliensis*

*C_{50}-based carotenoids are known in some bacteria.

Chemically, terpenoids are generally lipid-soluble and are located in the cytoplasm of the plant cell. Essential oils sometimes occur in special glandular cells on the leaf surface, whilst carotenoids are especially associated with chloroplasts in the leaf and with chromoplasts in the petal. Terpenoids are normally extracted from plant tissues with light petroleum, ether or chloroform and can be separated by chromatography on silica gel or alumina using the same solvents. There is, however, often difficulty in detection on a microscale, since all (except carotenoids) are colourless and there is no sensitive universal chromogenic reagent for them. Reliance often has to be placed on relatively non-specific detection on TLC plates with conc.H_2SO_4 and heating.

Isomerism is common among terpenoids, and pairs of isomeric forms may be isolated from plants; one such pair are the monoterpenes geraniol and nerol

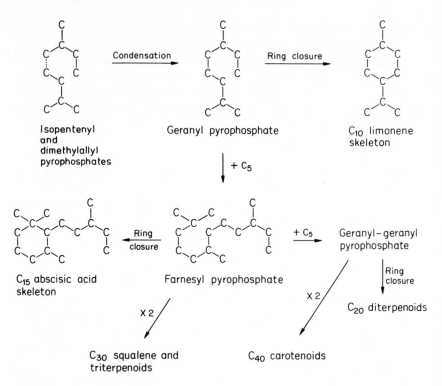

Fig. 3.1 The path of terpenoid biosynthesis in plants (for simplicity, only the carbon skeletons are shown).

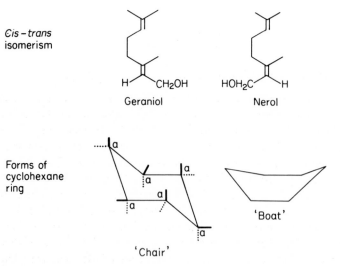

Fig. 3.2 Conformation of terpenoids.

(see Fig. 3.2). In addition, terpenoids are mostly alicyclic compounds and because the cyclohexane ring is usually twisted in the so-called 'chair' form (see Fig. 3.2), different geometric conformations are possible, depending on the substitution around the ring. The stereochemistry of the cyclic terpenoids is therefore highly involved and often difficult to determine. From the practical point of view, it must be remembered that both isomerization and structural re-arrangement within the molecule may occur quite readily under relatively mild conditions and artifact formation is always possible during isolation procedures.

A considerable number of quite different functions have been ascribed to plant terpenoids. Their growth-regulating properties are very well documented; two of the major classes of growth regulators are the sesquiterpenoid abscisins and the diterpenoid-based gibberellins. The important contribution of carotenoids to plant colour is well known and it is now certain that these C_{40} terpenoids are also involved as accessory pigments in photosynthesis. The importance of mono- and sesquiterpenes in providing plants with many of their distinctive smells and odours is also familiar to most scientists. Less is generally known of the role of terpenoids in the more subtle interactions between plants and animals, e.g. as agents of communication and defence among insects, but this is now an area of active research. Finally, it should be mentioned that certain non-volatile terpenoids have been implicated as sex hormones among the fungi.

Reviews of many aspects of plant terpenoids are included in Volume 4 of a comprehensive treatise on plant biochemistry (Stumpf, 1980). The introductory text on the chemistry of terpenes by Newman (1972) can also be recommended.

3.2 ESSENTIAL OILS

3.2.1 Chemistry and distribution

The mainly terpenoid essential oils comprise the volatile steam-distillable fraction responsible for the characteristic scent, odour or smell found in many plants. They are commercially important as the basis of natural perfumes and also of spices and flavourings in the food industry. Plant families particularly rich in essential oils include the Compositae, *Matricaria*, Labiatae, e.g. the mints, *Mentha* spp., Myrtaceae, *Eucalyptus*, Pinaceae, *Pinus*, Rosaceae, 'attar' of roses, Rutaceae, *Citrus* oils and Umbelliferae, anise, caraway, cumin, dill, etc. Other classes of chemical substance may be present with terpenes in essential oils. For example, sulphur-based smells are common in plants of the Cruciferae and in *Allium*, Liliaceae; the detection of these substances are dealt with elsewhere (see Chapter 4, p. 172). Also, terpenes are often associated in odour fractions with aromatic compounds such as the phenylpropanoids; methods of identification of these compounds have already been discussed earlier in Chapter 2 (p. 52).

Chemically, the terpene essential oils can be divided into two classes, the mono- and sesquiterpenes, C_{10} and C_{15} isoprenoids, which differ in their boiling point range (monoterpenes b.p. 140–180°C, sesquiterpenes b.p. >200°C). First of all, with regard to the monoterpenes, these substances can be further divided into three groups depending on whether they are acyclic (e.g. geraniol), monocyclic (e.g. limonene) or bicyclic (e.g. α- and β-pinene). Within each group, the monoterpenes may be simple unsaturated hydrocarbons (e.g. limonene) or may have functional groups and be alcohols (e.g. menthol), aldehydes or ketones (e.g. menthone, carvone).

Also included with monoterpenes on biosynthetic grounds are monoterpene lactones (better known as iridoids, see Bate-Smith and Swain, 1966) and tropolones. A typical monoterpene lactone is nepetalactone, the principal odour constituent of catmint, *Nepata cataria*, Labiatae, a plant which has a peculiar attraction for the domestic cat because of its odour. Other iridoids such as loganin are of interest because they are intermediates in the biosynthesis of the indole alkaloids (see Chapter 5). A typical tropolone is γ-thujaplicin; these substances have a restricted distribution in certain fungi and are also found as heartwood constituents in the Cupressaceae (Erdtman and Norin, 1966). The structures of a range of typical monoterpenoids are collected in Fig. 3.3.

Simple monoterpenes are widespread and tend to occur as components of the majority of essential oils. Some compounds are regularly found together in leaf oils, especially α- and β-pinene, limonene, Δ^3-carene, α-phellandrene and myrcene. Flower and seed oils tend to have more specialized monoterpenes present. However, whatever the tissue, complex mixtures tend to be the rule rather than the exception. While there may be ten or fifteen easily detectable components, there may be many other terpenoids present in trace amounts.

Like the monoterpenes, the sesquiterpenes fall chemically into groups according to the basic carbon skeleton; the common ones are either acyclic (e.g. farnesol), monocyclic (e.g. γ-bisabolene) or bicyclic (e.g. β-selinene, carotol). However, within each group there are many different compounds known; indeed, according to a recent estimate, there are several thousand sesquiterpenoids with well-defined structures, belonging to some 200 skeletal types. Two derived sesquiterpenoids deserve special mention because of their growth-regulating properties; abscisic acid and xanthinin. Abscisic acid, a sesquiterpene carboxylic acid related in structure to the carotenoid violaxanthin (see Section 3.5), is best known as the principal hormone controlling dormancy in seeds of herbaceous plants and in buds of woody plants (Addicott, 1983). Xanthinin, which occurs in cocklebur, *Xanthium pennsylvanicum*, has a less well defined role in plant physiology as an auxin-antagonist. It, however, represents an important class of sesquiterpenes which are also

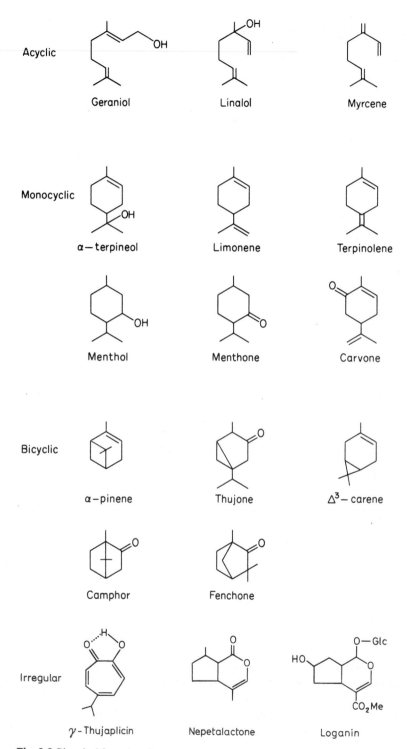

Acyclic

Geraniol   Linalol   Myrcene

Monocyclic

α−terpineol   Limonene   Terpinolene

Menthol   Menthone   Carvone

Bicyclic

α−pinene   Thujone   Δ³− carene

Camphor   Fenchone

Irregular

γ-Thujaplicin   Nepetalactone   Loganin

Fig. 3.3 Chemical formulae of monoterpenes.

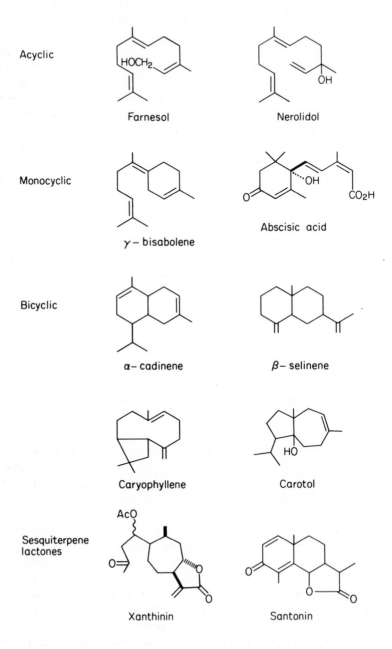

Acyclic

Farnesol

Nerolidol

Monocyclic

γ – bisabolene

Abscisic acid

Bicyclic

α – cadinene

β – selinene

Caryophyllene

Carotol

Sesquiterpene lactones

Xanthinin

Santonin

Fig. 3.4 Formulae of sesquiterpenoids.

lactones and which have a wide distribution in the Compositae (Seaman, 1980). Other properties possessed by these sesquiterpene lactones include their occasional bitter or pungent taste and their ability to act as allergens (Rodriguez *et al.*, 1976). The chemical formulae of the various sesquiterpenes mentioned above are shown in Fig. 3.4.

For their isolation from plant tissues, mono- and sesquiterpenes are nowadays separated by extraction into ether, petroleum or acetone. The classic procedure for essential oils is separation from fresh tissue by steam distillation. This step is now often omitted, because of the danger of artifact formation at the raised temperatures involved. Terpenes may either undergo rearrangement (e.g. dehydration in the case of tertiary alcohols) or poly-merization. The volatility of the simple terpenes means that they are ideal subjects for separation by GLC. Many have fragrant odours and indeed can often be recognized in plant distillates directly, if present as the major constituent.

The major references to the volatile terpenes are the two series of volumes *The Essential oils* edited by E. Guenther (1948–52) and *The Terpenes* by J. L. Simonsen (1947–52). These are now somewhat out of date. The mono- and sesquiterpenes identified as volatile constituents in most common fruits and vegetables have been exhaustively listed by Nursten (1970) and Johnson *et al.* (1971). More recent general reviews are those of Charlwood and Banthorpe (1978) and Loomis and Croteau (1980).

3.2.2   Recommended techniques for mono- and sesquiterpenes

(a) *Gas liquid chromatography*

This is undoubtedly the most important technique for study of essential oils, since it yields in one operation both qualitative and quantitative analyses. This is particularly important when a similar set of compounds occur throughout a particular plant group, since it is the quantitative variations that are most significant. Certainly, GLC is an indispensable tool for chemo-taxonomic studies of essential oils in leaf or bark, as in the gymnosperms. A typical separation of terpenes in pine resin by GLC is illustrated in Fig. 3.5.

For the identification of volatile terpenes in any plant material, it is essential to combine the use of GLC with other procedures, and especially with TLC. TLC is useful, for example, for monitoring fractions separated by preparative GLC; on the other hand, if a preparative GLC apparatus is not available, large-scale separations can be carried out on TLC, with the TLC fractions subsequently being monitored by GLC. For confirming identity, IR spectra of separated oils were at one time determined routinely, but it is now more usual to measure the mass spectra, since most terpenes give characteristic frag-mentation patterns. In critical cases, both IR and mass spectra should be determined.

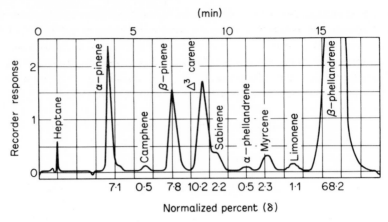

Fig. 3.5 Gas liquid chromatography separation of terpenes in pine resin. Separation on 10% oxydipropionitrile column on 60–80 mesh acid washed Chromosorb W (from Smith, 1964).

Preliminary fractionation of a crude ether or petroleum plant extract by column chromatography on silicic acid is sometimes advantageous, since it avoids contamination of the GLC column by high boiling impurities that may be present in such crude extracts. However, satisfactory results have been obtained in this laboratory by direct application of dry ether extracts from leaf or powdered seed.

A wide variety of column stationary phases have been employed for essential oils. The most popular non-polar phases are probably apiezon L and silicone SE 30; among polar phases, diethylene glycol adipate polyester and Carbowax 400 are widely used. It is important that the support material (e.g. Chromosorb W) should be freed from traces of iron, base or acid, since terpenes are sensitive to such impurities.

Temperature programming is necessary in the GLC of a crude oil preparation, in order to successfully separate the monoterpenes, the sesquiterpenes and their oxygenated derivatives. On non-polar columns, hydrocarbons elute according to the boiling point but, on other types of column, it is not always possible to predict the relative retention times. Some examples of relative retention times are shown in Table 3.2. These are given to illustrate the need for using more than one column for purposes of terpene separation and identification. Thus, pairs of related compounds (e.g. Δ^3-carene and α-phellandrene) may not separate on one column (Apiezon N) but they will on others (polyethylene glycol). In the detailed analysis of individual oil constituents of a particular plant material, it is normal practice to use a range of GLC columns for identification purposes. In surveying the variation in essential oil composition within populations of a single species, it is necessary to restrict the analysis to a single column, so that it is important first to find the stationary phase which yields the best separation. In chemotaxonomic work, it is worth noting that it is possible to obtain analyses directly on small pieces of plant material placed directly in the inlet port of the GLC apparatus. This procedure has been used successfully on fragments of herbarium material (Harley and Bell, 1967), which in spite of years in storage on paper sheets still yield sufficient essential oil for determination of their profiles.

Table 3.2 Relative retention times of terpenes in gas liquid chromatography

	RR_Ts on Column*		
Terpene	10% *Apiezon* N	15% *Polyethylene glycol*	15% *Polyethylene glycol bis-propionitrile*
α-Pinene	42	29	30
Camphene	50	41	44
β-Pinene	63	55	54
Δ³-Carene	82	73	67
Myrcene	60	82	88
α-Phellandrene	82	82	86
Limonene	100	100	100
β-Phellandrene	97	106	116
p-Cymene	100	175	232

*RR_Ts relative to limonene, from isothermal runs at 65°C on a 300 cm column (from von Rudloff, 1966).

When complex mixtures of oils are encountered, as in flavour analysis, combined GLC–MS is now routinely used for the resolution and identification of monoterpenoids. The addition of a computer for storing data and library searching has greatly facilitated these operations. Recent developments in such separations include the replacement of packed by open columns coated with the stationary phase, in what is termed capillary chromatography (Croteau and Ronald, 1983).

(b) *Thin layer chromatography*

As already mentioned, TLC can be used with advantage in combination with GLC for the analysis of terpenes, since the two techniques are complementary. It is even possible, in the absence of a GLC apparatus, to analyse essential oils using TLC as the only separation technique (e.g. Hörhammer *et al.*, 1964). Even when GLC is available, TLC is useful at all stages for separation and analysis of these terpenes. When dealing with the less volatile sesquiterpenoids, it may even become the method of choice.

Silica gel is the most widely used absorbent, with solvents such as benzene, chloroform, benzene–chloroform (1:1) and benzene–ethyl acetate (19:1). For the analysis of oxygen-containing terpenes (e.g. carvone), silica gel layers should not be activated prior to use, since the moisture present aids the separation. Terpene alcohols are best separated on paraffin-impregnated plates in 70% methanol. Activated silica gel plates are first immersed in 5% paraffin in petroleum for 1 min and then allowed to dry before use; the chromatographic solvent, 70% methanol, must also be saturated with paraffin oil. Another modification, to separate terpenes according to the number of double bonds, involves TLC on silica gel plates spread as a slurry with 2·5% aqueous AgNO$_3$ instead of with water. The solvent system to employ with the AgNO$_3$ treated plates is methylene dichloride–chloroform–ethyl acetate–*n*-propanol (45:45:4·5:4·5).

General methods of detection include spraying with 0·2% aqueous KMnO$_4$, 5% antimony chloride in chloroform, conc.H$_2$SO$_4$ or vanillin–H$_2$SO$_4$. The latter reagent is prepared fresh by adding 8 ml ethanol with cooling to 0·5 g vanillin in 2 ml conc.H$_2$SO$_4$. The plates are heated after spraying at 100–105°C until full development of colours has occurred. More selective agents are available for detecting terpenes with double bonds (bromine vapour) and those with ketonic groupings (2,4-dinitrophenyl-hydrazine). The responses of some of the common terpenes to a range of detection agents are indicated in Table 3.3.

Table 3.3 Detection of monoterpenes on thin layer chromatography plates

	*Response to test**			
Terpene	UV	*Bromine*	2,4-DNP	*conc.*H$_2$SO$_4$
Limonene	−	+	−	brown
α-Pinene	−	+	−	brown
Pulegone	+	+	+	yellow
Geraniol	−	+	−	purple
Carvone	+	+	+	pink
p-Cymene	+	−	−	−
α-Terpineol	−	+	−	green
1,8-Cineole	−	−	−	green

*Key: UV, examine in short UV light; bromine, spray with 0·05% fluorescein in water, expose plate to bromine vapour, yellow spots on a red background; 2,4-DNP, spray with 0·4 g 2,4-DNP in 100 ml 2 M HCl, yellow spots on white background; conc.H$_2$SO$_4$, spray with conc.H$_2$SO$_4$ and heat plate at 100°C for 10 min.

Further details of the TLC of essential oils may be found in Stahl (1969). Procedures applicable to the separation of essential oils in some seventy plant species are tabulated in this work.

(c) *Tropolones*

These are conveniently separated by chromatography on paper or cellulose plates. Zavarin *et al.* (1959) have used paper impregnated with phosphoric acid and iso-octane–toluene as solvent, while Wachtmeister and Wickberg (1958) employed paper impregnated with ethylenediamine tetracetic acid and light petroleum as solvent. The tropolones were detected with 1% aqueous $FeCl_3$.

(d) *Iridoids*

These monoterpene lactones most frequently occur in plants combined with sugar as glucosides and they therefore require special techniques for their analysis. PC, for example, is widely used in their detection. Weiffering (1966) has described a simple procedure, based on the Trim–Hill colour test (Trim and Hill, 1951), for surveying plant material for some of the common iridoids. Fresh tissue (1 g), or herbarium material, if necessary, is cut into small pieces and placed in a test tube with 5 ml 1% aqueous HCl. After 3–6 h, 0·1 ml of the macerate is decanted into another tube containing 1 ml of the Trim–Hill reagent (made up from 10 ml acetic acid, 1 ml 0·2% $CuSO_4$. $5H_2O$ in water and 0·5 ml conc.HCl). When the tube is heated for a short time in a flame, a colour is produced if certain iridoids are present. Asperulin, aucubin and monotropein give blue colours, harpagide a red-violet. Certain iridoids (catalpin, loganin) cannot be detected by this method and respond only to general tests for glucosides, e.g. benzidine–trichloroacetic acid (Duff *et al.*, 1965).

For more detailed identification, 100 ml dried tissue is ground with an equal weight of sand and treated with 10 ml 2% aqueous lead acetate. After filtering, adding dil. H_2SO_4 and centrifuging, the supernatant is treated with 50 mg

Table 3.4 R_Fs and colours of iridoids

	R_F ($\times 100$) in		Colours with	
Iridoid	BAW	IsoPrOH–water	*antimony chloride*	*anisaldehyde–H_2SO_4*
Asperulin	51	90	blue	blue
Aucubin	38	78	brown	red-violet
Catalpol	32	79	brown	orange
Harpagide	34	—	violet-brown	red
Loganin	63	93	red	orange
Monotropein	33	70	blue	blue

Solvents: n-BuOH–HOAc–H_2O (4:1:5) and isoPrOH–H_2O (3:2).
Sprays: 15% antimony chloride in chloroform and anisaldehyde–conc.H_2SO_4–ethanol (1:1:18); to develop colours heat papers for 2–5 min at 100°C.

sand and 50 mg celite in a Petri dish and the water present is allowed to evaporate off, with heating. The residue is extracted overnight with 4 ml 96% ethanol. The extract is centrifuged and concentrated to 0·1 ml. Aliquots are then examined by PC (Table 3.4) for the presence of specific iridoids. Detection reagents recommended are antimony chloride and anisaldehyde–H_2SO_4. It is also possible to use the Trim–Hill colour test on paper, if it is made up with toluene *p*-sulphonic acid instead of HCl and the paper is heated, after spraying, in the presence of acetic acid vapour (Bate-Smith, 1964).

Preparative separation of iridoids from plant extracts can be achieved by column chromatography on cellulose CF-11 eluted with *n*-butanol saturated with water or on silica gel eluted with chloroform–methanol in varying proportions. HPLC can be carried out on a μBondapak C_{18} column using water–methanol (1:1) as the mobile phase.

(e) *Sesquiterpene lactones*

A general procedure for the detection of these lactones in plants has been outlined by Mabry (1970). Dried leaves (20 g) are ground in a Waring blender with 100 ml chloroform. The slurry is filtered and the extract taken to dryness *in vacuo*. The residue is dissolved in 25 ml 95% ethanol and 25 ml 4% aqueous lead acetate added. The solution is filtered and the filtrate concentrated. The water–oil mixture is extracted with chloroform and the extract dried and evaporated. The residue is analysed directly by both TLC and NMR.

TLC is conducted on silica gel G in chloroform–ether (4:1), benzene–acetone (4:1), chloroform–methanol (99:1), benzene–methanol (9:1), benzene–ether (2:3) or light petroleum–chloroform–ethyl acetate (2:2:1). The lactones are detected as brown spots by placing the developed plates in a chamber containing iodine crystals. Alternatively, the lactones appear as green, brown, yellow, red or blue spots on plates sprayed with conc.H_2SO_4 and heated for 5 min at 100–110°C. The colour produced can be diagnostic of certain structural features in the lactone (Geissman and Griffin, 1971). Xanthinin and cumambrin B, for example are chemically related and both give a red colour λ_{max} 540 nm). Other spray reagents, namely 1% methanolic resorcinol–5% phosphoric acid (1:1) (Drozdz and Blosyk, 1978) and vanillin–H_2SO_4 (Picman *et al.*, 1980) have been developed for the selective detection of these lactones.

HPLC may be useful in conjunction with TLC for monitoring the purification of sesquiterpene lactones. In isolating the bitter lactones of chicory root, for example, we have found that isocratic elution from a Partisil column with chloroform–methanol (19:1) monitored at 254 nm to give excellent resolution.

Not all lactones show such strong UV absorbance as lactucin and lactupicrin in chicory and lower wavelengths may have to be adopted in other cases.

For example, in separating the lactones of *Parthenium* species on an Ultra-sphere-ODS column by gradient elution with acetonitrile-water, Marchand *et al.* (1983) detected the lactones in the eluates at 215 nm. For large-scale separation, HPLC is possible but, in general, standard chromatography on thick layers or columns of silica gel is usually quite satisfactory. The lactones are then identified mainly on the basis of melting point, rotation, NMR and mass spectra.

(f) *Abscisic acid*

Special methods have been devised for the detection and estimation in plants of the sesquiterpene growth inhibitor, abscisic acid. A typical procedure (Little *et al.*, 1972) starts with extraction of fresh tissue with 80% methanol, evaporation of methanol from the filtered extract, acidification and extraction into ether. The crude abscisic acid is then removed from this ether extract into aqueous satd. $NaHCO_3$ and recovered, after washing with ether and acidifying, into ether. The residue is chromatographed on Whatman 3MM paper in isopropanol–ammonia–water (8:1:1) and the zone at R_F 0·5–0·8 eluted with methanol. This is then further chromatographed on silica gel GF_{254} plates in benzene–ethyl acetate–acetic acid (14:6:1) and benzene–chloroform–formic acid (2:10:1), the plates being developed thrice in the latter solvent. The abscisic acid, detected by its quenching of the fluorescence of the silica gel plate, can finally be purified by sublimation at 0·1 mm up to 220°C.

The simplest method of identification is by GLC on 5% QF1 on 50–60 mesh Chromosorb W; the retention of the acid, as its methyl ester, is approximately 9 min. GLC identification can be neatly confirmed by irradiating an isolated sample with UV light for 4 h. This produces an equilibrium mixture of the natural *cis*- and the artificial *trans*-isomers, which give two peaks on GLC instead of one as originally (Lenton *et al.*, 1971). Confirmation of identity is preferably based on the use of combined GLC–MS (Gaskin and MacMillan, 1968). The UV spectrum of abscisic acid is also a useful characteristic (λ_{max} 263 nm in acid, 240 nm in alkali) in its identification.

In the purification of abscisic acid from plants, contamination with phenolics sometimes occurs. In such cases, crude extracts can be absorbed on to a PVP (Polyclar T) column. Elution with water removes the growth inhibitor, leaving the phenolics adsorbed on the column (Lenton *et al.*, 1971).

The earlier procedures for quantitative analysis of abscisic acid were based on spectropolarimetry (Milborrow, 1967), which can be applied to μg samples. Because of its UV absorbance, abscisic acid can now be readily determined following HPLC by monitoring the eluate at 254 nm. A reversed-phase system is needed (e.g. an ODS Hypersil column) and elution with a gradient ranging from 10% to 80% methanol in 0·1 M acetic acid (Horgan, 1981). GLC–MS using a deuterated internal standard and selected ion monitoring has been employed to determine abscisic acid levels on the 100 pg scale in wilting *Eucalyptus* leaves (Netting *et al.*, 1982). Immunoassay methods are also available, with a detection limit of around 30 pg (Weiler, 1983).

3.2.3   Practical experiment

(a) *Essential oils of umbellifer seeds*

The fruits of many Umbelliferae have characteristic odours or smells due to the presence of relatively large amounts of essential oils in the resin canals. As a result, the fruits are widely used in cooking as a flavouring (e.g. caraway) or in medicine for treatment of mild chronic conditions (e.g. dill). The compounds in the essential oils are mainly monoterpenes, especially aldehydes (e.g. cuminaldehyde) or ketones (e.g. carvone). Occasionally, the odour principle may be of slightly different structure; for example, carrot odour is due to 2-nonenal, a simple hydrocarbon derivative. Simple TLC procedures can be used to identify the major components of many umbellifer fruits and the ready accessibility of fruits from grocers or seed merchants make this an attractive experiment for students to gain experience of detecting monoterpenes in plants. These simple procedures can also be used to identify unknown seed samples, since different species give different TLC patterns; this can be quite important in pharmacognosy (see Betts, 1964).

(b) *Procedure*

(1) Crush a few seeds to a fine powder, using sand and a pestle and mortar, cover the powder with ether and leave to extract for at least 0·5 h.
(2) Concentrate the ether extract and spot the concentrate on to silica gel plates (in duplicate). Develop the plate in benzene, benzene–chloroform (1:1) or hexane–chloroform (3:2) (40–60 min).
(3) Examine one of the dried plates in UV light for dark absorbing spots and mark these in.
(4) Spray one plate with vanillin–H_2SO_4 reagent and heat at 100°C for 10 min to develop the colours. Spray the second plate with 2,4-DNP reagent and note orange spots on a pale yellow background.

Typical results with eight umbellifer species are given in Table 3.5 and Fig. 3.6. Note that fixed oil is present in all samples, giving a purple streak at approx. R_F 0·60, which should not inferfere with terpene detection. Besides the principal components indicated in Table 3.5, most fruits contain other compounds of higher or lower R_F. The higher R_F compounds are monoterpene hydrocarbons such as limonene (R_F 80), α-phellandrene and α-pinene. Due to their volatility, it is only usually possible to detect these substances in traces.

Table 3.5 Umbellifer fruits and their essential oil components*

Common name	Latin name	Approx. R_F ($\times 100$)	Component	Means of detection
Anise	*Pimpinella anisum*	71	anethole	} orange DNP
		39	anisaldehyde	
Caraway	*Carum carvi*	43	carvone	orange DNP
Coriander	*Coriandrum sativum*	26	linalol	mauve, vanillin
Cumin	*Cuminum cyminum*	58	cuminaldehyde	orange DNP
Dill	*Anethum graveolens*	42	carvone	orange DNP
Fennel	*Foeniculum vulgare*	72	anethole	} orange DNP
		38	anisaldehyde	
Indian dill	*Anethum sowa*	57	dillapiole	brown, vanillin
		41	carvone	orange DNP
Parsley	*Petroselinum crispum*	64	apiole	brown, vanillin
			myristicin	brown, vanillin

*Measured in $CHCl_3$–benzene (1:1) on silica gel.

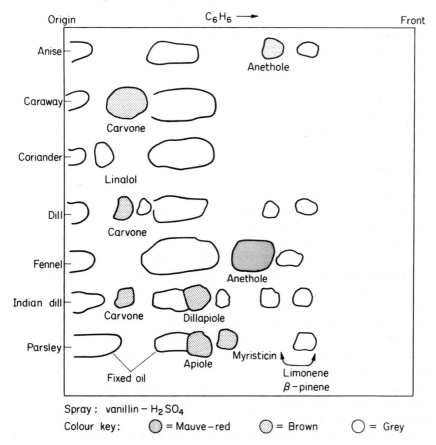

Fig. 3.6 Thin layer chromatography separation of essential oils in umbellifer seeds.

The lower R_F substances are mainly sesquiterpenoids. The above experiments can be extended by developing a third plate and testing it with fluorescein and bromine (Table 3.3). It is also valuable to provide several 'unknown' seed samples, to be given to students to determine their components. A few marker solutions of pure terpenes in ether should also be chromatographed, alongside the seed samples. Finally, it should be stressed that there is occasionally inter-varietal variation in essential oil components and different results may be obtained, depending on the source of the seed.

A useful procedure for the TLC of essential oils, followed by simple confirmatory chemical tests on a microscale, is outlined by Ikan (1969). He also gives details of how to separate the components of clove oil and citrus oils by column chromatography on silica gel.

3.3 DITERPENOIDS AND GIBBERELLINS

3.3.1 Chemistry and distribution

The diterpenoids (see Fig. 3.7) comprise a chemically heterogenous group of compounds, all with a C_{20} carbon skeleton based on four isoprene units; most are of very limited distribution. Probably the only universally distributed diterpene is the acyclic parent compound of the series, phytol, which is present as the ester attachment in the molecule of chlorophyll (see Chapter 5). Three classes of diterpenoids will be considered here: resin diterpenes, toxic diterpenes and the gibberellins.

The resin diterpenes include compounds such as abietic and agathic acids, found in both modern and fossil plant resins (Thomas, 1970). These resin compounds have a protective function in nature in that they are exuded from wood of trees or the latex of herbaceous plants. Abietic acid is widespread in gymnosperm resins, especially in *Pinus*. The various 'copal' resins of legume trees contain a range of different diterpenes (Ponsinet *et al.*, 1968), the dicyclic terpene hardwickic acid being but one example. A group of toxic diterpenes are the grayanotoxins, typified by grayanotoxin-1 (see Fig. 3.7), which occur in the leaves of many *Rhododendron* and *Kalmia* species. They are responsible for the poisonous nature of the foliage of these plants.

The third class of diterpenoids to be mentioned are the gibberellins, a group of hormones which generally stimulate growth and which are known to be widespread in plants. Gibberellic acid (abbreviated as GA_3) is the most familiar gibberellin, but in fact over sixty compounds in the series have been found to date. Chemically, they are closely related and thus difficult to separate and distinguish. The only really satisfactory method of determination is by GLC–MS.

The chemistry and natural distribution of diterpenes has been reviewed by Hanson (1968, 1972). There are many modern reviews of the plant gibberellins (e.g. MacMillan, 1983) and a monograph on their physiology and biochemistry has recently appeared (Crozier, 1983).

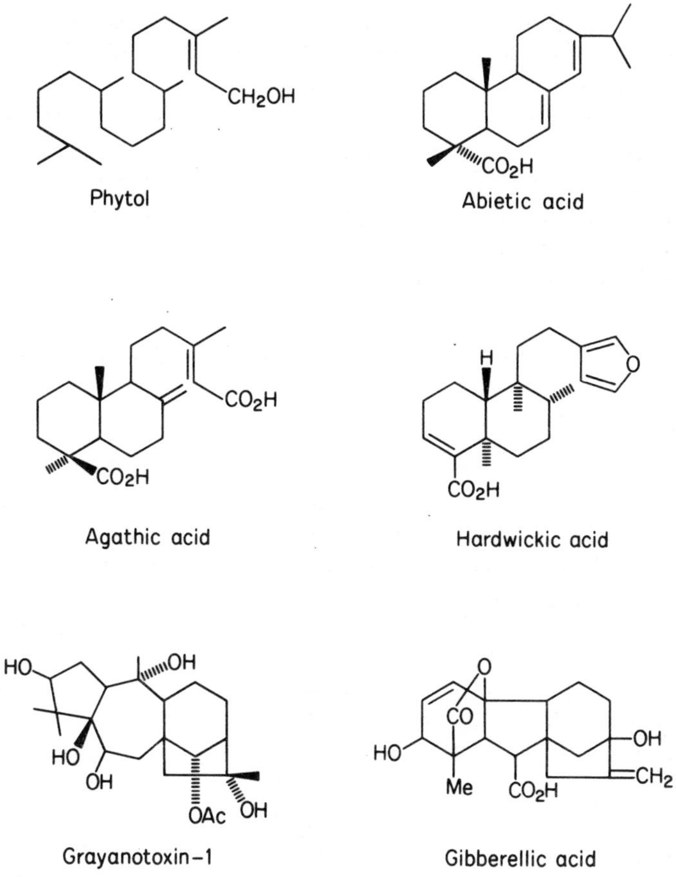

Fig. 3.7 Structure of some diterpenoids.

3.3.2 Recommended techniques

(a) *Diterpenes*

In general, these compounds are separated by GLC and TLC using the same methods as for the lower terpenes. However, diterpenes are less volatile than the sesquiterpenes and slightly different GLC techniques may be required in some cases. Further identification is largely based on IR and mass spectra.

GLC, for example, was used by Aplin *et al.* (1963) for surveying the distribution of eleven diterpene hydrocarbons in the Podocarpaceae. These authors extracted dried leaf tissue (20–40 g) in a Soxhlet with ether for 12–18 h. After concentration, the residue was dissolved in benzene and chromatographed on an alumina column (de-activated with 5% acetic acid) packed with light petroleum (b.p. 60–80°C). Elution with the latter solvent gave the diterpenes in the initial fractions and these were then submitted to GLC on two different columns. The first was a 350 cm column of embacel with 3% E301 silicone oil, which was operated at 160°C. The second was a 120 cm column of presiliconized Gas Chrom P (100–120 mesh) coated with 1% E 301 silicone oil at 140°C. Typical retention times on the first column (in min) were: rimuene 16·0, cupressene 17·8, isophyllocladene 20·9, isokaurene 23·0, phyllocladene 25·0 and kaurene 26·3. Thus, closely related pairs of isomers were satisfactorily separated and characterized by this procedure.

Similar procedures are used for the TLC of diterpenes as for other terpenoids. Demole and Lederer (1958) for example, separated phytol (R_F 35), isophytol (50), geranyl–linalol (44) and phytyl acetate (66) on silica gel in *n*-hexane–ethyl acetate (17:3). More recently, TLC on silica gel–AgNO$_3$ (10:1) with light petroleum as solvent has been exploited (Norin and Westfelt, 1963). Detection is by the standard procedures (conc.H$_2$SO$_4$, antimony chloride reagent, 0·2% KMnO$_4$ etc.).

When screening a series of plants for diterpenoids, a simple TLC procedure would normally be used, but HPLC might be advantageous, particularly in the case of oxygenated derivatives. Thus, Ganjian *et al.* (1980) were able to detect three biologically active diterpenes in a single dried leaf weighing 45 mg of *Rabdosia umbrosus* by HPLC. The methanol extract was injected into a Zorbax ODS C$_{18}$ column (25×0·46 cm), which was eluted with methanol–water (9:11) and monitored at 230 nm. For preparative work, diterpenoids are routinely purified by column chromatography on either silica or alumina (Croteau and Ronald, 1983).

(b) *Gibberellins*

In view of the importance of these substances as plant hormones, much effort has been devoted to their isolation and estimation in plant tissues. There are, however, sixty-two known gibberellins and no single TLC system is yet available for their resolution. Identification by TLC alone is therefore not to be relied on and other methods (see below) must be used in conjunction with TLC.

At least thirty solvents have been applied to the separation of gibberellins on layers of inactive or activated silica gel (Kaldewey, 1969). Two generally useful ones, on activated plates, are benzene–butanol–acetic acid (70:25:5) and benzene–acetic acid–water (50:19:31, upper layer); the time of development is approximately 1 h. TLC-electrophoresis on silica gel G at 3·8–5V

cm^{-1} has also been found useful (Schneider *et al.*, 1965). The standard method of detection is by spraying plates with H_2SO_4–water (7:3) and then heating at 120°C; gibberellins appear as UV fluorescent yellow-green spots.

For GLC, gibberellins must first be converted to their methyl esters or trimethylsilyl ethers. They may be separated on columns of 5% SE-30, 5% SE-52 and 5% OV-22, all on DMCS-treated Chromosorb W (Perez and Lachman, 1971). Alternatively, columns of 2% QF-1 or 2% SE-33 may be used (Durley *et al.*, 1971).

For reliable gibberellin identifications, combined GLC–MS is the technique of choice. In this procedure, the hormones are separated gas-chromatographically as their trimethylsilyl ethers, and then the mass spectra are automatically determined directly on the separated components as they emerge from the GLC column. GLC–MS can be applied to crude extracts or to 5 μg samples of impure gibberellins and known compounds can be identified without recourse to authentic samples (which are often inaccessible), as long as reference mass spectra are available (Binks *et al.*, 1968). For gibberellin conjugates (e.g. gibberellic acid glucoside or glucose ester), it is best to identify them by GLC–MS of the permethyl ethers (Rivier *et al.*, 1981).

The HPLC of plant gibberellins has been developed in recent years, but there are difficulties in detecting them because, generally, they do not absorb in UV. One solution is to separate them as their benzyl esters (Reeve and Crozier, 1978); another is to depend on their end absorption between 200 and 210 nm, when it is necessary to avoid solvents that are not transparent at these wavelengths. In the latter case, separations have been achieved on reverse-phase C_{18} silica columns using methanol–0·8% aqueous H_3PO_4 and detection at 205 nm (MacMillan, 1983).

The quantitative determination of gibberellins in plants is difficult and no ideal procedure that has been tested sufficiently is yet available. One approach is to employ an elaborate 6-stage clean-up procedure, involving several HPLC separations and finally GLC–MS with selective ion monitoring (e.g. Yamaguchi *et al.*, 1982). Another is to use a very sensitive analytical procedure, such as immuno-assay, which will work on relatively crude extracts. The latter procedure is expensive, since animals are needed for producing the antisera, but it has the advantage that it can be applied to quite small amounts of plant tissue (Weiler, 1983).

When identifying such a well known substance as gibberellic acid itself, many of the more complicated procedures mentioned above can be avoided. Simple methods can be applied, but careful comparison with authentic material at all stages is essential. The procedures used by Sircar *et al.* (1970) for identifying GA_3 in petals of *Cassia fistulosa* at a level of 5 mg kg^{-1} fresh weight are illustrative. The material was isolated by extraction with acetone and purified by bicarbonate fractionation and column chromatography on silica gel with chloroform as eluent. GA_3 was identified by R_F comparison on paper and TLC, IR and mass spectra and by spectrofluorimetry. Finally, identity was confirmed by bioassay using the lettuce hypocotyl, α-amylase and dwarf rice leaf sheath tests.

3.4   TRITERPENOIDS AND STEROIDS

3.4.1   Chemistry and distribution

Triterpenoids are compounds with a carbon skeleton based on six isoprene units and which are derived biosynthetically from the acyclic C_{30} hydrocarbon, squalene. They have relatively complex cyclic structures, most being either alcohols, aldehydes or carboxylic acids. They are colourless, crystalline, often high melting, optically active substances, which are generally difficult to characterize because of their lack of chemical reactivity. A widely used test is the Liebermann–Burchard reaction (acetic anhydride–conc.H_2SO_4), which produces a blue-green colour with most triterpenes and sterols.

Triterpenoids can be divided into at least four groups of compounds: true triterpenes, steroids, saponins and cardiac glycosides. The latter two groups are essentially triterpenes or steroids which occur mainly as glycosides. There are also the steroidal alkaloids, but these are covered in this book along with the other alkaloids (see Chapter 5).

Many triterpenes are known in plants and new ones are regularly being discovered and characterized (e.g. Das and Mahota, 1983). So far, only a few are known to be of widespread distribution. This is true of the pentacyclic triterpenes α- and β-amyrin and the derived acids, ursolic and oleanolic acids (see Fig. 3.8). These and related compounds occur especially in the waxy coatings of leaves and on fruits such as apple and pear and they may serve a protective function in repelling insect and microbial attack. Triterpenes are also found in resins and barks of trees and in latex (*Euphorbia, Hevea,* etc.).

Certain triterpenes are notable for their taste properties, particularly their bitterness. Limonin, the lipid-soluble bitter principle of *Citrus* fruits, is a case in point. It belongs to a series of pentacyclic triterpenes which are bitter, known as limonoids and quassinoids. They occur principally in the Rutaceae, Meliaceae and Simaroubaceae and are also of chemotaxonomic interest (Waterman and Grundon, 1983). Another group of bitter triterpenes are the cucurbitacins, confined mainly to the seed of various Cucurbitaceae but detected also in several other families including the Cruciferae (Curtis and Meade, 1971).

Sterols are triterpenes which are based on the cyclopentane perhydrophenanthrene ring system. At one time, sterols were mainly considered to be animal substances (as sex hormones, bile acids, etc.) but in recent years, an increasing number of such compounds have been detected in plant tissues. Indeed, three so-called 'phytosterols' are probably ubiquitous in occurrence in higher plants: sitosterol (formerly known as β-sitosterol), stigmasterol and campesterol. These common sterols occur both free and as simple glucosides. A less common plant sterol is α-spinasterol, an isomer of stigmasterol found in spinach, alfalfa and senega root. Certain sterols are confined to lower plants; one example is ergosterol, found in yeast and many fungi. Others occur mainly in lower plants but also appear occasionally in higher plants, e.g. fucosterol, the main steroid of many brown algae and also detected in the coconut.

Phytosterols are structurally distinct from animal sterols, so that recent discoveries of certain animal sterols in plant tissues are most intriguing. One of the most remarkable is of the animal estrogen, estrone, in date palm seed and pollen (Bennett *et al.*, 1966). Pomegranate seed is another source, but the amounts present are very low; according to Dean *et al.* (1971), there is only 4 μg estrone kg^{-1} tissue. Less remarkable, perhaps, is the detection of cholesterol as a trace constituent in several higher plants, including date palm, and in a number of red algae (Gibbons *et al.*, 1967). Finally, the occurrence of insect moulting hormones, the ecdysones, in plants must be mentioned, since they provide a fascinating insight into the way plants may have evolved in order to protect themselves from insect predation. Ecdysones were discovered in plants in 1966 (Nakanishi *et al.*, 1966) and have subsequently been found in a range of plant tissues, with high concentrations being present in a number of ferns (e.g. bracken, *Pteridium aquilinum*) and gymnosperms (e.g. *Podocarpus nakaii*).

Saponins are glycosides of both triterpenes and sterols and have been detected in over seventy families of plants (Tschesche and Wulff, 1973). They are surface-active agents with soap-like properties and can be detected by their ability to cause foaming and to haemolyse blood cells. The search in plants for saponins has been stimulated by the need for readily accessible sources of sapogenins which can be converted in the laboratory to animal sterols of therapeutic importance (e.g. cortisone, contraceptive estrogens, etc.). Compounds that have so been used include hecogenin from *Agave* and yamogenin from *Dioscorea* species. Saponins are also of economic interest because of their occasional toxicity to cattle (e.g. saponins of alfalfa) or their sweet taste (e.g. glycyrrhizin of liquorice root). The glycosidic patterns of the saponins are often complex; many have as many as five sugar units attached and glucuronic acid is a common component.

The last group of triterpenoids to be considered are the cardiac glycosides or cardenolides; here again there are many known substances, with complex mixtures occurring together in the same plant. A typical cardiac glycoside is oleandrin, the toxin from the leaves of the oleander, *Nerium oleander,* Apocynaceae. An unusual structural feature of oleandrin (see Fig. 3.8) and many other cardenolides is the presence of special sugar substituents, sugars indeed which are not found elsewhere in the plant kingdom. Most cardiac glycosides are toxic and many have pharmacological activity, especially as their name implies on the heart. Rich sources are members of the Scrophulariaceae,

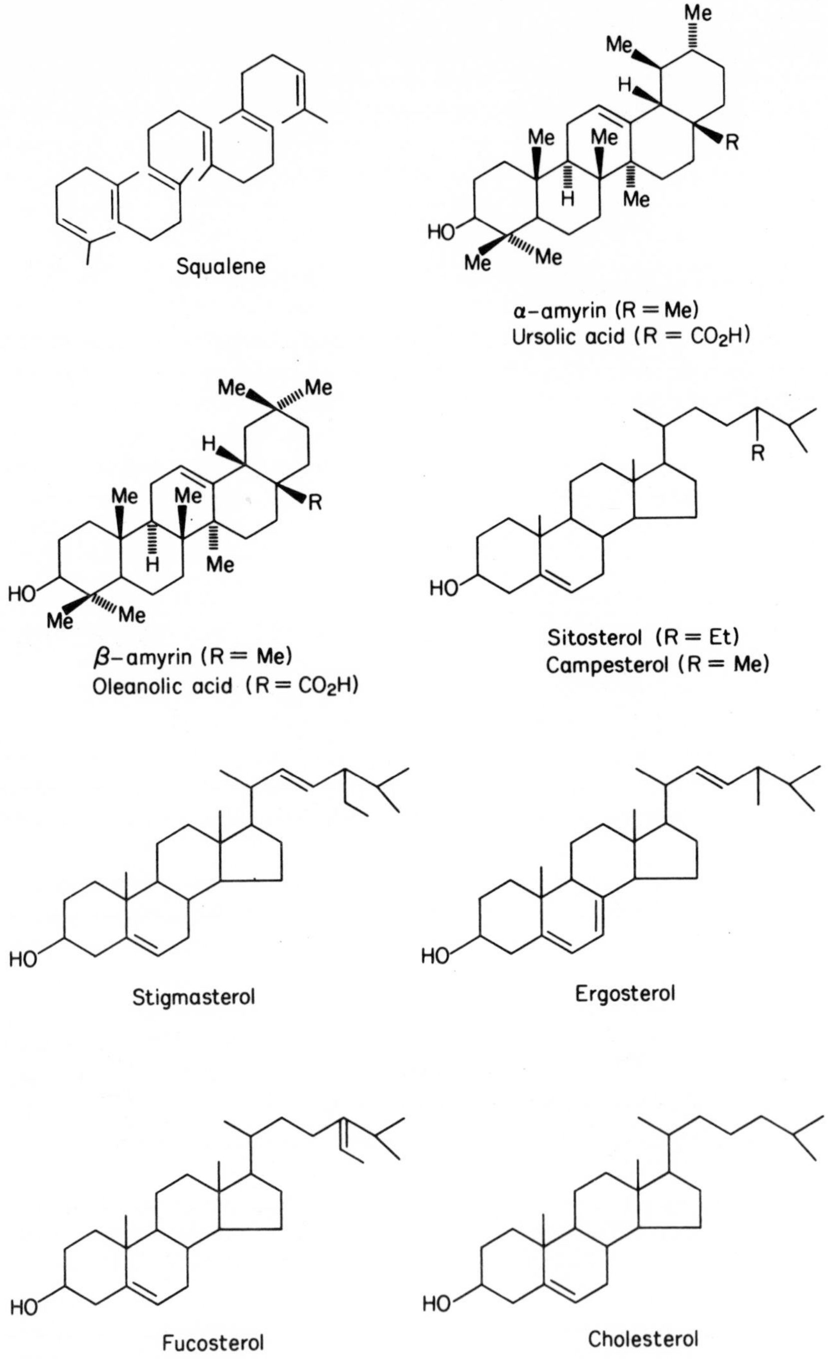

Squalene

α–amyrin (R = Me)
Ursolic acid (R = CO$_2$H)

β–amyrin (R = Me)
Oleanolic acid (R = CO$_2$H)

Sitosterol (R = Et)
Campesterol (R = Me)

Stigmasterol

Ergosterol

Fucosterol

Cholesterol

Fig. 3.8 Structures of some triterpenoids.

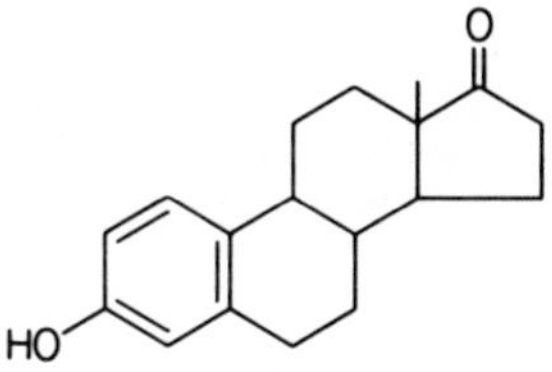

Estrone

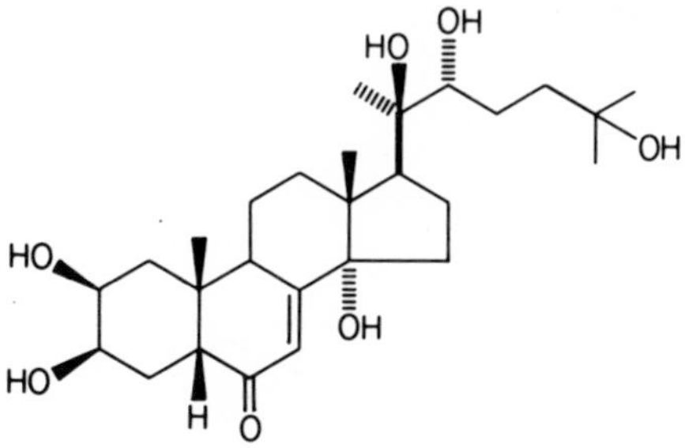

Ecdysterone

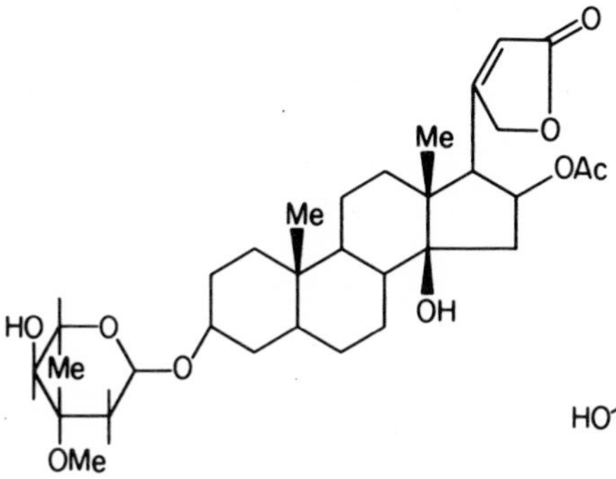

Oleandrin

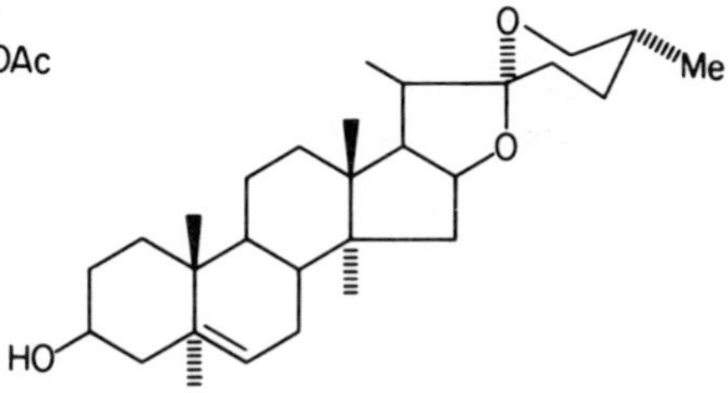

Diosgenin

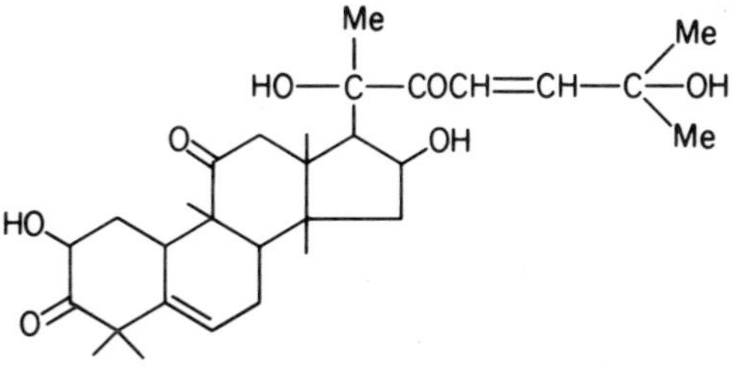

Cucurbitacin D

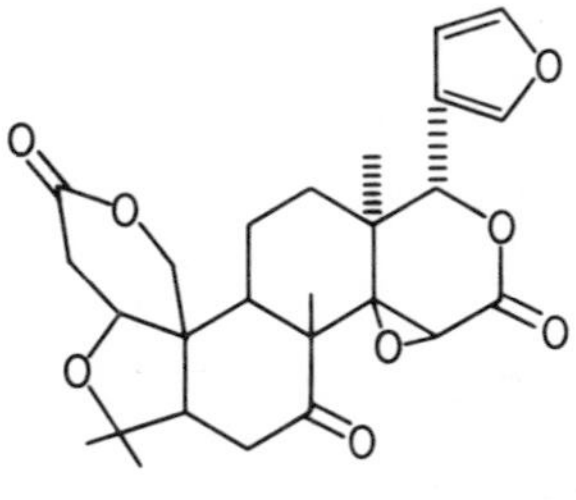

Limonin

Fig. 3.8 Structures of some triterpenoids *(continued)*.

Digitalis, Apocynaceae, *Nerium*, Moraceae and Asclepiadaceae, *Asclepias*. Special interest has been taken in the cardiac glycosides of *Asclepias*, because they are absorbed by Monarch butterflies feeding on these plants and are then used by the butterflies as a protection from predation by blue Jays (Rothschild, 1972). The butterfly is unharmed by these toxins, which on the other hand act as a violent emetic to the bird.

3.4.2 Recommended techniques

(a) *General*

All types of triterpenoid are separated by very similar procedures, based mainly on TLC and GLC. Identities are confirmed by melting point, rotation, GLC–MS, IR and NMR spectroscopy. PC is hardly ever used, although at one time it was very important for steroid separations (Bush, 1961). Separations on paper are occasionally valuable for distinguishing different glycosidic triterpenoids.

TLC is practically always carried out on layers of silica gel. Argentative TLC (on plates of silica gel treated with $AgNO_3$) is employed for separating triterpenoids according to the number of isolated double bonds present in the molecule. More than fifty detection reagents have been listed by Lisboa (1969) and Neher (1969) in recent reviews of steroid TLC. Of these, the most popular is probably the Carr–Price reagent, i.e. 20% antimony chloride in chloroform. A range of colours are produced, visible in both daylight and UV, on heating sprayed plates for 10 min at 100°C. The Liebermann–Burchard reaction has been adapted for TLC purposes and is also popular. Plates are sprayed with a mixture of 1 ml conc. H_2SO_4, 20 ml acetic anhydride and 50 ml chloroform and then heated at 85–90°C for 15 min. Again, a range of colours are formed with different triterpenoids and the sensitivity is quite good (2–5 μg). Other detection tests employed are those applicable to terpenoids in general, e.g. H_2SO_4 alone or diluted with water and alcohol, and which have already been mentioned in earlier sections in this chapter. Finally, the simplest of all laboratory solvents, namely water, can be used as a spray for detecting steroids on TLC plates (Gritter and Albers, 1962).

HPLC is occasionally used in triterpenoid analysis, having its widest application with the more highly substituted triterpene derivatives such as the ecdysones and limonoids. Reverse-phase partition techniques are most widely employed (Croteau and Ronald, 1983). Procedures for the HPLC of phytosterols are reviewed by Heftmann (1983). One example of the successful utilization of HPLC in sterol analysis was in the isolation of 22-dehydrocampesterol, a sterol previously only known in marine organisms, from a plant source, namely the seeds of *Brassica juncea*. Separation from the co-occurring brassicasterol was achieved on a Lichrosorb RP-18 column, which was eluted with methanol–water (98:2) as the mobile phase (Matsumoto *et al.*, 1983).

(b) *Triterpenes*

In surveying plants for their triterpenes, dried tissue should be first defatted with ether and then extracted with hot methanol. The concentrated methanol extracts can then be examined directly; in addition, acid hydrolysis should be carried out, to liberate aglycones if any glycosides are present, and the hydrolysate also examined. TLC is carried out on silica gel, in solvents such as hexane–ethyl acetate (1:1) and chloroform–methanol (10:1), with detection by antimony chloride in chloroform. Some triterpene mixtures are not readily separated by such procedures, α- and β-amyrin, for example, will only separate well if chromatographed in *n*-butanol–2 M NH₄OH (1:1). Similarly, betulinic, oleanolic and ursolic acids require special solvents, such as petroleum (b.p. 100–130°C)–dichloroethylene–acetic acid (50:50:0·7) (R_Fs 75, 50 and 20 respectively) or petroleum–ethyl formate–formic acid (93:7:0·7) (R_F 87, 70 and 20).

For GLC of triterpenes, liquid phases such as SE-30, OV-1 (methylsiloxane polymer), QF-1 and DEGS (diethylene glycol succinate) may be employed. 1–3% solutions of these phases are applied to a solid support such as Chromosorb W, previously washed with HCl and silanized with dimethyl-dichlorosilane in chloroform. Relatively high temperatures (220–250°C) are necessary, with gas flow rates of 50–100 ml min^{-1}. Relative retention times for a selection of eight common triterpenes are given in Table 3.7. Note that trihydroxytriterpenes (e.g. primulagenin) are only satisfactorily resolved on the column when applied as the trimethylsilyl ethers. Again, there are some mixtures which do not resolve easily. For example, α- and β-amyrin and taraxerol have very similar retention times on SE-30 columns, and it is necessary to use 2% XE 60 (a nitrile silicone) as liquid phase and inject them as the trimethylsilyl ethers in order to achieve good resolutions.

Table 3.7 Relative retention times by gas liquid chromatography of triterpenoids

Triterpenes	RR_T*	Sterols	RR_T†
Cholestane	1·0	Cholestane	1·0
Taraxerol	3·14	Cholesterol methyl ester	2·21
β-Amyrin	3·23	Campesterol	2·97
α-Amyrin	3·65	Stigmasterol	3·22
α-Amyrin acetate	4·74	Sitosterol	3·63
Oleanolic acid methyl ester	4·80	Cycloeucalenol	3·98
Ursolic acid methyl ester	5·25	Cycloartenol	4·47
Taraxasterol	5·75	24-Methylenecycloartanol	4·90
Primulagenin as the trimethylsilyl ether	6·57		

*1·5% SE-30 on Gas Chrom P (80–100 mesh) on a 150 cm×4 mm column at 240°C (actual retention time cholestane 2·8 min) (data from Ikekawa, 1969).
†3% OV-17 on Gas Chrom. Q (100–120 mesh) at 255°C (data from Atallah and Nicholas, 1971).

(c) *Phytosterols*

TLC and GLC procedures are similar to those for triterpenes. Occasionally, complex mixtures of sterols are found in a particular plant tissue and more elaborate procedures are needed for separation and identification. For example, sitosterol, cholesterol and stigmasterol are not easy to separate when present together. They will, however, separate if chromatographed as their acetates on Anasil B plates by continuous development for 2 h with hexane–ether (97:3). For the separation of the common sterols from their dihydro derivatives (e.g. sitosterol from sitostanol), argentative TLC is called for. The plates can be prepared by spraying fresh silica gel layers with concentrated aqueous methanolic $AgNO_3$ and then activating them for 30 min at 120°C. Separation is in chloroform and detection with H_2SO_4–water (1:1) (Ikan and Cudzinovski, 1965).

For the detection of oestrogens in plant tissues, relatively complex isolation procedures are necessary, including column chromatography on alumina (Bennett *et al.*, 1966). Final purification of oestrone can be achieved by TLC in cyclohexane–ethyl acetate (1:1) and in methylene dichloride–acetone (7:3). After spraying with 50% H_2SO_4 and heating, oestrone has a characteristic orange colour, with a green fluorescence in UV.

Procedures for the GLC of steroids have been mainly worked out for substances occurring in animals (Wotiz and Clark, 1969), but the methods work perfectly well with the phytosterols. The subject is briefly reviewed by Ikan (1969), who also provides simple practical details for the preparation of columns and so on. Relative retention times of seven common phytosterols are given in Table 3.7 to illustrate the sort of separations that can be achieved. Many workers prefer to separate these sterols as their more volatile trimethylsilyl ethers.

(d) *Saponins and sapogenins*

The formation of persistent foams during plant extraction or during the concentration of plant extracts is reliable evidence that saponins are present. Indeed, if large quantities of saponin occur in a plant, it is difficult to successfully concentrate aqueous alcoholic extracts, even when using a rotary evaporator. A simple test for saponins is therefore, to shake up an aqueous alcoholic plant extract in a test tube, and note if a persistent foam is formed above the liquid surface. Saponins can also be tested for in crude extracts by their ability to haemolyse blood cells. However, it is usually preferable to confirm such simple tests by TLC and by spectral measurements.

To test for sapogenins, dried tissue should be hydrolysed with molar HCl for 2–6 h, neutralized and the solid matter dried and extracted with petroleum. This extract is taken to dryness, and the residue dissolved in chloroform and the IR spectrum determined (Brain *et al.*, 1968). The same solution is then concentrated and subjected to TLC on silica gel in solvents such as acetone–hexane (4:1), chloroform–carbon tetrachloride–acetone (2:2:1) or any of those shown in Table 3.8. Sapogenins are then detected as pink to purple spots by spraying the plates with antimony chloride in conc. HCl and heating at 110°C for 10 min. The different sapogenins are not easily separated from each other by TLC (Table 3.8). For separating diosgenin from yamogenin, for example, it is necessary to carry out continuous development with methylene dichloride–ether (4:1) for 8 h (Bennett and Heftmann, 1966).

Saponins are much more polar than the sapogenins because of their glycosidic attachments and they are more easily separated by PC or by TLC on cellulose. However, TLC on silica gel is successful in solvents such as butanol saturated with water or chloroform–methanol–water (13:7:2, lower layer) (Kawasaki and Miyahara, 1963).

Table 3.8 R_F values of sapogenins

	R_F ($\times 100$) *in solvent*		
Compound	1	2	3
Diosgenin	55	55	34
Tigogenin	56	55	29
Smilagenin	62	61	—
Yamogenin	55	55	—
Hecogenin	41	32	19
Gitogenin	16	21	11

Key to solvents: 1, $CHCl_3$–Me_2CO (4:1); 2, $CHCl_3$–EtOAc (1:1); 3, hexane–Me_2CO (4:1). For separation of diosgenin and yamogenin, see text.

(e) *Cardiac glycosides*

Techniques to be employed depend on the structural complexity of the glycosides present in any one plant source. Thus, the cardiac glycosides of *Strophanthus* species can be separated by one-dimensional TLC on silica gel in the upper layer of ethyl acetate–pyridine–water (5:1:4). On the other hand, separation of the twenty-eight or so cardenolides in the foxglove, *Digitalis purpurea* requires two-dimensional TLC in ethyl acetate–methanol–water (16:1:1) and chloroform–pyridine (6:1) (Sjoholm, 1962). The simpler mixtures (up to fifteen components) in *Nerium oleander* and *Scilla maritima* can be fractionated by partition TLC using methyl ethyl ketone–toluene–water–methanol–acetic acid (40:5:3:2·5:1). For separating the cardiac glycosides of *Asclepias* species and of the butterflies that ingest them in the larval state, multiple development on silica gel plates is recommended in either ethyl acetate–methanol (97:3) twice or chloroform–methanol–formamide (90:60:1) four times (Brower *et al.*, 1982).

3.4.3 Practical experiments

(a) *Triterpenoids in seeds*

Sterols and sapogenins can be surveyed in plant seeds by simple screening procedures outlined in the previous section under sapogenins (p. 126). For example, it is possible to detect diosgenin in seed of the spice fenugreek, *Trigonella foenumgraecum*, Leguminosae, using as little as 5 g powdered seed (Hardman and Fazli, 1972). This is a good experiment to try out, since fenugreek seed is available from spice suppliers.

One or two practical points should be mentioned. TLC of the final chloroform extract in hexane–acetone (1:4) separates the sterols (R_F 50 and above) from sapogenins (R_F 45 and below). However, some of the diosgenin is converted during the acid hydrolysis (carried out to release the free sapogenins) to a diene derivative and this appears on TLC plates as a spot with high R_F (*ca.* 90). After spraying the plates with the antimony chloride reagent and heating, diosgenin appears as a red spot, which has yellow-green fluorescence in UV light.

Sterols are widely distributed in seeds and sitosterol may be present in sufficient amount for it to be isolated in crystalline form (e.g. in seed of chickpea, *Cicer arietinum*). The general procedure can be used for its isolation. When mixtures of several sterols occur, screening is probably best carried out by GLC. Procedures for screening crucifer seeds for their sterols by GLC are described by Knights and Berrie (1971).

(b) *Hecogenin from agaves*

This sapogenin can be isolated by straightforward extraction and saponification from fresh agaves (Ikan, 1969). This sapogenin is important as a source, by chemical transformations, of therapeutically important steroids, and the isolation is carried out commercially in Mexico. A 1 kg agave sample is extracted with 1 l. 95% EtOH at 100°C for 12 h. This is filtered hot, the solvent removed under reduced pressure and the residue is hydrolysed by refluxing for 2 h with 300 ml M ethanolic HCl. The reaction mixture is filtered and diluted with 400 ml ether. This solution is washed with water, with 5% aqueous NaOH and again with water and evaporated. The residue is hydrolysed again by refluxing with 10% alcoholic KOH for 30 min and, after cooling, hecogenin is extracted into ether, which is then evaporated to dryness. The product can be purified *via* acetone (norite) and should melt at 256–260°C. It forms a 2,4-dinitrophenylhydrazone (orange needles), m.p. 281–282°C.

3.5.1 Chemistry and distribution

Carotenoids, which are C_{40} tetraterpenoids, are an extremely widely distributed group of lipid-soluble pigments, found in all kinds of plants from simple bacteria to yellow-flowered composites. In animals, one particular carotenoid β-carotene is an essential dietary requirement, since it provides a source, through hydration and splitting of the molecule, of vitamin A, a C_{20} isoprenoid alcohol. Carotenoids, through dietary intake, also provide many brilliant animal colours, as in the flamingo, starfish, lobster and sea urchin. In plants, carotenoids have two principal functions: as accessory pigments in photosynthesis and as colouring matters in flowers and fruits. In flowers, they mostly appear as yellow colours (daffodil, pansy, marigold) while in fruits, they may also be orange or red (rose hip, tomato, paprika).

Although there are now over 300 known carotenoids (Isler, 1971), only a few are common in higher plants and problems of identification can often be resolved, in the first instance, by reference to these common substances. Well known carotenoids are either simple unsaturated hydrocarbons based on lycopene or their oxygenated derivatives, known as xanthophylls. The chemical structure of lycopene (see Fig. 3.9) consists of a long chain of eight isoprene units joined head to tail, giving a completely conjugated system of alternate double bonds, which is the chromophore giving it colour. Cyclization of lycopene at one end gives γ-carotene while cyclization at both ends provides the bicyclic hydrocarbon β-carotene. β-Carotene isomers (e.g. α- and ε-carotene) only differ in the positions of the double bonds in the cyclic end units. The common xanthophylls are either monohydroxycarotenes (e.g. lutein, rubixanthin), dihydroxy (zeaxanthin) or dihydroxyepoxy (violaxanthin).

Most of the rarer carotenoids have more complicated structures and may be more highly hydroxylated (e.g. with keto groups), more highly unsaturated (e.g. with allenic or acetylenic groups) or more extended with additional isoprene residues (giving C_{45} or C_{50} carotenoids). Just one example of a rarer carotenoid is the phenolic compound, 3,3′-dihydroxyisorenieratene (see Fig. 3.9), reported in a *Streptomyces* species. Combined forms of carotenoid occur, especially in flowers and fruits of higher plants and they are usually xanthophylls esterified with fatty acid residue, e.g. palmitic, oleic, or linoleic acids. Glycosides are normally very rare; in higher plants, the best known is the water-soluble crocin, the gentiobiose derivative of an unusual C_{20}–carotenoid crocetin, the yellow pigment of meadow saffron, *Crocus sativa*. Glycosides of C_{40} carotenoids with rhamnose or glucose as the sugars have been found quite recently in various algae (Herzberg and Liaan-Jensen, 1971).

When isolating a carotenoid from a new higher plant source, the chances are fairly high that it will be β-carotene, since this is by far the most common of all

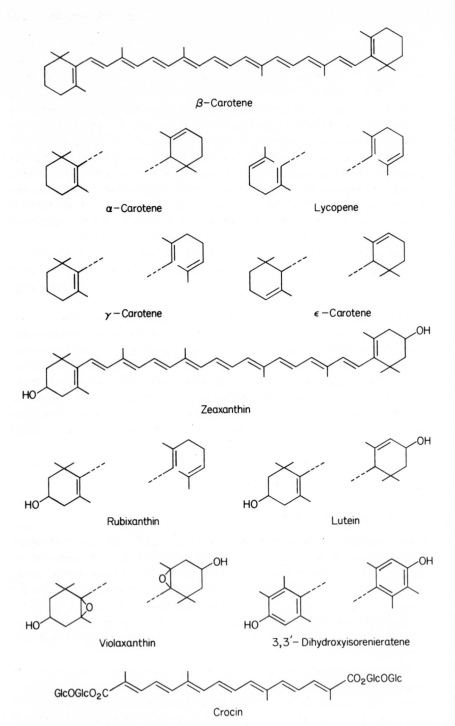

Fig. 3.9 Structures of some common carotenoids.

these pigments. In quantitative terms, however, it is not as important as certain xanthophylls. The annual production of carotenoids in plants has been estimated at 108 tons year^{-1} and this refers mainly to the synthesis of fucoxanthin (widespread in marine algae) and of lutein, violaxanthin and neoxanthin. These three latter compounds occur, with β-carotene, universally in the leaves of higher plants. Traces of α-carotene, cryptoxanthin or zeaxanthin may also occur in the lipid-soluble fraction of leaf extracts. In flowers too, carotenoid mixtures are the rule rather than the exception. The pigments are often highly oxidized with epoxides being common and carotenes only being present in traces. The amount of oxidation, however, varies from species to species and in the corona of certain narcissi, the red colour is due almost entirely to high concentrations (up to 15% of the dry weight) of β-carotene. Finally, when studying fruit pigments, it should be remembered that acyclic carotenoids tend to accumulate in some cases (e.g. lycopene in tomato) and that rather specific compounds may be synthesized by particular plant groups (e.g. rubixanthin by *Rosa* species, capsanthin by *Capsicum* species and so on).

The major reference to the biochemistry of plant carotenoids is Goodwin (1980). There are five chapters on carotenoids in the same author's edited two-volume *Chemistry and Biochemistry of Plant Pigments* (1976) and these should all be consulted for further information. The chapter in the second volume by B. H. Davies on analytical methods is especially valuable and should be perused by anyone contemplating carotenoid identification. The excellent monograph edited by O. Isler (1971) contains a complete list of all carotenoids known up to that date, together with comprehensive data on spectral properties.

3.5.2 Recommended techniques

(a) *Extraction and purification*

Carotenoids are unstable pigments. They are easily oxidized, particularly when exposed to air on TLC plates, and they may also undergo *trans–cis* isomerism during handling. It is necessary to keep these facts in mind during isolation. Most experiments can, in fact, be done in the laboratory under normal working conditions without incurring too much pigment loss. However, solutions of carotenoids should be kept in the dark as much as possible and, ideally, they should be stored at low temperatures under nitrogen gas. Peroxide-free solvents must always be employed.

Fresh plant tissue is extracted in a blender with either methanol or acetone and, after filtration, the carotenoids are taken into ether from such extracts, water being added if necessary to produce two layers. The combined ether extracts are then dried and evaporated under reduced pressure at temperatures below 35°C, preferably in the presence of nitrogen.

Since carotenoids may be present in esterified form, a saponification step is necessary. The residue from the original extract is dissolved in the minimum volume of ethanol, to which 60% aqueous KOH is added, 1 ml for every 10 ml of ethanolic solution. This is kept in the dark in the presence of nitrogen and may either be boiled for 5–10 min or left overnight at room temperature. After dilution with water, the pigments are taken back into ether and the ether extracts carefully washed, dried and concentrated. Occasionally, large amounts of sterols may be present at this stage; these can be conveniently removed by leaving a light petroleum solution at −10°C overnight, when most of the sterol is precipitated and can be removed by centrifugation.

At this stage, carotenoids may be directly analysed by TLC or PC. It is, however, often convenient when mixtures of pigments occur, to divide the hydrocarbons from the xanthophylls by phase separation. This is done by shaking a light petroleum solution with an equal volume of 90% methanol, when the dihydroxycarotenes go into the lower methanol phase, leaving the monohydroxycarotenes and the carotenes themselves in the upper petroleum phase. A second treatment of the upper layer with 90% methanol will then separate these two latter classes, with the second methanol fraction containing the monohydroxycarotenes. The pigments in the two methanol fractions are recovered by returning them into ether.

Similar separation of carotenoids according to the differing polarities can also be achieved by column chromatography, using sucrose as adsorbent and eluting with 0·5% *n*-propanol in petroleum. Pigment zones are recovered either by extruding the column and cutting out and eluting each fraction or by sequentially eluting the pigments from the column. A range of other adsorbents (e.g. Al_2O_3, MgO) and solvents have been employed for this purpose.

(b) *Chromatography*

Apart from column chromatography (see above) which is almost essential for large-scale isolation of carotenoids, the two techniques otherwise employed are TLC and PC. There is, however, no single support and solvent system which can be applied universally to all carotenoids. The choice depends largely on the relative polarities of the compounds to be separated and this is the reason why phase separation is a convenient preliminary to TLC. Six systems which can be used with the common carotenes and xanthophylls are shown in Table 3.9. For details of other systems, see Davies (1976) and Bolliger and König (1969).

Detection on TLC plates is no problem, since carotenoids are coloured, but it should be remembered that the colours fade with time, particularly if silica gel is the adsorbent. If the presence of colourless carotenoid precursors such as phytofluene is suspected, plates should be examined in UV light, since these

Table 3.9 Thin layer chromatography of carotenoids and xanthophylls

Pigments	R_F ($\times 100$) *in system*		
	1	2	3
Hydrocarbons			
α-Carotene	66	80	88
β-Carotene	49	74	84
γ-Carotene	11	41	45
ϵ-Carotene	70	84	—
Lycopene	01	13	15
	R_F ($\times 100$) *in system*		
	4	5	6
Xanthophylls			
Lutein	10	35	56
Zeaxanthin	05	24	55
Violaxanthin	05	21	84
Cryptoxanthin	54	75	07
Capsanthin	06	16	—
Neoxanthin	—	—	95

Key to systems: 1. activated MgO, petroleum (b.p. 90–110°C)–C_6H_6 (1:1); 2. activated MgO, petroleum (b.p. 90–110°C)–C_6H_6 (1:9); 3. silica gel–Ca (OH)$_2$ (1:6); petroleum–C_6H_6 (49:1); 4. sec magnesium phosphate, petroleum (b.p. 40–60°C)–C_6H_6 (9:1); 5. silica gel, CH_2Cl_2–EtOAc (4:1); 6. Kieselguhr G impregnated with 8% solution of triglyceride, acetone–MeOH–H_2O (3:15:2).

compounds can then be detected by their fluorescence. Ether is a suitable solvent for eluting the pigments from TLC plates, but to avoid ether-soluble contaminants, the TLC adsorbent needs to be prewashed with ether, before being used as a slurry to spread the plate. In order to maximize recovery when eluting carotenoids after TLC, the coloured zones should be scraped off the plate *before* the solvent has completely dried off.

Whilst carotenoid R_Fs measured on TLC plates vary from run to run, those measured on paper are more reliable and can be made reproducible within ±0·01 units (Jensen and Liaaen-Jensen, 1959; Jensen, 1960). This is one reason why PC is still used in the characterization of these natural pigments. Circular chromatography is carried out on filter paper impregnated with silica, kieselguhr or alumina. Carotenoid mixtures are supplied as a thin arc near the centre of a paper circle of diameter 16 cm and this is developed for 20–30 min in *n*-hexane or *n*-hexane–acetone. For example, using Whatman Chromedia AH81 (7·5% Al_2O_3 content) and SG81 (22% SiO_2 content) papers, Valadon and Mummery (1972) separated α-, β- and γ-carotene and lycopene (R_Fs 45, 35, 20 and 10 on SG81 in *n*-hexane) and also violaxanthin, rhodoxanthin, lutein, zeaxanthin and neoxanthin (R_Fs 32, 62, 60, 54 and 00 on AH81 in *n*-hexane—acetone, 4:1).

If an HPLC apparatus is available, it may be profitably used alongside TLC and PC for carotenoid separations. A range of columns and solvent systems can be used (Croteau and Ronald, 1983); lycopene, α- and β-carotene in tomato fruits can be separated, for example, on a Partisil-5 ODS C_{18} column, with elution by chloroform—acetonitrile (2:25). HPLC has been critically compared with both TLC and PC in its ability to separate particular carotenoid mixtures (Fiksdal *et al.*, 1978). HPLC is probably the ideal method to use for quantitative analysis. Stransky (1978) has described a simple isocratic elution of a Nucleosil 50-S column with iso-octane—ethanol (9:1) for estimating chloroplast carotenoids in the picomole range. An analysis is possible in the pigment extracted from only a milligram of spinach leaf and the procedure takes 15 min, whereas a comparable TLC method would require 4 h for completion.

(c) *Spectral properties*

Conclusive identification of a carotenoid is based on co-chromatography with an authentic sample in at least two solvents and on similar visible spectral comparison, again in at least two solvents. The spectra of carotenoids are quite characteristic between 400 and 500 nm, with a major peak around 450 nm and usually two minor peaks either side (see Fig. 3.10 for the visible spectrum of β-carotene). The exact positions of the three maxima vary from pigment to pigment and are sufficiently different to provide a means of identification (Table 3.10). Also, there are overall spectral shifts according to the solvent used and this is the reason why measurement in more than one solvent is recommended. The spectral maxima of most common carotenoids are given in Table 3.10, recorded in both petroleum and chloroform. A useful spectral test for 5,6-epoxides, such as violaxanthin, is to add 0·1 M HCl to an ethanolic solution; after 3 min, a spectral shift to lower wavelength (18—40 nm) will be observed.

IR spectroscopy is not useful for most carotenoids, but it is valuable for detecting certain structural features, such as keto- or acetylenic groups, in the rarer pigments. NMR and MS are important methods in the structural analysis of new pigments (see Isler, 1971), the latter procedure being particularly valuable since only 0·1 mg samples are needed and the fragmentation pattern often gives a clear indication of structure.

(d) *Crocin*

This is the only carotenoid likely to be encountered in higher plants which is water-soluble and which will not partition into ether during a preliminary

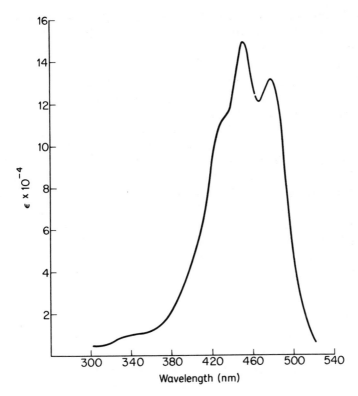

Fig. 3.10 Visible spectrum of β-carotene.

extraction. It could be mistaken for a yellow flavonoid (see Chapter 2, p. 71), since it is mobile on paper chromatograms in *n*-butanol–acetic acid–water (R_F 27) and in water (R_F 36) (Harborne, 1965). However, the spectrum is typically carotenoid (λ_{max}, 411, 437 and 458 nm in EtOH). On hydrolysis with 2 M HCl for 30 min at 100°C, it furnishes glucose and the dicarboxylic acid crocetin (λ_{max} 403, 427 and 452 in EtOH, λ_{max} 415, 433 and 462 nm in chloroform). Crocetin has an R_F (×100) on silica gel in 20% acetic acid in chloroform of R_F 90 (β-carotene 100) and in toluene–ethyl formate–formic acid (5:4:1) of 54 (β-carotene 85).

(e) *Authentic markers*

β-Carotene, canthaxanthin and crocin (from saffron) are used as food colourants and are commercially available. Lutein, violaxanthin and neoxanthin can be isolated from any higher plant leaf. Lycopene is readily obtained from tomatoes and capsanthin from paprika (for isolation procedures, see Ikan, 1969).

Table 3.10 Visible spectra of some carotenoids*

| Pigment | Spectral maxima (nm) in | |
	Petroleum or n-hexane	chloroform
Hydrocarbons		
α-Carotene	422,444,473	— ,454,485
β-Carotene	425†,451,482	— ,466,497
γ-Carotene	437,462,494	447,475,508
ε-Carotene	419,444,475	418,442,471 ‡
Lycopene	446,472,505	456,485,520
Xanthophylls		
Lutein	420,447,477	428,456,487
Violaxanthin	— ,443,472	424,452,482
Zeaxanthin	423,451,483	429,462,494
Neoxanthin	415,437,466	421,447,477
Rubixanthin	432,462,494	439,474,509
Fucoxanthin	425,450,478	— ,457,492
Cryptoxanthin	425,451,483	433,463,497

*Data from Davies (1976). Other solvents for spectral measurements are benzene, ethanol and carbon disulphide.
†Shoulder.
‡Measured in ethanol, not chloroform.

3.5.3   Practical experiments

(a) *Leaf carotenoids*

The four carotenoids that are universally present in leaf tissue can be isolated and identified by a simpler procedure than that outlined in the previous section. The following method is partly based on the experiments of Maroti and Gabnai (1971). Fresh leaf discs (0·25–0·5 g) are ground with sand and $MgCO_3$ in the presence of cold acetone (1 ml) and light petroleum (b.p. 60–80°C) (2 ml). The combined extract is applied as a streak on TLC plates of cellulose–silica gel G (2:1), which have been activated by heating. The plates are developed (40 min) with benzene–petroleum–ethanol–water (10:10:2:1). The β-carotene moves to the front, the chlorophylls have R_F 49–53 and the xanthophylls R_F 24–40. The two carotenoid bands are eluted with ether from the plates before they are completely dry. The β-carotene band is purified on a silica gel G plate developed with benzene–petroleum (b.p. 90–110°C) (1:1), in which it has R_F 82. It can be eluted and its identity confirmed by spectral measurement (Table 3.10) and by co-chromatography with authentic material (for solvents, see Table 3.9).

The xanthophyll band is re-run on a silica gel G plate in methylene dichloride–ethyl acetate (4:1) and the three bands of decreasing R_F, namely lutein, violaxanthin and neoxanthin are eluted with ether. This identity can be confirmed by spectral measurements. Remember that violaxanthin and neo-xanthin, which are epoxides, can be distinguished by the hypsochromic shift in their spectra when 0·1 M HCL is added to an ethanolic solution.

In our laboratory, we have run the above experiment in a simpler fashion by using a single TLC separation on a microcrystalline cellulose plate developed with petroleum (b.p. 40–60°C)–acetone–n-propanol (90:10:0·45) for about 1 h. It has the disadvantage that, if overloaded, the carotenoids may overlap with the chlorophylls, but lutein and β-carotene are usually clearly separated and give good spectra.

(b) *Petal carotenoids*

Most yellow flower colours among ornamental garden plants are due to carotenoids and complex mixtures have been isolated in many cases. Valadon and Mummery (1967, 1971) have made a special study of carotenoid pigments in the flowers of the members of the Compositae and a good way of gaining experience in the separation of carotenoid mixtures is to apply the general procedure outlined in the previous section (see pp. 131–134) to one or more of these plants. Various marigolds, chrysanthemums, gerberas and hawkweeds are available in most gardens, while dandelions and senecios are flourishing weeds in most temperate parts of the world.

The general procedure should be carried out up to the separation from sterols by cooling at −10°C in petroleum solution. The final carotene and xanthophyll fractions should then be separately chromatographed on columns of MgO–celite (1:2) eluted with hexane containing increasing amounts of ether. The various bands, after elution, are then identified from their position on the column and from their spectral and chromatographic properties. In the case of the yellow or orange flowers of the African marigold, *Tagetes erecta*, there are at least nine pigments with several others in trace amounts. In the hydrocarbon fraction, it should be possible to identify the following in order of elution: β-carotene (*ca.* 1% total carotenoid), 5,6-monoepoxycarotene (*ca.* 40%), 5,6-diepoxycarotene (*ca.* 25%) and cryptoxanthin (*ca.* 3%). The xanthophyll fraction should similarly give: 5,6-epoxylutein, lutein, zeaxan-thin, chrysanthemaxanthin and flavoxanthin (all as approximately 2–5% of total carotenoid).

(c) *Fruit carotenoids*

Occasionally, it is possible to isolate a single carotenoid in a relatively pure state from certain plant sources. This is true of capsanthin from pods of the red pepper or paprika, *Capsicum annuum*. The following procedure is taken from Ikan (1969). Finely ground dried pods (100 g) are extracted for 4 h with 200 ml petroleum (b.p. 40–60°C). After filtration, the extract is diluted with 600 ml ether, 100 ml 30% methanolic KOH added and the mixture stirred for 8 h. The ether layer is collected, washed, dried and concentrated to 20 ml. When diluted with 60 ml petroleum and left to stand for 24 h in the refrigerator, crystals of almost pure capsanthin should separate. Final purification, if desired, can be achieved by column chromatography on $CaCO_3$, with benzene–ether (1:1) as eluent. Capsanthin has λ_{max} 486 and 520 nm in benzene, 475 and 504 nm in *n*-hexane, 503 and 542 nm in carbon disulphide. It gives a deep blue colour when treated in chloroform solution with either conc.H_2SO_4 or antimony chloride.

REFERENCES

General references

Addicott, F.T. (1983) *Abscisic Acid*, Praeger, New York.

Bate-Smith, E.C. and Swain, T. (1966) The asperulosides and the aucubins. in *Comparative Phytochemistry* (ed. T. Swain), Academic Press, London, pp. 159–74.

Bolliger, H.R. and Koenig, A. (1969) Carotenoids and chlorophylls. in *Thin Layer Chromatography* (ed. E. Stahl), George Allen and Unwin, London, pp. 266–72.

Bush, I.E. (1961) *The Chromatography of Steroids*, Pergamon Press, Oxford.

Charlwood, B.V. and Banthorpe, D.V. (1978) The biosynthesis of monoterpenes. *Progr. Phytochem.*, **5**, 65–126.

Croteau, R. and Ronald, R.C. (1983) Terpenoids. in *Chromatography, Fundamentals and Applications: Part B*, Elsevier, Amsterdam, pp. 147–190.

Crozier, A. (ed.) (1983) *Biochemistry and Physiology of Gibberellins*, Praeger, New York.

Das, M.C. and Mahota, S.B. (1983) Triterpenoids. *Phytochemistry*, **22**, 1071–96.

Davies, B.H. (1976) Carotenoids. in *The Chemistry and Biochemistry of Plant Pigments*, 2nd edn (ed. T.W. Goodwin), Academic Press, London, pp. 489–532.

Goodwin, T.W. (ed.) (1976) *The Chemistry and Biochemistry of Plant Pigments*, 2nd edn, Academic Press, London.

Goodwin, T.W. (1980) *The Biochemistry of the Carotenoids*, Vol. 1: *Plants*, 2nd edn, Chapman and Hall, London.

Guenther, E. (1948–52) *The Essential Oils*, D. van Nostrand, New York.

Hanson, J.R. (1968) Recent advances in the chemistry of the tetracyclic diterpenes. *Prog. Phytochem.*, **1**, 161–89.

Hanson, J.R. (1972) in *Chemistry of Terpenes and Terpenoids* (ed. A.A. Newman), Academic Press, New York, pp. 155–206.

Heftmann, E. (1983) Steroids. in *Chromatography, Fundamentals and Applications: Part B*, Elsevier, Amsterdam, pp. 191–222.

Ikekawa, N. (1969) Gas liquid chromatography of polycyclic triterpenes. *Meth. Enzymol.*, **15**, 200–18.

Isler, O. (ed.) (1971) *Carotenoids*, Birkhauser-Verlag, Basle.

Lisboa, B.P. (1969) Thin layer chromatography of steroids. *Meth. Enzymol.*, **15**, 3–157.

Loomis, W.D. and Croteau, R. (1980) Biochemistry of terpenoids. in *Biochemistry of Plants*, Vol. 4, *Lipids, Structure and Function*, Academic Press, New York, pp. 364–419.

MacMillan, J. (1983) Gibberellins in higher plants. *Biochem. Soc. Trans.*, **11**, 528–33.

Neher, R. (1969) TLC of steroids and related compounds. in *Thin Layer Chromatography* (ed. E. Stahl), George Allen and Unwin, London, pp. 311–62.

Newman, A.A. (ed.) (1972) *Chemistry of Terpenes and Terpenoids*, Academic Press, London.

Rodriguez, E., Towers, G.H.N. and Mitchell, J.C. (1976) Biological activities of sesquiterpene lactones. *Phytochemistry*, **15**, 157–80.

Seaman, F.C. (1980) Sesquiterpene lactones as taxonomic characters in the Asteraceae. *Bot. Rev.*, **48**, 121–595.

Simonsen, J.L. (1947–52) *The Terpenes*, Cambridge University Press, Cambridge.

Stumpf, P.K. (ed.) (1980) *The Biochemistry of Plants*, Vol. 4, *Lipids, Structure and Function*, Academic Press, New York.

Thomas, B.R. (1970) Modern and fossil plant resins. in *Phytochemical Phylogeny* (ed. J.B. Harborne), Academic Press, London, pp. 59–80.

Tschesche, R. and Wulff, G. (1973) Chemie und Biologie der Saponine. *Fortschr. Chemie Org. Naturst.*, **30**, 461–606.

Waterman, P.G. and Grundon, M.F. (eds) (1983) *Chemistry and Chemical Taxonomy of the Rutales*, Academic Press, London.

Wotiz, H.H. and Clark, S.J. (1969) Gas liquid chromatographic methods for the analysis of steroids and sterols. *Meth. Enzymol.*, **15**, 158–99.

Supplementary references

Aplin, R.T., Cambie, R.C. and Rutledge, P.S. (1963) *Phytochemistry*, **2**, 205.

Attalah, A.N. and Nicholas, H.J. (1971) *Phytochemistry*, **10**, 3139.

Bate-Smith, E.C. (1964) *Phytochemistry*, **3**, 623.

Bennett, R.D. and Heftmann, E. (1966) *J. Chromatog.*, **21**, 488.

Bennett, R.D., Ko, S.T. and Heftmann, E. (1966) *Phytochemistry*, **5**, 231.

Betts, T.J. (1964) *J. Pharm. Pharmacol.*, **16**, Suppl., 131T.

Binks, R., MacMillan, J. and Pryce, R.J. (1968) *Phytochemistry*, **8**, 271.

Brain, K.R., Fazli, F.R.Y., Hardman, R. and Wood, A.B. (1968) *Phytochemistry*, **7**, 1815.

Brower, L.P., Seiber, J.N., Nelson, C.J., Lynch, S.P. and Tuskes, P.M. (1982) *J. Chem. Ecol.*, **8**, 579.

Curtis, P.J. and Meade, P.M. (1971) *Phytochemistry*, **10**, 3081.

Dean, P.D.G., Exley, D. and Goodwin, T.W. (1971) *Phytochemistry*, **10**, 2215.

Demole, E. and Lederer, E. (1958) *Bull. Soc. Chim. France*, 1128.

Drozdz, B. and Blosyk, E. (1978) *Planta Med.*, **33**, 379.

Duff, R.B., Bacon, J.S.D., Mundie, C.M., Farmer, V.C., Russell, J.D. and Forrester, A.C. (1965) *Biochem. J.*, **96**, 1.

Durley, R.C., MacMillan, J. and Pryce, R.J. (1971) *Phytochemistry*, **10**, 1891.

Erdtman, H. and Norin, T. (1966) *Fortschr. Chemie Org. Naturst.*, **24**, 206.

Fiksdal. A.. Mortensen. J.T. and Liaaen-Jensen. S. (1978) *J. Chromatog..* **157,** 111.

Ganjian, I., Kubo, I. and Kubota, T. (1980) *J. Chromatog.*, **200,** 250.

Gaskin, P. and MacMillan, J. (1968) *Phytochemistry*, **7,** 1699.

Geissman, T.A. and Griffin, T.S. (1971) *Phytochemistry*, **10,** 2475.

Gibbons, G.F., Goad, L.J. and Goodwin, T.W. (1967) *Phytochemistry*, **6,** 677.

Gritter, R.J. and Albers, R.J. (1962) *J. Chromatog.*, **9,** 392.

Harborne, J.B. (1965) *Phytochemistry*, **4,** 647.

Hardman, R. and Fazli, F.R.Y. (1972) *Planta Med.*, **21,** 131.

Harley, R.M. and Bell, M.G. (1967) *Nature* (Lond.), **213,** 1241.

Herzberg, S. and Liaaen-Jensen, S. (1971) *Phytochemistry*, **10,** 3121.

Horgan, R. (1981) *Prog. Phytochem.*, **7,** 137.

Hörhammer, L., Hamidi, A.E. and Richter, G. (1964) *J. Pharm. Sci.*, **53,** 1033.

Ikan, R. (1969) *Natural Products, A Laboratory Guide*, Academic Press, London.

Ikan, R. and Cudzinovski, M. (1965) *J. Chromatog.*, **18,** 422.

Jensen, A. (1960) *Acta Chem. Scand.*, **14,** 2051.

Jensen, A. and Liaaen-Jensen, S. (1959) *Acta Chem. Scand.*, **13,** 1863.

Johnson. A.E.. Nursten. H.E. and Williams. A.A. (1971) *Chem. Ind.* (Lond.). **556,** 1212.

Kaldewey. H. (1969) in *Thin Layer Chromatography* (ed. E. Stahl). George Allen and Unwin, London, pp. 471–93.

Kawasaki, T. and Miyahara, K. (1963) *Chem. Pharm. Bull. (Tokyo)*, **11,** 1546.

Knights, B.A. and Berrie, A.M.M. (1971) *Phytochemistry*, **10,** 131.

Lenton, J.R., Perry, V.M. and Saunders, P.F. (1971) *Planta*, **96,** 271.

Little, C.H.A., Strunz, G.M., France, R.L. and Bonga, J.M. (1972) *Phytochemistry*, **11,** 3535.

Mabry, T.J. (1970) in *Phytochemical Phylogeny* (ed. J.B. Harborne), Academic Press, London, pp. 269–300.

Marchand, B., Behl, H.M. and Rodriguez, E. (1983) *J. Chromatog.*, **265,** 97.

Maroti, I. and Gabnai, E. (1971) *Acta Biol. Szeged.*, **17,** 67.

Matsumoto, T., Shimizu, N., Shigemoto, T., Itoh, T., Iida, I. and Nishioka, A. (1983) *Phytochemistry*, **22,** 789.

Milborrow, B.V. (1967) *Planta*, **76,** 93.

Nakanishi, K., Koreeda, M., Sasaki, S., Chang, M.L. and Hus, H.Y. (1966) *Chem. Commun.*, 915.

Netting, A.G., Milborrow, B.V. and Duffield, A.M. (1982) *Phytochemistry*, **21,** 385.

Norin. T. and Westfelt. L. (1963) *Acta Chem. Scand.*, **17,** 1828.

Nursten, H.E. (1970) in *The Biochemistry of Fruits and Their Products* (ed. A.C. Hulme), Academic Press, London, pp. 239–68.

Perez, A.T. and Lachman, W.H. (1971) *Phytochemistry*, **10,** 2799.

Picman, A.K., Ranieri, R.L., Towers, G.H.N. and Lam, S. (1980) *J. Chromatog.*, **189,** 187.

Ponsinet, G., Ourisson, G. and Oehlschlager, A.C. (1968) *Recent Adv. Phytochem.*, **1,** 271.

Reeve, D.R. and Crozier, A. (1978) in *Isolation of Plant Growth Substances* (ed. J.A. Hillman), Cambridge University Press, Cambridge, p. 41.

Rivier, L., Gaskin, P., Albone, K.S. and MacMillan, J. (1981) *Phytochemistry*, **20,** 687.

Rothschild. M. (1972) in *Phytochemical Ecology* (ed. J.B. Harborne). Academic Press. London, pp. 1–12.

Rudloff, E. von (1966) *Phytochemistry*, **5,** 331.

Schneider, G., Sembdner, G. and Schreiber, K. (1965) *J. Chromatog.*, **19,** 358.

Sircar, P.K., Dey, B., Sanyal, T., Ganguly, S.N. and Sircar, S.M. (1970) *Phytochemistry,* **9,** 735.

Sjoholm, I. (1962) *Svensk Farm. Tidskr.*, **66,** 321.

Smith, R.H. (1964) *Phytochemistry*, **3,** 259.

Stahl, E. (ed.) (1969) *Thin Layer Chromatography,* George Allen and Unwin, London.

Stransky, H. (1978) *Z. Naturforsch.*, **33c,** 836.

Trim, A.R. and Hill, R. (1951) *Biochem. J.*, **50,** 310.

Valadon, L.R.G. and Mummery, R.S. (1967) *Phytochemistry*, **6,** 983.

Valadon, L.R.G. and Mummery, R.S. (1971) *Phytochemistry*, **10,** 2349.

Valadon, L.R.G. and Mummery, R.S. (1972) *Phytochemistry*, **11,** 413.

Wachtmeister, C.A. and Wickberg, B. (1958) *Acta Chem. Scand.*, **112,** 1335.

Weiffering, J.H. (1966) *Phytochemistry*, **5,** 1053.

Weiler, E.W. (1983) *Biochem. Soc. Trans.*, **11,** 485.

Yamaguchi, I., Fujisawa, S. and Takahashi, N. (1982) *Phytochemistry*, **21,** 2049.

Zavarin, E., Smith, R.M. and Anderson, A.B. (1959) *J. Org. Chem.*, **24,** 1318.

CHAPTER FOUR

Organic Acids, Lipids and Related Compounds

4.1 Plant acids
4.2 Fatty acids and lipids
4.3 Alkanes and related hydrocarbons
4.4 Polyacetylenes
4.5 Sulphur compounds

4.1 PLANT ACIDS

4.1.1 Chemistry and distribution

A unique feature of plant metabolism, when compared to that of animals and micro-organisms, is the ability of plants to accumulate organic acids in the cell vacuole, sometimes in considerable amount. For example, the expressed sap of lemon fruits has a pH of 2·5, due to the presence of as much as 58 mg ml^{-1} of citric acid. Indeed, the acidity of practically all edible fruits is due to such accumulation, but the phenomenon is not confined to fruit tissue. Leaves of many plants also have this ability and members of one plant family, the Crassulaceae, are notable for the diurnal variation in the amounts of the leaf acids, mainly citric, malic and isocitric acids (Kluge and Ting, 1978). Because of their importance in intermediary metabolism and plant respiration, a multitude of methods have been devised for determining these organic acids and for identifying them in plant extracts.

The simple organic acids that accumulate in plants fall conveniently into two groups: the tricarboxylic (Krebs) cycle acids and others (Table 4.1). The nine tricarboxylic acid cycle acids occur, of course, in catalytic amounts in all plant tissues, but only two of these, citric and malic, regularly accumulate in such tissues. Citric acid is a major fruit constituent in the orange, lemon, strawberry, blackcurrant and gooseberry, whereas malic acid is the dominant

142

acid in the grape, apple, plum and cherry (Ulrich, 1970). The other tricarboxylic acid cycle acids are apparently less common, although there are plants known in which they may be an abundant or major component (Table 4.1).

Table 4.1 Some major plant acids

Acid	Typical source(s)	Concentration present*
KREBS CYCLE ACIDS		
Citric	citrus fruits	58 mg ml^{-1} juice
Malic	apple fruit, leaves of Crassulaceae	15 mEq 100 g^{-1} FW
Isocitric	blackberry	10 mEq 100 g^{-1} FW
Cis-Aconitic	sugar cane juice	1·5% DW
Succinic	lucerne shoots	
Fumaric	sunflower stems	40 mg 100 g^{-1} FW
	also *Fumaria* species *Myrrhis odorata*, etc.	
Oxalacetic } α-Ketoglutaric }	barley grain	
OTHER ACIDS		
Formic	stinging nettle hairs	
Acetic	free and esterified in plant volatiles	
Monofluoracetic	*Dichapetalum cymosum* and related species	
Oxalic (as calcium salt)	rhubarb petioles	11 mEq 100 g^{-1} FW
	spinach leaves	300 mEq 100 g^{-1} DW
Tartaric	grape	11 mEq 100 ml^{-1} juice
	in leaves of many plants	
Malonic	*Phaseolus coccineus* stem	2 mg g^{-1} FW
	in many legumes and umbellifer leaves	
Shikimic } Quinic }	widespread, especially in woody plants	
Ascorbic (as lactone)	universal (as vitamin C)	

*FW = fresh weight, DW = dry weight, mEq = milliequivalents.

Of the other plant acids, acetic acid may be considered the most important, since it serves as a universal precursor of fatty acids, lipids and many other organic plant products. Acetyl coenzyme A, together with malonyl coenzyme A, are the active forms involved in biosynthesis of fatty acids from acetate. Acetic acid also occurs in trace amounts both free and combined (e.g. as ethyl acetate) in the essential oils of many plants and it is also the acyl group in a number of phenolic acylated glycosides. Acetyl derivatives of anthocyanins, the acyl group being attached through sugar, occur in grapes and in wine.

The presence of an acetic acid analogue, monofluoracetic acid, in the South

African plant 'Gifblaar' or *Dichapetalum cymosum* should also be mentioned. This substance inhibits the tricarboxylic acid cycle at very low concentrations and is hence toxic to all living organisms. The fatal dose in man is $2-5$ mg kg^{-1} body weight. Another toxic plant acid is oxalic, present in *Oxalis* shoots, rhubarb petioles, spinach and *Begonia* leaves. It is not normally poisonous to men or animals eating these plants, since it is mainly present as the highly insoluble calcium salt, which passes through the mammalian system in un-changed form. Indeed, calcium oxalate is so insoluble that it sometimes crystallizes out in cell vacuoles and a characteristic of many plant cells is the presence of so-called 'raphides' or long needles of calcium oxalate. 'Raphides' provide an anatomical feature in plants which is of some taxonomic interest (Gibbs, 1963).

Three other plant acids must be mentioned because of their widespread occurrence, namely ascorbic, shikimic and quinic. L-Ascorbic acid or vitamin C, an organic acid in lactone form, is universal in plants. It is an essential dietary requirement in man and is, fortunately, widely distributed in food plants. The concentration varies from tissue to tissue; in fruits, there may be 0·01% (apples), 0·2% (blackcurrant) or as much as 1% fresh weight (rose hips). Shikimic and quinic acids are two cyclohexane carboxylic acids of interest because they are precursors of aromatic compounds in plants. Shikimic acid is on the direct pathway between sedoheptulose and the aromatic amino acid phenylalanine and is not usually present in any quantity. Small amounts of it are found in many plants, especially woody ones (Boudet, 1973). By contrast, quinic acid occurs in quantities at least ten-fold those of shikimic and it contributes significantly to acidity in unripe apples and in a number of other fruits.

Organic acids are classified chemically according to the number of carboxylic acid groups (see Fig. 4.1) or according to whether other functional groups are present (e.g. as hydroxy acids, keto acids, etc.). The simplest monocarboxylic acid is formic, HCO_2H, the next homologue being acetic, CH_3CO_2H. Higher homologues in the series are usually only found in plants in trace amounts in the essential oil fraction. Isovaleric acid, for example, $(CH_3)_2CH.CH_2CO_2H$, the rancid principle of butter-fat is found in hop oil, tobacco and in various species of *Valeriana*.

The simplest dicarboxylic acid is malonic acid, with succinic being the next higher homologue. Unsaturated derivatives of succinic acid are fumaric and maleic acids, two geometric isomers, with fumaric being the *trans-* and maleic the *cis*-form. Monohydroxysuccinic acid is known as malic acid, while dihydroxysuccinic acid is tartaric acid. The keto acid corresponding to succinic is oxaloacetic, the next higher homologue of which is α-ketoglutaric, a key compound in amino acid biosynthesis (see p. 176).

The best known tricarboxylic acid is citric. Together with its isomer isocitric acid and its dehydration product aconitic acid, it participates in the tricarboxylic acid cycle. While citric acid has one centre of asymmetry, isocitric has two and can thus exist in four optical isomeric forms; only one of these is known naturally.

Organic acids are water-soluble, colourless liquids or relatively low melting solids. The majority are non-volatile, the exceptions being the simple mono-carboxylic acid series based on formic acid. They are generally chemically

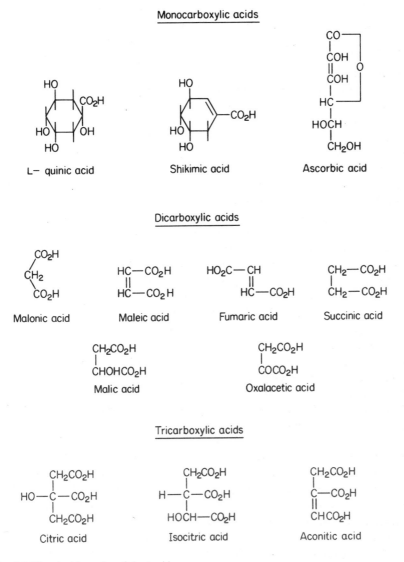

Fig. 4.1 Chemical formulae of plant acids.

stable, although α-keto acids readily undergo decarboxylation and may have to be isolated as derivatives. Acids are easily recognized by their taste in solution and by the low pH of crude aqueous plant extracts, when they occur in quantity. They are universally detected by their effect on acid–base indicators such as bromcresol green or bromothymol blue.

4.1.2   Recommended techniques

(a) *Purification*

For preliminary surveys, hot alcoholic extracts of fresh plants should be concentrated, filtered and examined for acids directly by PC. If oxalic acid is present in excessive amounts, it may interfere with the separation of the other acids on paper. The bulk of it can be removed as the insoluble calcium salt, by adding $Ca(OH)_2$ solution and centrifuging off the precipitate.

For more detailed studies, fractionation of crude extracts is preferable. This can be done by passing the concentrated aqueous extract through a cation exchange column to remove cations and amino acids. Passage through a weak basic ion exchange resin in the hydroxyl form will lead to the absorption of the organic acid anions. After washing, the organic acids are then released with 0.1 M HCl. Strongly basic columns should be avoided, since sugars present in the crude extracts may react with the resin, producing acids such as lactic and glycolic as artifacts.

(b) *Paper chromatography*

For analysing simple mixtures of the common plant acids, PC is still a useful procedure. Streaking occurs unless the acids are kept in the un-ionized form, so that it is imperative to either use a strong organic acid (e.g. formic) in the solvent or to chromatograph the acids as their ammonium salts by using a basic solvent. Two-dimensional PC may be carried out with n-propanol–1 M NH_4OH (7:3 or 3:2) followed by n-butanol–formic acid–water (10:3:10); equilibration with the lower layer of the solvent mixture is important in the separation in the second direction. Another good pair of solvents are 95% ethanol–1 M NH_4OH (19:1) and n-butanol–formic acid–water (4:1:5) (Carles *et al.*, 1958). After development, the paper must be dried thoroughly to remove all traces of formic acid. The paper is then sprayed with bromothymol blue (0.04 g in 100 ml 0.01 M NaOH) or similar indicator mixture at a suitably adjusted pH. The acids appear as blue spots on a yellow background; the contrast of the background can be adjusted by fuming the paper with ammonia vapour. The colours tend to fade and records should be made immediately. Some typical R_F values for common acids are given in Table 4.2.

Table 4.2 Paper chromatographic separation of common plant acids

Acid	R_F ($\times$ 100) *in*	
	n-butanol–formic–water (4:1:5)	*n*-PrOH–1 M NH₄OH (7:3)
Malic	45	06
Tartaric	22	03
Citric	37	00
Succinic	74	13
Fumaric	89	11
Malonic	63	09
Lactic	67	40

Many other solvents have been proposed for separating particular mixtures of organic acids and other means of detection are possible (Carles *et al.*, 1958). Tricarboxylic acids can be detected by reaction with pyridine–acetic anhydride (7:3), when citric and isocitric give yellow, aconitic brown-yellow and fumaric brown colours.

A simple one-dimensional separation of organic acids may be achieved (Blundstone, 1962) by using *n*-butyl formate–formic acid–water (10:4:1) as solvent. If sodium formate (0·05% w/v) and bromophenol blue (0·02% w/v) are added to the solvent mixture, the organic acid spots will immediately become apparent during the paper development. Separation takes about 4 h and typical R_Fs ($\times$ 100) are tartaric 21, citric 24, oxalic 40, succinic 56 and lactic 64).

The two cyclohexane carboxylic acids quinic and shikimic will separate well in the butanol–formic acid solvent (4:1:5), with R_Fs of 22 and 46 respectively. These acids, after separation, can be determined quantitatively by a microbiological assay involving an *Aerobacter aerogenes* mutant (Cookman and Sondheimer, 1965). Quinic and its related keto acid dehydroquinic will not, however, separate well on paper; it is then necessary to employ GLC with a 4% QF-1 on Chromosorb W column, applied to the trimethylsilyl ethers (Shyluk *et al.*, 1967).

Keto acids are usually converted to the more stable 2,4-dinitrophenylhydrazones before being separated by chromatography. Due to the fact of *cis–trans* isomerism in these compounds, most acids give not one but two 2,4-DNP spots (Isherwood and Niavis, 1956). A useful technique for confirming the identity of a keto acid is to hydrogenate the 2,4-DNP, when the corresponding amino acid is formed (Towers *et al.*, 1954). α-Ketoglutaric acid, for example, gives glutamic acid, which can be identified by procedures described in a later chapter (p. 178).

(c) *Thin layer chromatography*

Cellulose layers may be used, with similar solvents as on paper. Silica gel plates should be developed with mixtures such as methanol–5 M NH_4OH (4:1) or benzene–methanol–acetic acid (79:14:7). The latter mixture will separate, for example, maleic (R_F 07) from fumaric (R_F 23). Polyamide layers are useful for separating the acids of the tricarboxylic acid cycle. The solvent to employ is di-isopropyl ether–petroleum–carbon tetrachloride–formic acid–water (5:2:20:8:1). Treating the plate with iodic acid solution is sometimes valuable, since some acids are modified or decomposed in a characteristic way (Knappe and Rohdewald, 1964).

Acid-treated silica gel layers (obtained by adding 5 ml propionic acid to water (60 ml) when preparing the slurry, for plate spreading) have been suggested for the separation of keto acid 2,4-DNPs; the solvent is petroleum–ethyl formate (13:7). These compounds will also separate on untreated silica gel in benzene–acetic acid (19:1) or *n*-butanol–ethanol–1% NH_4OH (6:1:3). For further details of TLC procedures for organic acids, see Peereboom (1969).

(d) *Gas liquid chromatography*

Separation of the trimethylsilyl ethers may be conducted on a 0·3% OV-17 column (60–80 mesh Chromosorb G support) at 160°C (Horii *et al.*, 1965). For quantitative analysis, GLC of the methyl esters has been proposed by Mazliak and Salsac (1965). The liquid phase is butanediol succinate (20%) on a silanized (60–80 mesh) Chromosorb W support, with a 3 m × 6 mm column operating at 120–200°C. Results obtained on the acids of apple pulp and chestnut leaves were in good agreement with analyses based on silica gel or ion exchange column separations.

(e) *Ascorbic acid*

This water-soluble vitamin can be separated equally satisfactorily on paper or by TLC on silica gel. For PC, it is advisable to add traces of KCN to the solvent, in order to prevent traces of copper present in the paper from oxidizing the ascorbic acid. R_F values (×100) in *n*-butanol–acetic acid–water (4:1:5) and phenol–1% acetic acid are 37 and 35 respectively. It gives an immediate black colour when the paper is sprayed with ammoniacal $AgNO_3$. Dehydroascorbic acid, the oxidation product of ascorbic, has R_Fs 27 and 38 in the above solvents and gives a brown colour with ammoniacal $AgNO_3$.

TLC on silica gel G or GF_{254} can be carried out in water (R_F 96) or ethanol (R_F 22). Use of ethanol–10% acetic acid (9:1) separates ascorbic (R_F 50) from the dehydro-derivative (R_F 73) and from isoascorbic acid (R_F 54). Ascorbic acid can be detected on plates of silica gel GF_{254} as a dark blue spot in UV light of wavelength 254 nm; the limit of detection is 3 μg. For quantitative determination, in direct plant extracts, ascorbic is first oxidized with 2,6-dichlorophenol—indophenol to dehydroascorbic, which is then treated with 2,4-dinitrophenylhydrazine in aqueous HCl. The 2,4-DNP derivative is then purified by TLC on silica gel in chloroform—ethyl acetate—acetic acid (60:35:5) and the concentration of the red 2,4-dinitrophenylhydrazone determined spectrophotometrically. Other chromatographic solvents and detection methods are also available for the identification of this important vitamin (Bolliger and Koenig, 1969). HPLC techniques have been developed additionally for vitamin C analysis and reverse-phase systems (e.g. μBondapak C_{18}) eluted with slightly acidic solvents and monitored at 254 nm are recommended (van Niekerk, 1982).

In a recent survey of foliar ascorbic acid levels in 213 angiosperm species (Jones and Hughes, 1983), the following simpler procedure was employed. Freshly collected, weighed leaf samples (between 0·5 and 2·0 g) were ground in a mortar with some sand and 6% w/v metaphosphoric acid. The extract was made up accurately to a suitable volume (between 10 and 20 ml) and filtered. A portion of the filtrate (5 ml) was then titrated against standard 2,6-dichlorophenolindophenol of which 1 ml $\equiv$ 0·2 mg ascorbic acid. The dye was previously standardized by titration against a 0·02% solution of ascorbic acid in 6% metaphosphoric acid.

(f) *Monofluoracetic acid*

This organic acid probably occurs in a number of plants besides 'Gifblaar' (Table 4.1) and in view of its extreme toxicity, methods of analysis are especially important. The following procedure is that of Vickery *et al.* (1973). Plant tissue is digested with water at 90°C for 1 h, and the extract filtered, made alkaline and concentrated to *ca.* 10 ml. It is then acidified with dil. H_2SO_4 and distilled down to *ca.* 1 ml. This concentrate is used either for the standard red colour test with thioindigo (AOAC, 1965) or for PC. In the latter case, 40 μl of this solution which should contain approx. 1 mg ml^{-1} monofluoracetic, is spotted on paper and developed by ascent in ethanol—conc.NH$_4$OH—pyridine—water (95:3:3·1). Monofluoracetic acid appears as a blue spot on a brown background with an R_F of 65 (at 28°C), when the paper is sprayed with Nile blue reagent. This is prepared by dissolving 0·4 g of the dye in 100 ml ethanol, adding triethanolamine until the colour is purple and then diluting with a further 100 ml ethanol.

(g) *Further identification of acids*

Organic acids generally have no absorption in the UV and there are only a few characteristic bands in the IR spectrum, the main one being between 1610 and 1550 cm^{-1}. Confirmation of identity must therefore be based on physical properties and/or preparation of derivatives. The melting point or sublimation range may be a useful criterion for some acids (Palmer and Wyman, 1965). Keto acids are readily characterized as their 2,4-dinitrophenylhydrazones. Citric acid forms a pentabromacetone derivative, m.p. 73–74°C. Certain other acids can be recognized by their distinctive colours with specific reagents. Malic acid, for example, gives a yellow colour with 2,7-dihydroxynaphthalene and conc.H_2SO_4.

(h) *Authentic markers*

All the common plant acids are available commercially. This includes shikimic acid, until recently a rather rare substance, and also the somewhat unstable α-ketoglutaric acid (sometimes listed as α-oxoglutaric acid).

4.2   FATTY ACIDS AND LIPIDS

4.2.1   Chemistry and distribution

Fatty acids occur mainly in plants in bound form, esterified to glycerol, as fats or lipids. These lipids comprise up to 7% of the dry weight in leaves in higher plants and are important as membrane constituents in the chloroplasts and mitochondria. Lipids also occur in considerable amounts in the seeds or fruits of a number of plants and provide such plants with a storage form of energy to use during germination. Seed oils from plants such as the olive, palm, coconut and peanut are exploited commercially and used as food fats, for soap manufacture and in the paint industry. Plant fats, unlike animal fats, are rich in unsaturated fatty acids and there is evidence that some of these may be essential as a dietary requirement in man.

Lipids are defined by their special solubility properties and are extractable with alcohol or ether from living plant tissues. Such extraction removes certain other classes of lipid, such as leaf alkanes (see Section 4.3.2, p. 161) and steroids (see Chapter 3, p. 126), but leaves behind all the water-soluble components of the plant tissue. This section is concerned with the neutral lipids, the fats or triglycerides, and the polar lipids, the phospho- and glycolipids, of plants.

The general formulae for the three main classes of plant lipid are given in Fig. 4.2. Structural variation within each class is due to the different fatty acid residues that may be present. Triglycerides are 'simple' if the same fatty acid is present in all three positions in the molecule; one example is triolein, a triglyceride with three oleic acid residues. Much more common are 'mixed' triglycerides, in which different fatty acids are present in all three positions.

Triglycerides

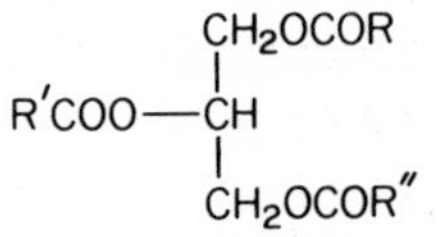

$$
\begin{array}{l}
CH_2OCOR \\
| \\
R'COO-CH \\
| \\
CH_2OCOR''
\end{array}
$$

Phospholipids

$$
\begin{array}{l}
CH_2OCOR \\
| \\
R'COO-CH \qquad O \\
| \qquad\qquad || \\
CH_2O-P-O-base \\
\qquad\quad | \\
\qquad\quad O^-
\end{array}
$$

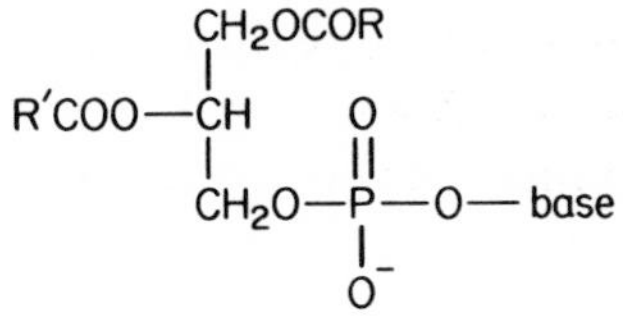

base = choline, $-CH_2CH_2\overset{+}{N}Me_3$
ethanolamine, $-CH_2CH_2NH_2$
serine, $-CH_2CH(NH_2)\,CO_2H$

Glycolipids

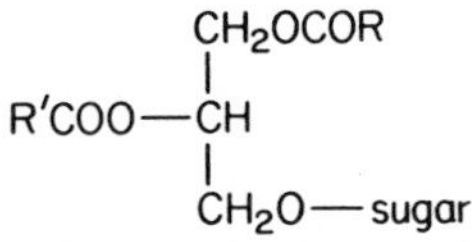

$$
\begin{array}{l}
CH_2OCOR \\
| \\
R'COO-CH \\
| \\
CH_2O-sugar
\end{array}
$$

sugar = galactose
galactosylgalactose
quinovose

(R, R', R'' = hydrocarbon chains of different fatty acids, see Fig. 4.3)

Fig. 4.2 General structures of plant lipids.

Phospholipids are complex in structure; all contain not only a phosphate group but also at least one other, usually basic, substituent. This basic residue may be choline, ethanolamine or serine. In addition, there are phospholipids with two or more glycerol residues or with inositol attached, instead of an organic base. Phosphatidylglycerol, for example, is a major fraction (*ca.* 20%) of the phospholipids in leaf tissue and is located particularly in the chloroplasts and mitochondria.

By contrast with the numerous types of phospholipids in plants, there are only a few glycolipids. Most important are monogalactosyl and digalactosyl diglycerides, highly surfactant molecules which play a role in chloroplast metabolism. Finally, there is one sulpholipid; this is a diglyceride with the sugar quinovose attached. Quinovose is a 6-deoxyglucose with a sulphonic acid residue in the 6-position. First discovered in green algae, this sulpholipid seems to be universal in plants as an essential component of the chloroplast.

Identification of lipids mainly requires the determination of their fatty acid components. Although numerous fatty acids are now known in plants, most lipids have the same few fatty acid residues, which makes their identification that much easier. The common fatty acids are either saturated or simple unsaturated compounds of C_{16} or C_{18}-chain length (see Fig. 4.3). Palmitic acid, a C_{16} acid, is the major saturated acid in leaf lipids and also occurs in quantity in some seed oils, e.g. peanut oil. Stearic acid, C_{18}, is less prominent in leaf lipids, but is a major saturated acid in seed fats in a number of plant families (Shorland, 1963).

Unsaturated acids based on C_{16} and C_{18} are widespread in both leaf and seed oils. Oleic acid comprises 80% of the fatty acid content of olive oil, 59% in peanut oil and is often accompanied by the di-unsaturated linoleic acid. The

Carbon chain length	Trivial name	Formula
	Saturated acids	
C_{14}	Myristic	$CH_3(CH_2)_{12}CO_2H$
C_{16}	Palmitic	$CH_3(CH_2)_{14}CO_2H$
C_{18}	Stearic	$CH_3(CH_2)_{16}CO_2H$
C_{20}	Arachidic	$CH_3(CH_2)_{18}CO_2H$
	Unsaturated acids	
C_{16}	Palmitoleic	$CH_3(CH_2)_5 CH{=}CH(CH_2)_7CO_2H$
	Oleic	$CH_3(CH_2)_7CH{=}CH(CH_2)_7CO_2H$
C_{18}	Linoleic	$CH_3(CH_2)_4CH{=}CH\ CH_2CH{=}CH(CH_2)_7CO_2H$
	Linolenic	$CH_3CH_2CH{=}CH\ CH_2CH{=}CH\ CH_2\ CH{=}CH(CH_2)_7CO_2H$
	Unusual acids	
C_{16}	Petroselinic	$CH_3(CH_2)_7CH{=}CH\ (CH_2)_5CO_2H$
C_{22}	Erucic	$CH_3(CH_2)_7CH{=}CH\ (CH_2)_{11}CO_2H$
C_{19}	Sterculic	$CH_3(CH_2)_7\underset{\underset{CH_2}{\diagdown\diagup}}{C}{=}C(CH_2)_7CO_2H$
	Cutin acids	
C_{24}	Lignoceric	$CH_3(CH_2)_{22}CO_2H$
C_{26}	Cerotic	$CH_2(CH_2)_{24}CO_2H$
C_{16}	10,16 – dihydroxy– palmitic	$HO(CH_2)_6CHOH(CH_2)_8CO_2H$

Fig. 4.3 Structures of some representative fatty acids.

tri-unsaturated linolenic acid is common, occurring in linseed oil to the extent of 52% of the total acid, with linoleic (15%) and oleic (15%). Unsaturated acids can exist in both *cis-* and *trans-*forms, but most natural acids in fact have the *cis-*configuration.

A number of rarer fatty acids are found as lipid components, many occurring characteristically in seed oils of just a few related plants. Three examples may be quoted here. Petroselinic acid, an isomer of the more common palmitoleic acid, occurs to the extent of 76% of the total acids in parsley seed, *Petroselinum crispum*; it appears to be present in many other Umbelliferae and also in the related Araliaceae in *Aralia*. Erucic acid is another unusual acid, found especially in the Cruciferae and the Tropaeoliaceae. It is present in high concentration in rape seed oil, *Brassica rapa*, and because of its alleged toxic properties, successful efforts have been made to breed rape varieties with low erucic acid content. The third unusual acid (see Fig. 4.3) is sterculic, with a unique cyclopropene ring in the middle of the molecule, which is found in *Sterculia* species, Sterculiaceae and in the related Malvaceae.

Cutin acids, although not fat components, are closely related in structure and may be synthesized from lipid fatty acids by chain elongation before being deposited in the leaf cutin. They are considered here briefly, because similar methods of identification apply to them as to the common fatty acids. The saturated cutin acids generally have a longer chain length than palmitic or stearic acids, the range being from C_{24} to C_{32}. The other main chemical feature of the cutin acids is the presence of hydroxy groups in their structures. 10,16-Dihydroxypalmitic acid, for example, is a major component in many plant cutins. In gymnosperms, 9,16-dihydroxypalmitic acid is found instead, while in ferns and lycopods, the 16-monohydroxy acid is present (Hunneman and Eglinton, 1972). The chemistry of plant cutins and suberins has been reviewed recently by Holloway (1982).

There are many excellent up-to-date reviews and books on the chemistry and biochemistry of lipids. The best general references to plant lipids are the book of Hitchcock and Nichols (1971) and the review chapter of Harwood (1980). The chemistry of the fatty acids is covered by Gunstone (1967, 1975). The fatty acid composition of plant seed oils is reviewed by Shorland (1963), but see also Harwood (1980). Specialized reviews are included in the continuing publication *Progress in Lipid Research*.

4.2.2   Recommended techniques

(a) *General*

Plant lipid analysis has developed considerably in recent years and most of the traditional procedures have been replaced by chromatographic techniques. The brief account here is based on that of Hitchcock and Nichols (1971).

Other reviews which may be consulted are those of Morris and Nichols (1967, 1970) and of Mangold (1969). A key paper on procedures for detailed analysis of plant lipids is that of Galliard (1968) who applied modern techniques to a complete survey of lipids in potato tuber tissue. The two main chromatographic techniques used are TLC for separation and purification of the lipids and GLC for identifying the fatty acids produced on saponification.

(b) *Extraction*

Fresh leaf tissue is extracted by maceration in 20 vols of cold isopropanol (this alcohol de-activates hydrolytic enzymes) and this is followed by re-extraction with chloroform–methanol (2:1). Seed tissue can be extracted directly with the latter solvent mixture or with petroleum. For tissues in which the lipids are very tightly bound such as cereals, extraction with chloroform–ethanol–water (200:95:5) is advisable. Direct extracts should be stored at $-5°C$ in the presence of a trace of antioxidant (0·005% butylated hydroxytoluene (BHT)) if they are not to be processed immediately.

(c) *Fractionation*

Before further analysis, it is frequently desirable at this stage to separate the lipids into neutral and polar fractions and to remove steroids and other contaminants. This can be done by column chromatography on silicic acid in ethereal solution. The neutral lipids will pass through, leaving the phospho- and glycolipids adsorbed; these can then be recovered by eluting the column with chloroform–methanol mixtures. A similar result can be obtained by preparative TLC on silica gel with chloroform as solvent; triglycerides move about halfway up the plate leaving the other lipids at the origin.

(d) *Thin layer chromatography*

The total lipids of plant tissues can be analysed by two-dimensional TLC, a typical separation being illustrated in Fig. 4.4. This was obtained from potato tuber (Galliard, 1968), but most other plant tissues show a range of similar components. The solvent in the first direction is chloroform–methanol–acetic acid–water (170:25:25:4), and that in the second, chloroform–methanol–7 M NH$_4$OH (65:30:4). In order to avoid decomposition of lipids during TLC, Galliard (1968) recommends adding BHT (5 mg%) to the first solvent and drying the plate, after the first run, at 50°C in an atmosphere of nitrogen.

For most purposes, one-dimensional TLC is satisfactory for separation of lipids prior to their analysis. Neutral lipids can be separated on silica gel plates, using isopropyl ether–acetic acid (24:1) and petroleum–ether–acetic acid (90:10:1) as consecutive solvents in one dimension. In this way, tri-

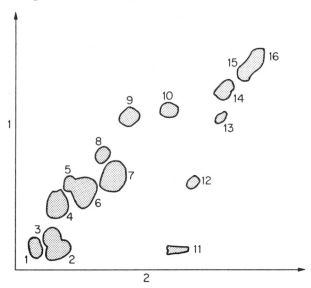

Fig. 4.4 Two-dimensional separation of lipids by thin layer chromatography on silica gel. Key: solvents 1 and 2 given in the text, spots detected with 50% H_2SO_4. Identification of spots: 1, origin; 2, phosphatidylserine; 3, phosphatidylinositol; 4, phosphatidylcholine; 5, sulpholipid; 6, digalactosyldiglyceride; 7, phosphatidylethanolamine; 8, phosphatidylglycerol; 9, glucocerebroside; 10, steryl glucoside; 11, phosphatidic acid; 12, unidentified galactolipid; 13, diphosphatidylglycerol; 14, monogalactosyl diglyceride; 15, steryl glycoside; 16, neutral lipids (triglycerides).

glycerides (R_F at about 70) separate from diglycerides (at 50) and monoglycerides (at 15). Hydrocarbons travel to near the front and phospholipids remain at the origin. Individual triglycerides in the band at R_F 70 can then be separated by argentation TLC, according to their degree of unsaturation. Silica gel plates are impregnated with 5% $AgNO_3$ and the plates developed with isopropanol–chloroform (3:197). Triglycerides then separate into several fractions, the R_F decreasing with increasing number of double bonds from one to six. For complex triglyceride mixtures, further separations may be necessary. Reversed phase partition chromatography may be employed, where the silica gel plate is impregnated with silicone oil and the solvent is acetonitrile–acetic acid–water (14:2:5).

Phospho- and glycolipids can be separated one-dimensionally on silica gel plates using solvents such as chloroform–methanol–acetic acid–water (170:30:20:7) and acetone–benzene–water (90:30:8). Phosphatidylglycerol is better separated from phosphatidylethanolamine by running the second solvent on plates previously impregnated with 0·15 M ammonium sulphate (Khan and Williams, 1977). Approximate R_F values of the different lipids in these two main systems are shown in Table 4.3 (from Harwood, 1980).

(e) *Detection*

Lipids may be detected without alteration by spraying plates with fluorescent dyes. A solution of 0·2% ethanolic 2',7'-dichlorofluorescein gives light green fluorescent spots in UV light, on a violet background. A spray of 0·5% ethanolic Rhodamine B or 6G produces yellow or blue-violet spots on a pink background. A non-specific, destructive but very sensitive spray for lipids is 25% H_2SO_4, followed by heating the plate at 230°C. Sterols give red-purple, glycolipids red-brown, sulpholipids bright red and other lipids pale brown colours. The limit of sensitivity is 1 μg. Lipids containing bases or sugars can be detected by reagents which react specifically with these residues. Thus, ninhydrin can be used to detect ethanolamine- and serine-containing phospholipids. A good sugar spray for glycolipids is 0·2% anthrone in conc.H_2SO_4, with heating at 70°C for 20 min to develop the colours; galactolipids give green and sulpholipids violet spots.

Table 4.3 TLC separations of plant lipids on silica gel

Lipid	R_F ($\times$100) *in**	
	$CHCl_3-MeOH-HOAc-H_2O$ (170:30:20:7)	$Me_2CO-C_6H_6-H_2O$ (90:30:8)
Triacylglycerol	93	92
Diacylgalactosylglycerol	90	82
Diacyldigalactosylglycerol	42	35
Diacylsulphoquinovosylglycerol	25	30
Phosphatidylcholine	35	10
Phosphatidylethanolamine	65	19
Phosphatidylinositol	15	—
Phosphatidylglycerol	46	22

*Data from Harwood (1980) measured on silica gel G plates activated at 110°C for 60 min and then developed at 20°C. Separation in the second solvent on $(NH_4)_2SO_4$-treated plates (see text) gives similar mobilities to untreated plates except that phosphatidylglycerol moves further, up to 62.

(f) *Determination of lipid components*

Acid or alkaline saponification of lipid yields fatty acids and glycerol and also, in the case of polar lipids, sugars or amines and phosphate. Acid hydrolysis is carried out with 2 M H_2SO_4 under nitrogen at 100°C for 6 h. After dilution with water, the fatty acids, and bases (if present) are extracted into chloroform. The aqueous residue is neutralized with $BaCO_3$, filtered and then chromatographed on paper in *n*-butanol–pyridine–water (7:3:1) for 40 h, along with marker solutions of sugars and glycerol. The presence of glycerol and galactose is detected with the alkaline $AgNO_3$ reagent. Amines in the chloroform layer may be detected by TLC (see Chapter 5, p. 187). Phosphate in the aqueous layer may be detected, as the sodium salt, by TLC on silica gel S-HR with methanol–1 M NH_4OH–10% trichloracetic acid–water (10:3:1:6). Phosphate appears as a blue spot, after spraying with 1% aqueous ammonium molybdate, followed by 1% stannous chloride in 10% HCl.

(g) *Gas liquid chromatography*

The fatty acids obtained after acid hydrolysis are converted to the methyl esters with ethereal diazomethane and then analysed by GLC. Alternatively, the fatty acid methyl esters can be obtained directly by transmethylation of the parent lipids by refluxing them for 90 min with methanol–benzene–H_2SO_4 (20:10:1). GLC is commonly carried out on columns of polyethylene glycol adipate (10% on 100–120 mesh Celite) at 200°C or of 3% SE 30 on 80–100 mesh Gas Chromosorb Q at 204°C. The peaks obtained are compared in retention times with standard fatty acid methyl esters. Typical separations on these two columns are illustrated in Fig. 4.5.

Some unsaturated acids cannot be identified by GLC alone. In such cases, it is necessary to carry out argentation TLC in order to determine the degree of

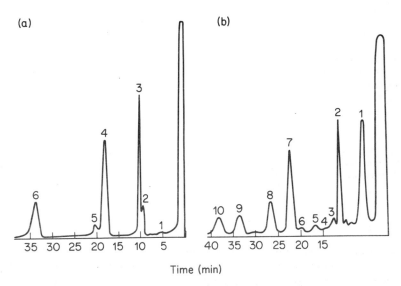

Fig. 4.5 Gas liquid chromatography separation of fatty acid methyl esters. Key: Column (a) SE-30 at 204°C, peak 1 is myristic, 2 palmitoleic, 3 palmitic, 4 oleic and linoleic, 5 stearic and 6, ricinoleic. Column (b) PEGA at 200°C, peak 1 is myristic, 2 palmitic, 3 palmitoleic, 4 C_{16} dienoic, 5 C_{16} trienoic, 6 stearic, 7 oleic, 8 linoleic, 9 linolenic and 10 C_{20} monoenoic acid. (From Hitchcock and Nichols, 1971.)

unsaturation. Hydrogenation to the corresponding saturated acid is useful, in order to determine the chain length. Finally, the positions of double bonds can be determined by cleaving with permanganate or by ozonolysis and identifying the fragments produced.

(h) *Confirmation of identity*

Individual lipids are further characterized by comparison with authentic samples and by spectroscopic procedures. UV spectroscopy is only useful for lipids with conjugated double bonds in their structures. These show absorption between 240 and 300 nm, while all other lipids lack absorption above 220 nm. IR spectroscopy is also of limited value. The methylene groups of fatty acids give rise to several characteristic bands, while unsaturated lipids show extra bands due to *cis*- or *trans*-ethylenic absorption.

The most valuable spectroscopic procedure is MS, which is used especially for identification of fatty acids (Odham and Stenhagen, 1972). Combined GLC–MS on the methyl ester trimethylsilyl ethers is the preferred method of analysis for the long-chain hydroxyacids present in cutins (Hunneman and Eglinton, 1972). Analyses of fats has been automated and procedures have been devised for linking TLC and GLC in the routine determination of lipids. Such a method is described by Pohl *et al.* (1970). The same authors have also devised a microscale procedure for TLC–GLC of the unsaturatd fatty acids (Pohl *et al.*, 1969).

The quantitative analysis of fatty acids can also be carried out routinely by straightforward GLC. For example, in a study of seed and leaf lipids in tobacco and other *Nicotiana* species, Koiwai *et al.* (1983) released the bound acids from small (10–50 mg) leaf or seed samples with 5% H_2SO_4 in methanol and the fatty acid methyl esters were analysed by GLC on a glass column packed with 5% BDS on Chromosorb W for leaf samples and with 5% DEGS on Gaschrom Q for seed samples. The column temperature was maintained at 200°C with a helium gas flow rate of 50 ml min^{-1} and the effluent monitored by a flame ionization detector.

(i) *High performance liquid chromatography*

Although HPLC has been applied to many lipid separations, there is still the major difficulty of finding a suitable detection system for a class of compound essentially lacking UV absorbance. In the case of the fatty acids, this can be overcome by derivatizing them and separating them as their phenacyl or *p*-bromophenacyl esters. In the case of the bound lipids, it is possible to measure their end absorption at about 195 nm or use a refractive index detector, but in general HPLC has not yet become a routine procedure in plant lipid studies. For further details of HPLC methods in lipid determinations, see Hammond (1982).

(j) *Authentic markers*

A range of phospholipids and neutral lipids (e.g. triolein) are available commercially. The firm of Koch-Light specialize in supplying phospholipids and related lipid enzymes.

4.2.3 Practical experiment

(a) *Procedure for determining the composition of peanut triglyceride*

(1) A weighed amount of peanuts (about four nuts) is macerated in a mortar with sand and 8 ml light petroleum (b.p. 40–60°C). After centrifugation, the residue is re-extracted with a further 8 ml of petroleum. The extracts are combined and taken to dryness in a weighed flask, evaporation being conducted in a stream of nitrogen. The lipid extract is weighed and the percentage of lipid in the original nuts can then be calculated.

(2) The lipid extract is then taken up in 5 ml cyclohexane and 3 μl spotted on a 5% $AgNO_3$–silica gel plate, along with marker solutions of triolein and tristearin, and the plate developed in isopropanol–chloroform (3:187).

(3) After drying, the plate is sprayed with 0·2% dibromo-R-fluorescein and examined in UV light. The separated triglycerides appear as yellow spots. The relative amounts of different unsaturated lipids can then be estimated from the size of the spots separated, the faster moving spots having mono-unsaturated fatty acid residues and the slower spots having di- and tri-unsaturated acids present.

(4) In order to identify the acids present, the remaining lipid extract is boiled for 10 min with 5 ml 2 M KOH in water–ethanol (1:1). This is cooled by adding 5 ml water and extracted with petroleum (2×20 ml) to remove unhydrolysed lipid.

(5) The hydrolysate is then acidified with 10 M H_2SO_4 and the fatty acid extracted with petroleum (3×15 ml).

(6) These extracts are then washed, dried and taken to dryness under nitrogen. Boron trifluoride–methanol complex (5 ml) is added and the solution is boiled for 2 min. After cooling by adding 15 ml water, the methyl esters are taken into petroleum (2×15 ml) and the extracts washed, dried and evaporated to dryness.

(7) The methyl esters are taken up in a minimum amount of petroleum and an aliquot injected into the inlet port of a GLC apparatus, which has a column of 10% diethyleneglycol adipate on 100–120 mesh Celite and which is operated at temperatures between 180 and 200°C. The relative retention times (RR_T) of the peaks emerging are measured, taking the RR_T of palmitic as 1·0. A standard mixture of methyl esters of palmitic, stearic, oleic and linoleic acids should be run concurrently; these acids emerge

from the GLC column in the order indicated above. Finally, the concentrations of the different acids in the peanut oil can be determined from measurement of the peak areas (see Chapter 1, p. 13).

4.3   ALKANES AND RELATED HYDROCARBONS

4.3.1   Chemistry and distribution

Long-chain hydrocarbons are best known as constituents of petroleum, which is at least partly derived from fossilized plant matter. In living plants, hydrocarbons are universally distributed in the waxy coatings on leaves and other plant organs. The alkane fraction is commonly a mixture of hydrocarbons of similar properties. The qualitative pattern is relatively similar from plant to plant, but there are considerable quantitative variations. Alkanes also occur in fungi and other lower plant groups, and the pattern is generally like that found in higher plants (Weete, 1972). Biosynthetically, these hydrocarbons are related to the fatty acids (see p. 152) and, in the simplest instances, are formed from them by chain elongation and decarboxylation. The function of alkanes in the cuticle waxes of plants is a protective one, the water-repellent properties providing a means of controlling water balance in the leaf and stem. Their universal presence in leaf coatings may also be to provide a measure of disease resistance to the plant.

Alkanes are saturated long-chain hydrocarbons with the general formula $CH_3(CH_2)_nCH_3$. They are usually present in the range of C_{25} to C_{35} carbon atoms, i.e. general formula with $n = 23$ to 33. Odd numbered members of the series, C_{25}, C_{27}, etc., predominate over the even numbered members of the series, often to the extent of 10:1. The major constituents of waxes are thus C_{27}–C_{31} alkanes. Examples are *n*-nonacosane, $C_{29}H_{60}$ and *n*-hentriacontane, $C_{31}H_{64}$, the major constituent of candelilla wax from *Euphorbia* species.

There are also a considerable number of alkane derivatives in waxes, formed by the introduction of unsaturation, branching of the chain, or oxidation to alcohol, aldehyde or ketone. Branching most commonly occurs near the end of the carbon chain. Two types may be mentioned; isoalkanes, with general formula $(CH_3)_2CH(CH_2)_n$-CH_3 and anteisoalkanes, general formula $(CH_3)(C_2H_5)CH(CH_2)_n$-CH_3. Branched alkanes are by no means universally present and rarely occur in any quantity. Olefinic alkanes, or alkenes, have a similar distribution in that they occur fairly frequently but in relatively low amount. Exceptionally high amounts of alkenes have been detected in rye pollen, rose petals and sugar cane.

Hydrocarbon alcohols are fairly common in plants, ceryl alcohol CH_3-$(CH_2)_{24}CH_2OH$ being a regular consituent in many cuticular waxes.

By contrast, aldehydes and ketones are infrequent. Two taxonomically interesting β-diketones might be mentioned: the compound $CH_3(CH_2)_{10}$-

COCH$_2$CO(CH$_2$)$_{14}$CH$_3$ found in waxes of certain *Eucalyptus* species and the related hentriacontan-14,16-dione present as a major wax constituent of cereals and grasses.

Detailed knowledge on the plant hydrocarbons has only accumulated significantly in the last twenty years and this has been entirely due to the application of one particular technique, GLC, to their separation and estimation. A key paper on leaf waxes is that of Eglinton *et al.* (1962), who first applied this technique to a survey of alkanes in the Crassulaceae. Of the many more recent papers that have been published, those of Herbin and Robins (1968; 1969) on *Aloe*, of Nagy and Nordby (1972) on citrus fruit hydrocarbons and of Evans *et al.* (1980) on *Rhododendron* leaf waxes may be mentioned. General reviews of plant wax constituents include those of Douglas and Eglinton (1966), Martin and Juniper (1970) and Baker (1982).

4.3.2   Recommended techniques

(a) *Extraction*

As a general precaution against contamination of plant samples, it is essential to employ clean glassware and redistilled solvents and also to avoid contact with stopcock grease or plastic tubing. Another source of possible contamination is contact with 'Parafilm', a thermoplastic sealing material widely employed in phytochemical laboratories. Indeed, this material, when washed with benzene or hexane is very useful since it will give a solution containing the standard range of *n*-alkanes, which can then be used for GLC comparison with alkanes from a plant extract (Gaskin *et al.*, 1971).

Extraction of plant waxes is simply carried out by dipping unbroken leaves or stems into ether or chloroform for very short periods of time (e.g. 30 s). This removes the surface alkanes without attacking cytoplasmic constituents. A filtration at this stage may be desirable to remove any dirt. An alternative procedure is to Soxhlet-extract dried powdered leaf for several hours in hexane. Such an extract will be considerably contaminated with leaf lipids and some fractionation will be necessary before the hydrocarbons are obtained pure. For example, steam-distillation may be desirable, in order to remove any essential oils from such an extract.

(b) *Purification*

It is common practice to fractionate the direct wax extract either in order to remove undesirable components (such as lipids) or to separate the hydrocarbon classes according to polarity. In many cases, all that is necessary is to pass the crude extract through an alumina column (e.g. Alcoa F-20 grade Al$_2$O$_3$) and elute with light petroleum. The first fraction contains the alkanes, the later fractions containing the alcohols, aldehydes or ketones. The purity of the alkane fraction can be checked by IR spectroscopy; oxygenated impurities if present will be apparent by IR absorption between 600 and 3500 cm^{-1}. A more thorough approach to the purification of alkanes is to saponify the crude wax with methanolic KOH and then remove ketonic materials by reacting them with 2,4-dinitrophenylhydrazine in aqueous 2 M HCl.

If lipids are present, these may be separated by TLC of the wax extract on silica gel in chloroform–benzene (1:1) followed by detection with a Rhodamine B fluorescein spray. The hydrocarbons have R_F ca. 90 and the fatty acid esters have R_F ca. 50. Alternatively, separation from lipids may be achieved by argentative TLC with light petroleum as solvent. Argentative TLC (see p. 155) may also be employed for fractionating alkane types; in 2% ether in petroleum (b.p. 30–60°C) saturated, mono-unsaturated and polyunsaturated hydrocarbons separate in order of decreasing R_F.

If the leaf waxes contain β-diketones as happens in many grass species, these can be separated by column chromatography on copper acetate–silica gel, previously prepared by mixing 80 g silica gel (200–400 mesh) with a solution of 25 g cupric acetate in 100 ml H$_2$O and then drying at 120°C. Other wax constituents are first eluted off with hexane–ether (9:1) and then the β-diketone–Cu complex formed on the column is eluted with chloroform–ethanol (17:3) at 50°C. The β-diketones are recovered by decomposing the complex with acid and are then identified by GLC–MS of the trimethylsilyl enol ethers (Tulloch, 1983).

(c) *Gas liquid chromatography*

The earlier procedures described by Eglinton *et al.* (1962) for the GLC of *n*-alkanes have been employed, with relatively little modification, up to the present time. These authors used a 120×0·5 cm column of 80–100 Celite coated with 0·5% Apiezon L grease, and obtained separations as indicated in Fig. 4.6. They found also that there was a linear relationship between the logarithm of the relative retention times and the alkane carbon number, in both the *n*- and isoalkane series. Identification was based in the first instance on the use of this linear relationship and on the intensification of appropriate peaks when genuine *n*-alkanes of known carbon number were added to the plant extract.

More recently, other types of column packing have been used. Two useful additional systems are 3% SE-30 on 100–120 Varaport 30 and 10% polyethylene glycol adipate. When a wide range of alkanes are being separated, a programmed temperature operation is desirable, such as one based on raising the temperature of the column from 70°C to 300°C at 6°C min^{-1}.

Alkenes can be tested for, using preparative GLC, by their reaction with bromine in carbon tetrachloride. Also, they have shorter relative retention

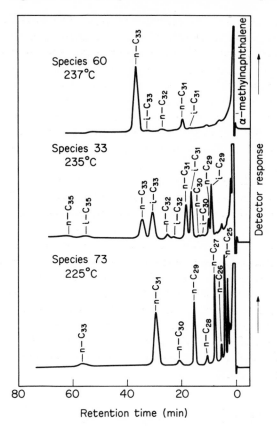

Fig. 4.6 Gas liquid chromatography of plant alkanes. Key: Species 60 is *Monanthes polyphylla*, species 33 is *Aeonium saundersii-Bolle* and species 73 is *Euphorbia aphylla*. Column 0·5% Apiezon L on 80–100 celite (from Eglinton *et al.*, 1962).

times (RR_Ts) than the corresponding alkanes on an Apiezon L column, but longer RR_Ts on a polyethylene glycol adipate column. Alkene peaks can also be made to disappear by catalytically hydrogenating the plant extract before GLC; on the subsequent GLC trace, the peaks of those alkanes related to the alkenes reduced will be intensified.

Confirmation of identification is most frequently done by mass spectral studies, and in many recent studies of plant alkanes, combined GC–MS is used. IR spectroscopy may also be employed for identification, but it is not very sensitive to impurities. The advantage of mass spectral studies is that they may well reveal trace amounts of isomers in what otherwise appear to be pure GLC fractions.

(d) *Ethylene*

The identification of this simple hydrocarbon ($CH_2{=}CH_2$) is of especial interest to plant physiologists since in recent years, it has been recognized to be an important natural growth regulator. Ethylene produced by plants is measured using a GLC apparatus set up for gas analysis. Since only very small amounts of ethylene are produced by plant tissues, it is essential that the GLC recorder used is operating at the maximum level of sensitivity. Before analysis, ethylene can be condensed in a liquid oxygen trap, or by passing it on to a column of silica gel impregnated with mercuric perchlorate (Phan, 1965).

Originally, GLC was carried out on a column of 30% silicone oil 550 coated on Firebrick (60–80 mesh) (Meigh *et al.*, 1960). More recently, Galliard *et al.* (1968) have employed a stainless steel column (150 cm × 3 mm) packed with 10% Triton x–305 on NAW Chromosorb G (80–100 mesh), operating isothermally at room temperature with a flame ionization detector. The same authors removed polar volatile compounds from the gas samples by employing a pre-column (17 cm × 4 mm) packed with 20% diglycerol on celite and fitted with a back-flushing device. Using this procedure, 0·03 p.p.m. ethylene can be detected.

More recently still, Muir and Richter (1972) have recommended GLC on a column of Porapak, at a temperature of 80°C. In a typical run on this support, the authors found oxygen emerging after 1·5 min, methane at 2 min and ethylene at 4 min. Although ethylene clearly separates from related hydrocarbons (e.g. methane, ethane, propylene) on most column stationary phases, it is as well to confirm its identification in natural plant vapours by GLC on at least two different types of column, e.g. on Porapak together with silicic acid (80–100 mesh) or alumina F_1 (80–100 mesh) as the contrasting phase. Conclusive identification of ethylene in a new plant source really requires GLC–MS analysis as well (Ward *et al.*, 1978).

4.4 POLYACETYLENES

4.4.1 Chemistry and distribution

Polyacetylenes (or acetylenic compounds) are an unusual group of naturally occurring hydrocarbons which all have one or more acetylenic groups in their structures. It is remarkable that while acetylene, $CH{\equiv}CH$, itself is a highly reactive, even dangerously explosive gas, the long-chain hydrocarbon derivatives are sufficiently stable to be isolated and characterized by standard phytochemical techniques. Indeed, over 650 polyacetylenes are now known as plant products. Only a few are simple hydrocarbon derivatives; most have additional functional groups and are either alcohols, ketones, acids, esters, aromatics or furans. Some typical structures found among the polyacetylenes are indicated in Fig. 4.7.

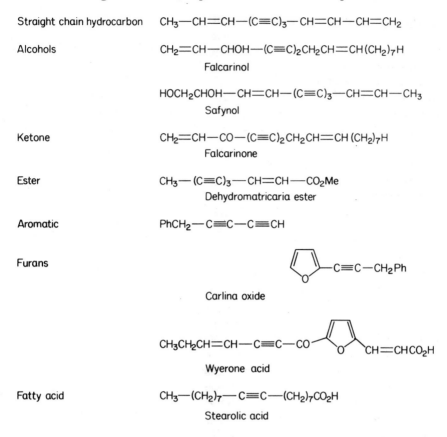

Straight chain hydrocarbon $CH_3—CH=CH—(C≡C)_3—CH=CH—CH=CH_2$

Alcohols $CH_2=CH—CHOH—(C≡C)_2CH_2CH=CH(CH_2)_7H$
Falcarinol

$HOCH_2CHOH—CH=CH—(C≡C)_3—CH=CH—CH_3$
Safynol

Ketone $CH_2=CH—CO—(C≡C)_2CH_2CH=CH(CH_2)_7H$
Falcarinone

Ester $CH_3—(C≡C)_3—CH=CH—CO_2Me$
Dehydromatricaria ester

Aromatic $PhCH_2—C≡C—C≡CH$

Furans
Carlina oxide

$CH_3CH_2CH=CH—C≡C—CO$
Wyerone acid

Fatty acid $CH_3—(CH_2)_7—C≡C—(CH_2)_7CO_2H$
Stearolic acid

Fig. 4.7 Structures of some representative polyacetylenes.

Polyacetylenes have a taxonomically interesting distribution pattern in higher plant families; they occur regularly in only five families, namely the Campanulaceae, Compositae, Araliaceae, Pittosporaceae and Umbelliferae. The former two and latter three families are especially closely linked in other ways (Sorensen, 1968). Acetylenic acids (e.g. stearolic acid) have a rather different distribution from the other polyacetylenes and are found in *Santalum*, Santalaceae and other families in the Santalales and also in certain Malvales, where they occur in association with cyclopropene fatty acids (Bu'Lock, 1966). Acetylenics are also found in the higher fungi, in two families of the Basidiomycetes, the Agaricaceae and Polyporaceae. The fungal compounds are slightly different in having a chain length mainly between C_8 and C_{14}, whereas the higher plant acetylenes are mostly C_{14} to C_{18} compounds.

In their biosynthesis, the polyacetylenes are probably formed from the corresponding fatty acid, via an olefinic intermediate by successive dehydrogenations, followed by other modifications (Bu'Lock, 1966). If acetylenes have an overall function, it is most likely as toxins in either plant–animal or plant–plant interactions. Thus, some are highly poisonous, e.g. those found in the roots of the water dropwort, *Oenanthe crocata* and of fool's parsley, *Aethusa cynapium*, while others in fungi have antibiotic activity. Also, two acetylenes, wyerone acid in broad bean, *Vicia faba*, and safynol in safflower, *Carthamus tinctorius*, have been implicated as natural phytoalexins and are toxic to micro-organisms which attack these plants.

A comprehensive review of the chemistry and chemotaxonomy of the poly-acetylenes is available (Bohlmann *et al.*, 1973).

4.4.2   Recommended techniques

(a) *Thin layer chromatography*

Fresh tissue is extracted by maceration with ether in the presence of activated alumina and left to extract for at least 24 h at 4°C in the dark. The decanted solution is dried over Na_2SO_4 and taken to dryness at 15°C. The samples, in ether, are then chromatographed on either Al_2O_3 or silica gel plates in benzene–chloroform (10:1), pentane–ether (9:1) or chloroform–methanol (9:1). On spraying the plates with 0·4% isatin in conc.H_2SO_4, and heating, acetylenes appear as brown or green spots. They will appear as yellow spots on plates sprayed with 1% $KMnO_4$ in 2% aqueous Na_2CO_3. If silica gel plates with fluorescent indicator are used (HF_{254}), acetylenes can be simply detected by their quenching action.

(b) *Spectroscopy*

The most characteristic property of polyacetylenes is their UV spectrum. Almost all compounds show a series of three or more very intense peaks in the region 200–320 nm (see Fig. 4.8) and UV spectroscopy is widely used as a means of preliminary detection. For example, in screening plants of the Campanulaceae (leaves and roots) for polyacetylenes, Bentlev *et al.* (1969) depended on UV monitoring of crude ether extracts for their detection. Thus, two compounds in the root of the bellflower, *Campanula glomerata* after pre-liminary purification on silica gel in ether–petroleum, had λ_{max} 252, 266 and 288 nm and 277, 294 and 313 nm respectively. UV spectra of some pure polyacetylenes found in the Umbelliferae are given in Table 4.4, to further illustrate the range and number of spectral maxima.

Unfortunately, UV spectroscopy is not a conclusive test for presence or absence of acetylenes, since just a few compounds fail to give a series of intense peaks, showing instead only a single broad band. In such cases, IR spectra may be measured, since there is a characteristic band at about 2200 cm^{-1} for

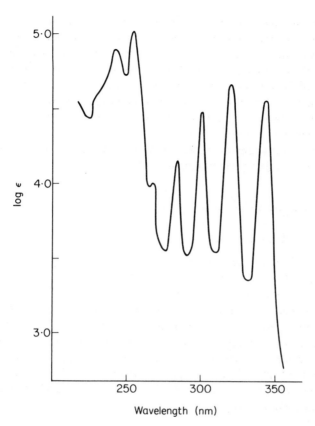

Fig. 4.8 Ultraviolet spectrum of a polyacetylene. Absorption spectrum of $CH_3(C\equiv C)_3CH=CH\text{-}CO_2Me$ (from Sorensen, 1968).

Table 4.4 Ultraviolet spectral maxima of some polyacetylenes of the Umbelliferae

Polyacetylene	UV maxima (nm)	Source
Falcarinol	229,240,254	root of domestic carrot
Falcarinolone	259,274,290	caraway, *Carum carvi*
Falcarinone	208,246,260,274,291	*Falcaria vulgaris*
Falcarindione	245,253,268,285,303	caraway, *Carum carvi*
Aethusin	210,248,264,280,294, 313,334	root of Fool's parsley, *Aethusa cynapium*

Data is from Bohlmann *et al.* (1961). Details of the distribution of polyacetylenes in the Umbelliferae can be found in Bohlmann (1971).

the acetylenic triple bond. Even so, substances with a triple bond adjacent to two conjugated double bonds cannot be clearly identified by either UV or IR measurements (Sorensen, 1968). Two other spectral procedures, NMR and MS, are now widely employed for structural identification of polyacetylenes. Even these techniques are not completely unequivocal in indicating acetylenic substituents.

Most polyacetylenes can be isolated in reasonably stable form and the carrot root compound, falcarinol, can even be purified by gas chromatography using a column temperature of 192°C and a liquid phase of 5% Dow II silicone oil on 60–80 mesh Chromosorb W (Crosby and Aharonson, 1967). Some, however, are much more unstable. A ketoaldehyde present in parsnip seed was obtained as a colourless oil, which resinified within 2 days, even when kept in the dark at −40°C (Jones *et al.*, 1966).

4.4.3   Practical experiments

(a) *The polyacetylenes of the carrot*

The domestic carrot contains four acetylenes, the major one being falcarinol (Bentley *et al.*, 1969). This compound is weakly toxic, producing neurotoxic symptoms on injection into mice (Crosby and Aharonson, 1967). However, the quantities in carrot are so low (2 mg kg^{-1} carrot) that it does not present a dietary hazard. The structure advanced by Crosby and Aharanson for their 'carotatoxin' was later revised by Bentley *et al.* (1969) to that of the already known polyacetylene, falcarinol. This compound can be easily isolated from carrot (and from roots of a number of related umbellifers) and detected by its UV spectrum.

(b) *Procedure*

(1) Three medium-sized fresh carrots are cut into small pieces, covered with ether and stored in the refrigerator for at least 24 h.

(2) The ether is then decanted, the extract dried over $MgSO_4$ and evaporated to dryness at below 35°C. The residue is taken up in a small volume of *n*-hexane and applied as a streak to a narrow (5×20 cm) and full size (20×20 cm) silica gel plate.

(3) The plates are developed in benzene–chloroform (10:1) for 1 h. The smaller plate is then sprayed with 0·4% isatin in conc.H_2SO_4 and heated at 110°C for 5 min. The most intense brown spot is due to falcarinol and the position of the major band on the unsprayed plate can be determined by comparison with the marker plate.

(4) The band is then scraped off, eluted with ether (3 ml) and the UV spectrum measured. Several intense peaks should be obtained, cor-

responding in position to those of falcarinol (Table 4.4). Extra peaks may also be present due to contamination with related acetylenes. If this is so, falcarinol can be purified by re-running it on silica gel in pentane–ether (9:1).

(c) *Other sources of polyacetylenes*

Another plant which can be used as a convenient source of acetylenics is the water dropwort, *Oenanthe crocata*, which grows abundantly near ditches and streams in temperate regions of the Northern hemisphere. The roots are poisonous and the plant has been described as the 'five finger death' because of their shape. Fresh roots can be collected, and after careful washing, extracted in the same way as in the carrot experiment above. A TLC plate of a concentrated extract developed in pure chloroform will show the presence of several acetylenes after spraying with isatin in conc.H_2SO_4. One of the major toxins, a ketone, has spectral maxima at 210, 249, 266, 291, 314 and 336 nm.

4.5 SULPHUR COMPOUNDS

4.5.1 Chemistry and distribution

In addition to the ubiquitous sulphur amino acids cysteine and methionine there are other sulphur-containing organic compounds in plants, mostly of restricted distribution. Many are volatile and have either an acrid taste or an obnoxious smell so that their presence is rather obvious during extraction and isolation. The most important undoubtedly are the glucosinolates (mustard oil glucosides) of the Cruciferae and the organic disulphides of *Allium*, since the flavours of mustard, radish, onion and garlic are due to these substances. A third group are the acetylenic thiophenes found in roots and leaves of members of the Compositae. There are also a few miscellaneous sulphur derivatives, such as the sulphur alkaloids of *Nuphar*, Nympheaceae, which will not be considered further here.

The glucosinolates are the *in vivo* precursors of the mustard oils themselves and only yield acrid volatile flavours after enzymic hydrolysis. This occurs whenever fresh crucifer tissue is crushed, since the glucosinolates are always accompanied by the appropriate hydrolytic enzyme, a thioglucosidase known as myrosinase. The products of enzymic release are the isothiocyanates, formed by hydrolysis and rearrangement of the parent thioglucosides according to the scheme shown in Fig. 4.9. Thus, when screening plants for mustard oil glycosides, it does not matter whether detection is *via* the glucosinolate, the isothiocyanate formed on hydrolysis or the enzyme myrosinase.

Glucosinolates

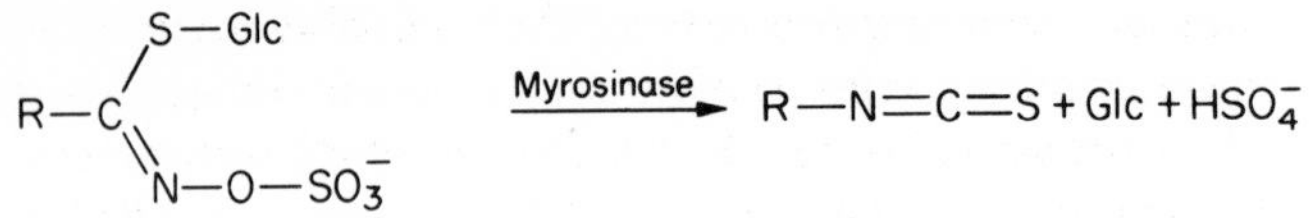

$$R-C \overset{\displaystyle S-Glc}{\underset{\displaystyle N-O-SO_3^-}{\Big\langle}} \quad \xrightarrow{\text{Myrosinase}} \quad R-N{=}C{=}S + Glc + HSO_4^-$$

R = CH_3-, glucocapparin
R = $CH_2{=}CH{-}CH_2-$, sinigrin
R = $MeS\,(CH_2)_3-$, glucoibervirin
R = $PhCH_2-$, glucotropaeolin
R = $p-HO-C_6H_4-$, sinapine

Sulphides

Me$-$S$-$S$-$Me dimethyldisulphide
$n-$Pr$-$S$-$S$-$Pr$-n$ di$-n-$propyldisulphide
$CH_2{=}CH{-}CH_2{-}S{-}S{-}CH_2{-}CH{=}CH_2$ diallyldisulphide
$CH_2{=}CH{-}CH_2{-}S{-}S{-}CH_3$ methylallyldisulphide
$CH_2{=}CH{-}S{-}CH{=}CH_2$ diallysulphide

Thiophenes

Fig. 4.9 Sulphur compounds of plants.

About seventy glucosinolates are known, the majority being aliphatic derivatives (e.g. glucocapparin, sinigrin) with the remainder having benzyl substituents (e.g. sinapine) of one sort or another. The basic structure is the same (see Fig. 4.9) and the sulphur-linked sugar is always glucose. Glucosinolates are almost certainly biosynthesized from amino acids, the pathway probably being related to that utilized for cyanogenic glucoside synthesis (see p. 202). Their function in plants is still being elucidated; they are known to have antibacterial properties and some are feeding attractants to caterpillars and aphids feeding on crucifers.

The glucosinolates are universally distributed in the Cruciferae and also occur in related families in the same plant order, the Rhoeadales, namely the Capparaceae, Resedaceae, Tovariaceae and Moringaceae. The absence of mustard oil glucosides from the Papaveraceae, a family often closely associated with the Cruciferae, is good chemotaxonomic evidence for the separation of the poppy family into another plant order (Kjaer, 1963; 1966). There are a number of miscellaneous occurrences in unrelated plant groups, notably of glucotropaeolin (giving benzyl isothiocyanate) in seeds of *Carica papaya*, Caricaceae, the classic source of the proteolytic enzyme, papain.

Sulphides, which are widely distributed in the species of *Allium* (Saghir *et al.*, 1968; Bernhard, 1970) are immediately recognizable by their pungent smells and lachrymatory properties. They probably occur in intact tissue bound as sulphur amino acids; during isolation, they can give rise by decomposition to alkyl thiols (e.g. methylmercaptan, MeSH), substances with even more repellent odours. The sulphides of *Allium* are mainly alkyl mono- or disulphides; the structures of some typical members are illustrated in Fig. 4.9.

Finally, there are the thiophenes, a remarkably diverse group of natural products (Ettlinger and Kjaer, 1968) occurring almost entirely in one family, the Compositae, and then in association with polyacetylenes (see p. 165). The first to be discovered α-terthienyl, found in marigold, *Tagetes* petal, is a simple trimer of thiophene itself, but most of the 100 or so thiophenes have side-chains containing acetylenic substitutions (e.g. methylphenyltriacetylene from *Coreopsis*). They are normally isolated along with the polyacetylenes and are purified and identified by very similar procedures.

4.5.2 Recommended techniques

(a) *Glucosinolates*

These are extracted with boiling alcohol from fresh tissue; it is important to avoid enzymic hydrolysis during this procedure. They are purified on columns of anion exchange resin (Amberlite IR-400) or of acid washed 'anionotropic' alumina (Kjaer, 1960). They can then be separated and isolated by PC in *n*-butanol–ethanol–water (4:1:4) or *n*-butanol–pyridine–water (6:4:3), detection being by means of ammoniacal $AgNO_3$. Alternatively, TLC on silica gel can be used, with solvents such as chloroform–methanol (17:3) and ethyl acetate–ethanol–water (9:1:2), when they can be detected as yellow spots with iodine vapour (Matsuo, 1970).

For surveying plant seeds for glucosinolates, Gmelin and Kjaer (1970) extracted defatted powdered seed samples (*ca*. 1 g) with 70% methanol. The concentrates were chromatographed on paper in the two solvents given above. The papers were then dipped in an acetone solution of $AgNO_3$ (0·75 g $AgNO_3$, 10 ml water and 390 ml acetone) and sprayed on both sides with 5% NaOH in methanol. The glucosinolates appeared as dark brown spots.

(b) *Isothiocyanates*

These are obtained by steam-distillation of crushed crucifer tissue, or by enzymic or acid hydrolysis of isolated glucosinolates. They have a characteristic pungent taste and smell. The non-volatile isothiocyanates can be separated on paper in similar solvents as for glucosinolates, detection being with ammoniacal $AgNO_3$. TLC is carried out on silica gel G in carbon tetrachloride–methanol–water (20:10:1). The volatile isothiocyanates are either separated by GLC or, more frequently, by PC as their thiourea derivatives. Thioureas are prepared by warming 0·3 g of the crude iso-thiocyanate oil with 1 ml 95% ethanol and 1 ml 25% NH_4OH. An exothermic reaction takes place and on cooling, the thiourea derivative(s) should crystallize out. The derivatives are then separated either on paper in water-saturated chloroform or *n*-butanol–ethanol–water (4:1:4) or on silica gel G plates in ethyl acetate–chloroform–water (3:3:4) (Wagner *et al.*, 1965). Thioureas give characteristic blue colours on treatment with Grote's reagent. This is prepared by adding two drops of bromine to a mixture of sodium nitroprusside (0·5 g), water (10 ml), hydroxylamine hydrochloride (0·5 g) and $NaHCO_3$ (1 g); after aeration and filtration, the solution is made up to 25 ml and is stable for up to 2 weeks (Grote, 1931).

In the quantitative analysis of glucosinolates in leaf tissues of Cruciferae, the glucosinolates should be first isolated by extraction into boiling 80% methanol and the aqueous concentrate then treated with myrosinase (prepared from Colman's mustard powder by the method of Wrede, 1941) for 12 h to convert them to the isothiocyanates. These are collected by ether extraction and analysed by GLC on a glass column packed with 5% FFAP on Chromosorb W using a sulphur detector (Feeny and Rosenberry, 1982). The identity of the individual glucosinolates present should be checked at the same time by PC, using the methods described above.

(c) *Sulphides*

These have been routinely separated in the volatile fraction of *Allium* bulbs or leaves, by GLC on 5% Carbowax 20 M on a Firebrick (100–120 mesh) support using a 3 m×3·2 mm column at 90°C. In order to improve the sensitivity, it is advantageous to use dual channel GLC, with both hydrogen flame and electron capture detectors (Saghir *et al.*, 1964; Bernhard, 1970). This procedure gives good separation of the various alkyl sulphides present.

(d) *Thiophenes*

These are isolated and purified by similar procedures as for polyacetylenes (see p. 166). TLC is carried out on silica gel or Al_2O_3 plates in pentane–ether (9:1) or benzene–chloroform (10:1). Plates are sprayed with 0·4% isatin in conc.H_2SO_4 and then heated. Thiophenes give intense purple, red, blue or blue-green colours and can be distinguished from simple polyacetylenes, which give brown or green colours. The presence of a thiophene in a plant extract has to be confirmed by isolation of the pure compound, followed by spectral studies. 2,5-Disubstituted thiophenes, for example, have a characteristic band at 838 cm^{-1} in the IR spectrum (measured in KBr) and also several distinctive signals in the NMR spectrum.

REFERENCES

General references

Baker, E.A. (1982) Chemistry and morphology of plant epicuticular waxes. in *The Plant Cuticle* (eds D.F. Cutler, K.L. Alvin and C.E. Price), Academic Press, London, pp. 139–66.

Bohlmann, F., Burkhardt, T. and Zdero, C. (1973) *Naturally Occurring Acetylenes*, Academic Press, London.

Douglas, A.G. and Eglinton, G. (1966) The distribution of alkanes. in *Comparative Phytochemistry* (ed. T. Swain), Academic Press, London, pp. 55–7.

Ettlinger, M.G. and Kjaer, A. (1968) Sulphur compounds in plants. *Recent Adv. Phytochem.*, **1**, 59–144.

Gunstone, F.D. (1967) *An Introduction to the Chemistry and Biochemistry of Fatty Acids and their Glycerides*, 2nd edn, Chapman and Hall, London.

Gunstone, F.D. (1975) Determination of the structure of fatty acids. in *Recent Advances in the Chemistry and Biochemistry of Plant Lipids* (ed. T. Galliard and E.I. Mercer), Academic Press, London, pp. 21–43.

Harwood, J.L. (1980) Plant acyl lipids: structure, distribution and analysis. in *The Biochemistry of Plants*, Vol. 4, *Lipids* (ed. P.K. Stumpf), Academic Press, New York, pp. 2–56.

Hitchcock, C. and Nichols, B.W. (1971) *Plant Lipid Biochemistry*, Academic Press, London.

Holloway, P.J. (1982) The chemical constitution of plant cutins. in *The Plant Cuticle* (eds D.F. Cutler, L.K. Alvin and C.E. Price), Academic Press, London, pp. 1–32.

Kjaer, A. (1960) Naturally derived isothiocyanates (mustard oils) and their parent glucosides. *Fortschr. Chemie Org. Naturst.*, **18**, 122–76.

Kjaer, A. (1963) The distribution of sulphur compounds. in *Chemical Plant Taxonomy* (ed. T. Swain), Academic Press, London, pp. 453–73.

Kjaer, A. (1966) The distribution of sulphur compounds. in *Comparative Phytochemistry* (ed. T. Swain), Academic Press, London, pp. 187–94.

Kluge, M. and Ting, I.P. (1978) *Crassulacean Acid Metabolism*, Springer-Verlag, Berlin.

Mangold, H.K. (1969) Aliphatic lipids. in *Thin Layer Chromatography* (ed. E. Stahl), George Allen and Unwin, London, pp. 363–420.

Martin, J.T. and Juniper, B.E. (1970) *The Cuticle of Plants*, Edward Arnold, London.

Shorland, F.B. (1963) The distribution of fatty acids in plant lipids. in *Chemical Plant Taxonomy* (ed. T. Swain), Academic Press, London, pp. 253–312.

Sorensen, N.A. (1968) The taxonomic significance of acetylenic compounds. *Recent Adv. Phytochem.*, **1**, 187–228.

Ulrich, R. (1970) Organic acids. in *The Biochemistry of Fruits and their Products*, Vol. I, (ed. A.C. Hulme), Academic Press, London, pp. 89–118.

Supplementary references

Association of Official Agricultural Chemists (A.O.A.C.) (1965) *Methods of Analysis*, 10th edn, A.O.A.C., Washington D.C., U.S.A., p. 399.

Bentley, R.K., Jenkins, J.K., Jones, E.H.R. and Thaller, V. (1969) *J. Chem. Soc.*, (C) 830.

Bernhard, R.A. (1970) *Phytochemistry*, **9**, 2019.

Blundstone, H.A.W. (1962) *Nature* (Lond.), **197**, 377.

Bohlmann, F. (1971) in *Biology and Chemistry of the Umbelliferae* (ed. V.H. Heywood). Academic Press, London, pp. 279–92.

Bohlmann F., Arndt, C., Bornowski, H. and Kleine, K.M. (1961) *Chem. Ber.*, **94**, 958.

Bolliger, H.R. and Koenig, A. (1969) in *Thin Layer Chromatography* (ed. E. Stahl), George Allen and Unwin, London, pp. 259–310.

Boudet, A. (1973) *Phytochemistry*, **12**, 363.

Bu'lock, J.D. (1966) in *Comparative Phytochemistry* (ed. T. Swain), Academic Press, London, pp. 79–95.

Carles, J., Schneider, A. and Lacoste, A.M. (1958) *Bull. Soc. Chim. Biol.*, **40**, 221.

Cookman, G. and Sondheimer, E. (1965) *Phytochemistry*, **4**, 773.

Crosby, D.G. and Aharonson, N. (1967) *Tetrahedron*, **23**, 465.

Eglinton, G., Gonzalez, A.G., Hamilton, R.J. and Raphael, R.A. (1962) *Phytochemistry*, **1**, 89.

Evans, D., Kane, K.H., Knights, B.A. and Math, V.B. (1980) in *Contributions Toward A Classification of Rhododendron* (eds J.L. Luteyn and M.E. O'Brien), The Botanic Garden, New York, pp. 187–246.

Feeny, P. and Rosenberry, L. (1982) *Biochem. System. Ecol.*, **10**, 23.

Galliard, T. (1968) *Phytochemistry*, **7**, 1907.

Galliard, T., Rhodes, M.J.C., Wooltorton, L.S.C. and Hulme, A.C. (1968) *Phytochemistry*, **7**, 1465.

Gaskin, P., MacMillan, J., Firn, R.D. and Pryce, R.J. (1971) *Phytochemistry*, **10**, 1155.

Gibbs, R.D. (1963) in *Chemical Plant Taxonomy* (ed. T. Swain), Academic Press, London, pp. 41–88.

Gmelin, R. and Kjaer, A. (1970) *Phytochemistry*, **9**, 591.

Grote, I.W. (1931) *J. Biol. Chem.*, **93**, 25.

Hammond, E.W. (1982) in *HPLC in Food Analysis* (ed. R. Macrae), Academic Press, London, pp. 167–86.

Herbin, G.A. and Robins, P.A. (1968) *Phytochemistry*, **7**, 239, 1325.

Herbin, G.A. and Robins, P.A. (1969) *Phytochemistry*, **8**, 1985.

Horii, Z., Matika, M. and Tamura, Y. (1965) *Chem. Ind.*, 1494.

Hunneman, D.H. and Eglinton, G. (1972) *Phytochemistry*, **11**, 1989.

Isherwood, F.A. and Niavis, C.A. (1956) *Biochem. J.*, **64**, 549.

Jones, E. and Hughes, R.E. (1983) *Phytochemistry*, **22**, 2493.

Jones, E.H.R., Safe, S. and Thaller, V. (1966) *J. Chem. Soc.* (C), 1220.

Khan, M.U. and Williams, J.P. (1977) *J. Chromatog.*, **140**, 179.

Knappe, E. and Rohdewald, I. (1964) *Z. Analyt. Chem.*, **211**, 49.

Koiwai, A., Suzuki, F., Mat Suzaki, T. and Kawashima, N. (1983) *Phytochemistry*, **22**, 1409.

Matsuo, M. (1970) *J. Chromatog.*, **49**, 323.

Mazliak, P. and Salsac, L. (1965) *Phytochemistry*, **4**, 693.

Meigh, D.F., Norris, K.H., Craft, C. and Lieberman, M. (1960) *Nature* (Lond.), **186**, 902.

Morris, L.J. and Nichols, B.W. (1967) in *Chromatography*, 2nd edn (ed. E. Heftmann), Reinholt, New York, pp. 466–509.

Morris, L.J. and Nichols, B.W. (1970) in *Progress in Thin Layer Chromatography and Related Methods* (eds A. Niederwieser and G. Pataki), Vol. 1, Science Publishers, Ann. Arbor, pp. 74–93.

Muir, R.M. and Richter, E.W. (1972) in *Plant Growth Substances 1970* (ed. D.J. Carr), Springer-Verlag, Berlin, pp. 518–25.

Nagy, S. and Nordby, H.E. (1972) *Phytochemistry*, **11**, 2865.

Odham, G. and Stenhagen, E. (1972) in *Biochemical Applications of Mass Spectrometry* (ed. G.R. Waller), Wiley-Interscience, New York, pp. 211–50.

Palmer, J.K. and Wyman, A.H. (1965) *Phytochemistry*, **4**, 305.

Peereboom, J.W.C. (1969) in *Thin Layer Chromatography* (ed. E. Stahl), George Allen and Unwin, London, pp. 650–4.

Phan, C.T. (1965) *Phytochemistry*, **4**, 353.

Pohl, P., Glasl, H. and Wagner, H. (1969) *J. Chromatog.*, **42**, 75.

Pohl, P., Glasl, H. and Wagner, H. (1970) *J. Chromatog.*, **49**, 488.

Saghir, A.R., Mann, L.K., Bernhard, R.A. and Jacobsen, J.V. (1964) *Proc. Am. Soc. Hort. Sci.*, **84**, 386.

Saghir, A.R., Mann, L.K., Ownbey, M. and Berg, R.Y. (1968) *Am. J. Bot.*, **53**, 477.

Shyluk, J.P., Youngs, C.G. and Gamborg, O.L. (1967) *J. Chromatog.*, **26**, 268.

Towers, G.H.N., Thompson, J.F. and Steward, F.C. (1954) *J. Am. Chem. Soc.*, **76**, 2392.

Tulloch. A.P. (1983) *Phytochemistry*, **22**, 1605.

Van Niekerk, P.J. (1982) in *HPLC in Food Analysis* (ed. R. Macrae), Academic Press, London, pp. 187–226.

Vickery, B., Vickery, M.L. and Ashu, J.T. (1973) *Phytochemistry*, **12**, 145.

Wagner, H., Horhammer, L. and Nufer, H. (1965) *Arzneimittel-Forsch.*, **15**, 453.

Ward, T.M., Wright, M., Roberts, J.A., Self, R. and Osborne, D.J. (1978) in *Isolation of Plant Growth Substances* (ed. J.R. Hillman), Cambridge University Press, Cambridge, p. 135.

Weete, J.D. (1972) *Phytochemistry*, **11**, 1201.

Wrede, F. (1941) in *Die Methoden der Fermentforschung* (eds E. Baumann and K. Myrback), Thieme, Leipzig, p. 1835.

Nitrogen Compounds

5.1 INTRODUCTION

Although only 2% of the dry weight of plants consists of the element nitrogen, compared to 40% for carbon, there are still a very large number of different nitrogen-containing organic substances known in plants. Nitrogen is first available to the plant in the form of ammonia, produced either from nitrogen fixation in the root (symbiotically in legumes) or from enzymic reduction of absorbed nitrate in shoot and leaf. Nitrogen first appears in organic form as glutamine, the key reactions being the transfer of ammonia to glutamic acid, catalysed by glutamine synthetase, followed by the transfer of nitrogen from glutamine to α-ketoglutarate, catalysed by glutamine α-ketoglutarate aminotransferase. These two enzymes operate together in what is often termed the GS–GOGAT pathway. The other amino acids are subsequently synthesized from the corresponding α-keto acids, the amino group being passed on from glutamic acid through the catalytic action of non-specific aminotransferases. Amino acids are also involved in the biosynthesis of practically all the other nitrogenous plant compounds, from the proteins (see Chapter 7) to the alkaloids, amines, cyanogenic glycosides, porphyrins, purines, pyrimidines and cytokinins.

In this chapter, which deals with methods of identifying nitrogen compounds, the amino acids will be considered first, since they play such a central role in nitrogen metabolism in plants. In the case of plant as distinct from animal systems, this poses special problems because of the existence of over 200 non-protein amino acids, in addition to the twenty protein amino acids common to all living systems.

By far the largest class of nitrogen compounds in plants, however, are the toxic alkaloids; some 5500 such substances are known. With so many structures, the problems of identifying an unknown alkaloid from a new plant source are very considerable. A much smaller group of toxic nitrogen compounds, also dealt with here, are the cyanogenic glycosides; these are characterized by their ability to release hydrogen cyanide on hydrolysis.

To the plant physiologist, the most important nitrogen compound is the universal growth hormone, auxin or indoleacetic acid and the identification of indoles is the subject of another section in this chapter. A further section deals with the purines and pyrimidines, the building blocks of the nucleic acids. Included under this heading are a group of plant hormones which control cell division, the cytokinins, which are purine derivatives and which require similar methods for their identification. A third section, on plant amines, may be of interest to plant physiologists, since certain polyamines such as spermine and spermidine have recently been implicated in plant growth processes.

The last group of nitrogen compounds covered in this chapter are the porphyrin pigments, the most important ones being the photosynthetic catalysts, the chlorophylls. Although complex in structure, the chlorophylls are synthesized, in the first instance, from a simple amino acid precursor, glutamate.

The chemical property which distinguishes nitrogen compounds from other organic substances is that they are usually basic; thus, they form salts with mineral acids. Also, they can be extracted from plant tissues using weak acidic solvents and can then be selectively precipitated from such extracts by addition of ammonia. Many nitrogen compounds are charged molecules (e.g. amino acids, amines, many alkaloids) so that electrophoresis can be used directly for their separation. The best known detection technique for nitrogen compounds is the spray reagent, ninhydrin; this is widely used for detecting amino acids, but other amino compounds also react with it. Another widely used reagent – for alkaloids – is that due to Dragendorff, namely a solution of potassium iodide and bismuth subnitrate. However, in spite of the availability of these and other reagents, there are often difficulties in detecting certain classes of nitrogen compound in plant extracts.

For a general account of nitrogen metabolism in plants, see Beevers (1976) or Miflin (1981). References to individual classes of nitrogenous compounds will be found in the respective sections.

5.2 AMINO ACIDS

5.2.1 Chemistry and distribution

The plant amino acids are conveniently divided into two groups, the 'protein' and 'non-protein' acids, although the division between the two groups is not entirely sharp and methods of identifying and separating both groups are

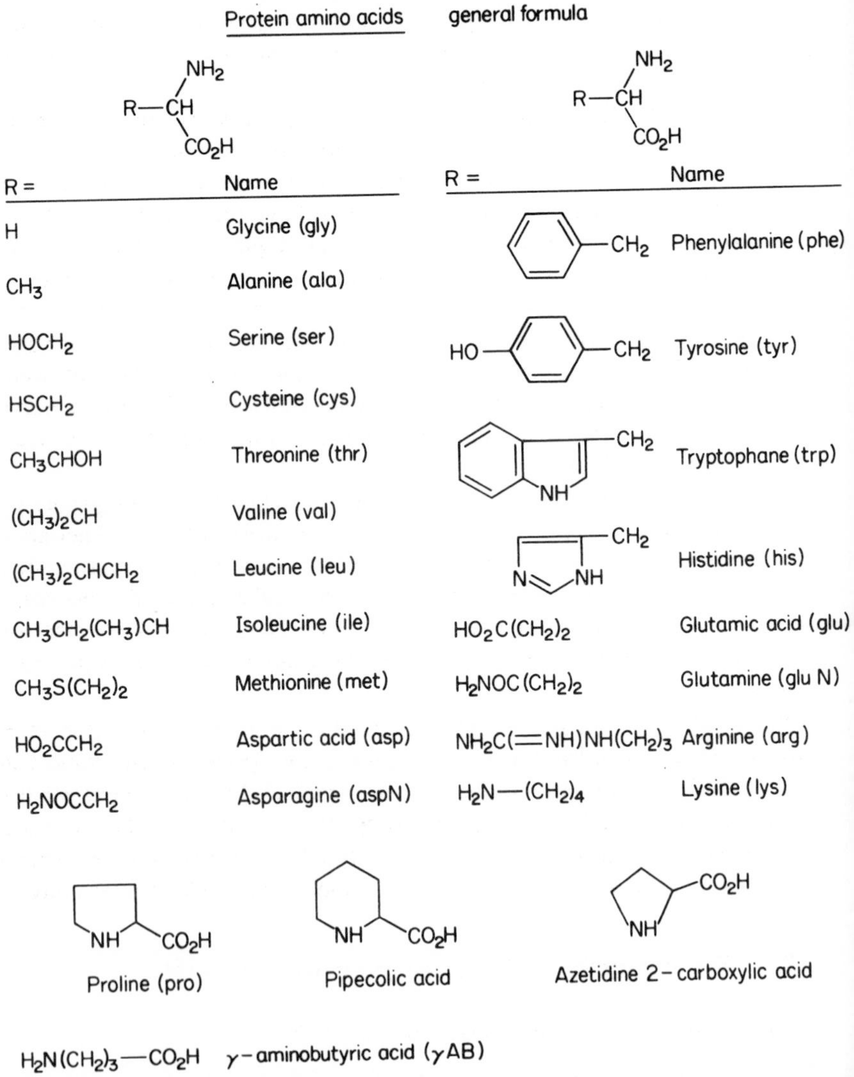

Protein amino acids		general formula	
$R-CH\begin{smallmatrix}NH_2\\CO_2H\end{smallmatrix}$		$R-CH\begin{smallmatrix}NH_2\\CO_2H\end{smallmatrix}$	
R =	Name	R =	Name
H	Glycine (gly)	⬡—CH_2	Phenylalanine (phe)
CH_3	Alanine (ala)		
$HOCH_2$	Serine (ser)	HO—⬡—CH_2	Tyrosine (tyr)
$HSCH_2$	Cysteine (cys)		
CH_3CHOH	Threonine (thr)	(indole)—CH_2	Tryptophane (trp)
$(CH_3)_2CH$	Valine (val)		
$(CH_3)_2CHCH_2$	Leucine (leu)	(imidazole)—CH_2	Histidine (his)
$CH_3CH_2(CH_3)CH$	Isoleucine (ile)	$HO_2C(CH_2)_2$	Glutamic acid (glu)
$CH_3S(CH_2)_2$	Methionine (met)	$H_2NOC(CH_2)_2$	Glutamine (glu N)
HO_2CCH_2	Aspartic acid (asp)	$NH_2C(\!=\!NH)NH(CH_2)_3$	Arginine (arg)
H_2NOCCH_2	Asparagine (aspN)	$H_2N\!-\!(CH_2)_4$	Lysine (lys)

Proline (pro)   Pipecolic acid   Azetidine 2-carboxylic acid

$H_2N(CH_2)_3\!-\!CO_2H$ γ-aminobutyric acid (γAB)

Fig. 5.1 Structure of amino acids.

essentially the same. The 'protein' amino acids are generally recognized to be twenty in number and are those found in acid hydrolysates of plant (and animal) proteins (see Fig. 5.1). They also occur together in the free amino acid pool of plant tissues at concentrations varying between 20 and 200 μg fresh weight; there are considerable quantitative variations from tissue to tissue, depending on the metabolic status of the plant in question. In general, glutamic and aspartic acids, and their acid amides glutamine and asparagine, tend to be present in larger amount than the others, since they represent a storage form of nitrogen. On the other hand, histidine, tryptophane, cysteine and methionine are often present in such low amounts in plant tissues that they cannot be readily detected.

Only one of the 'non-protein' amino acids is regularly present in plants – the more or less ubiquitous γ-aminobutyric acid. The remainder, of which over 200 structures are known (Fowden, 1981), are of more restricted occurrence. Their role in the plant is not entirely clear, although their presence (often in high concentration) in seeds and their subsequent metabolism during germination suggests they may be important as nitrogen storage materials. Most are structural analogues of one or other of the twenty 'protein' amino acids. For example, two analogues of proline are pipecolic acid, which has one more methylene group than proline, and azetidine 2-carboxylic acid, which has one less (see Fig. 5.1). Pipecolic acid is mainly found in certain legume seeds, while azetidine 2-carboxylic acid occurs characteristically in many members of the Liliaceae.

Amino acids are colourless ionic compounds, their solubility properties and high melting points being due to the fact that they are zwitterions. They are all water-soluble, although the degree of solubility varies, the aromatic amino acids (e.g. phenylalanine, tyrosine) being rather sparingly soluble. Since they are basic, they form hydrochlorides with conc.HCl and being acids, they can be esterified. The esters are more volatile than the free acids and can thus be separated by GLC.

'Neutral' amino acids (e.g. glycine, alanine) are those in which the amino groups are balanced by an equal number of acidic groups. Basic amino acids (e.g. lysine) have an additional free amino group and acidic amino acids (e.g. glutamic acid) an additional acidic group. Because of their different charge properties, amino acid mixtures can be divided into neutral, basic and acidic fractions by using either electrophoresis or ion exchange chromatography.

5.2.2 Recommended techniques – protein amino acids

(a) *Thin layer chromatography and paper chromatography*

The separation and quantitative estimation of the protein amino acids either as they occur in the free amino acid 'pool' or as they are obtained in different protein hydrolysates, is one of the most important and best known techniques of plant biochemistry. A great variety of different procedures are described in the literature, although the fundamental approach has changed little from that used in the very early days of PC.

In spite of the fact that it is possible to separate the twenty protein amino acids by one-dimensional chromatography (Hanes *et al.*, 1961), the normal practice is to employ two-dimensional separations. It is most frequently carried out on paper or on thin layers of silica gel G, cellulose or silica gel–cellulose mixtures. For paper, the best solvent pair is probably *n*-butanol–acetic acid–water (BAW) and phenol–water. The same pair may be used for TLC on silica gel G (Table 5.1), but for TLC on microcrystalline cellulose, replacement of BAW by chloroform–methanol–2 M NH_4OH (2:2:1) as the first solvent is recommended (Brenner *et al.*, 1969).

Table 5.1 R_Fs and ninhydrin colours of protein amino acids

Amino acid	R_F ($\times 100$) *in*		Ninhydrin colour
	BAW	PhOH–H_2O*	
Glycine	18	24	red-violet
Alanine	22	29	
Serine	18	20	
Cysteine†	10	04	
Threonine	20	26	violet
Valine	32	40	
Leucine	44	48	
Isoleucine	43	49	
Methionine	35	49	
Aspartic acid	17	06	blue-violet
Asparagine	14	—	orange-brown
Glutamic acid	24	10	
Glutamine	15	—	violet
Arginine	06	19	
Lysine	03	09	
Proline	14	50	yellow
Phenylalanine	43	55	
Tyrosine	41	47	grey-violet
Tryptophan	47	63	
Histidine	05	32	

*Measured on silica gel G plates (air-dried), by ascent (solvent migration 10 cm) BAW = *n*-BuOH–HOAc–H_2O (4:1:1); PhOH–H_2O = PhOH–H_2O (3:1, by wt.).
†As cysteic acid.

One advantage of using PC is that a concentrated aqueous alcoholic plant extract can be applied directly to the paper for separation. By contrast, TLC systems are sensitive to the presence of salts and sugars that may contaminate a crude plant extract, and purification on ion exchange resins is a normal pre-requisite for good separations (Brenner *et al.*, 1969). In such purifications, care must be taken to avoid losing material on the acidic and basic ion exchange resins which are used. One procedure which avoids the necessity for preliminary purification is conversion of the amino acids to their dinitrophenyl (DNP) derivatives and their subsequent separation on TLC. These derivatives are yellow compounds, formed by reaction with 2,4-dinitrofluoro-benzene in aqueous acetone. Details of the preparation and separation of dinitrophenylamino acids can be found in Brenner *et al.* (1969).

(b) *Development*

The standard reagent for amino acids is ninhydrin (triketohydrindene hydrate), which is commercially available in a form ready for spraying on to chromatograms or plates. Alternatively, it may be prepared fresh as a 0·1% solution in acetone. After spraying, the paper or plate is heated for 10 min at 105°C when most amino acids give purple or grey-blue colours; proline (and hydroxyproline) are distinctive in giving yellow colours (Table 5.1). Modification of the ninhydrin spray with cadmium acetate avoids the necessity for heating the plate to develop the colours. 112 ml of a solution of cadmium acetate (1 g) in water (100 ml), acetic acid (20 ml) and acetone (1 litre) is used for dissolving 1 g ninhydrin and the paper or plate is dipped in this reagent. On leaving overnight in the dark in a closed vessel containing H_2SO_4, the paper or plate shows the amino acids as dark red spots on a white background.

(c) *High voltage electrophoresis and thin layer chromatography*

The best procedure for achieving sharp separations of the common amino acids is to combine the use of electrophoresis and TLC (Blackburn, 1965; Bieleski and Turner, 1966). This is done as follows:

(1) The crude extract is applied as a narrow band (2·5 cm) near one corner of a plate spread with cellulose MN 300, which has been dried at room temperature. A marker spot of thionin (Michrome dye no. 215 from Edward G. Gurr) is placed at the opposite end of the plate.

(2) The plate is sprayed lightly with formic acid–acetic acid buffer pH 2·0. Using a wick of dialysis tubing and Whatman 3 MM paper (held in place by glass strips), the plate is developed horizontally in the same buffer at 1000 V (10–20 mA) in a Shandon-cooled plate electrophoresis tank for 25–35 min (thionin marker moves *ca.* 4–5 cm).

(3) The plate is blown dry and the bands which have separated are reduced to spots by dipping the plate (turned through 90°C) in distilled water and allowing it to develop to 2·5 cm. The plate is re-dried.

(4) The plate is then developed twice in the second direction with methyl ethyl ketone–pyridine–water–acetic acid (70:15:15:2) for 1·5 h and finally once in *n*-propanol–water–*n*-propyl acetate–acetic acid–pyridine (120:60:20:4:1) for 4 h.

(5) The plate is then developed with the ninhydrin–cadmium reagent (see above). A typical separation achieved by this procedure (from Kipps, 1972) is shown in Fig. 5.2.

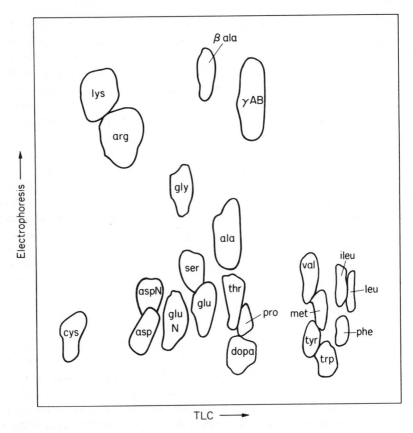

Fig. 5.2 Two-dimensional thin layer chromatography and electrophoretic separation of protein amino acids.

(d) *Gas liquid chromatography*

This technique cannot be applied directly to amino acids because they are so involatile. In recent years, successful separations by GLC have been obtained using temperature programming on their derivatives, for example, on the *N*-acetyl-*n*-amyl esters. These are prepared from the amino acids by successive treatment with amyl alcohol and anhydrous HBr and then with acetic anhydride. GLC is carried out on a column of Chromosorb W (60–80 mesh) coated with 1% polyethylene glycol (Carbowax 1546 or 6000), with temperatures between 125 and 155°C and flow rates of 60 to 240 ml min^{-1} (for details and references, see Ikan, 1969).

(e) *Quantitative measurements*

The high resolution provided by PC or TLC makes this an attractive method to use for quantitative determination. It is simple to stain the paper or plate with ninhydrin under standardized conditions, elute or scrape off the various coloured components and measure their concentration separately from their visible colour in the spectrophotometer. A good alternative is first to prepare the DNP derivatives of the amino acids in the crude extract, separate the derivatives by two-dimensional chromatography and, since they are yellow in colour, determine their concentration directly without further staining.

The above methods, however, do not compete in terms of convenience, accuracy and sensitivity with the separation and analysis (with ninhydrin) of amino acids on ion exchange resins, as originally devised by Moore and Stein (1956). These procedures are now available, completely automated, and most quantitative measurements on amino acids in plant tissues are routinely determined in one of the various models of automatic amino acid analyser available today.

(f) *Other techniques*

HPLC has yet to be widely used for separating amino acids, but it is a procedure that promises to be of value as soon as a suitably sensitive detector is developed. At present, most workers derivatize the amino acids before separation, usually making the phenylthiohydantoin compounds which can be detected by their intense UV absorbance. These amino acid derivatives can be separated by reverse-phase HPLC on a μBondapak C$_{18}$ column eluted with sodium acetate–methanol mixtures (Kuster and Niederwieser, 1983). GLC–MS is another technique that is occasionally used for quantitative analysis, in cases where high sensitivity and accurate identification are needed. For other methods of separating and quantifying amino acids, see Rosenthal (1982).

5.2.3 Recommended techniques – non-protein amino acids

(a) *Extraction*

Non-protein amino acids frequently occur in high concentration in seeds, particularly in those of the Leguminosae, and many surveys have been carried out on seed extracts. A typical extraction procedure, used by Dunnill and Fowden (1965), is as follows:

(1) Finely ground seed (1 g) is shaken with 75% ethanol (25 ml) for one day.
(2) The supernatant, after centrifugation, is applied to a small column (12×0·8 cm) of ZeoKarb 225 (H^+ form in 75% ethanol) to retain organic acids; it is washed with aqueous ethanol and then eluted with aqueous ethanol containing 2 M NH_4OH (25 ml).
(3) This eluate is concentrated and applied to a 3 MM sheet of filter paper for chromatography, the volume applied being equivalent to 0·25 g seed.

(b) *Chromatography*

Two-dimensional PC is still the most widely used technique for detecting non-protein amino acids in plants (see Fig. 5.3). Dunnill and Fowden (1965) recommend 75% phenol–water (ammonia vapour), followed by *n*-butanol–acetic acid–water (90:10:29) as one pair of solvents, ethyl acetate–pyridine–water (2:1:2) and *n*-butanol–3 M NH_4OH (top layer) as a second pair. One-dimensional PC in several solvents is sometimes sufficient to distinguish different non-protein amino acids. R_F values of six such acids which occur in *Vicia* species are shown in Table 5.2.

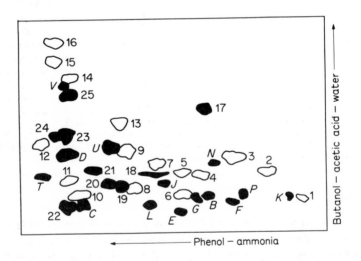

Fig. 5.3 Two-dimensional paper chromatography of non-protein amino acids. Key: This is a diagrammatic representation of the positions occupied on a 2-D chromatogram by protein amino acids (open spots) and by the various non-protein amino acids found in seed of the Cucurbitaceae (solid spots). Spot 17 is *m*-carboxyphenylalanine, spot 20 is citrulline and spot 21 N^4-methylasparagine (from Dunnill and Fowden, 1965).

(c) *Electrophoresis*

For screening plant tissues for acidic amino acids, it is important to use paper electrophoresis as well as PC. Bell and Tirimanna (1965) in surveying legume seeds, used a range of four buffers: formic acid–acetic acid–water (33:148:1819) pH 1·9, acetic acid–pyridine–water (5:0·5:95) pH 3·6 acetic acid–pyridine–water (0·2:5:95) pH 6·5 and 0·1 M Na_2CO_3 pH 11·6. For acid buffers, they used horizontal electrophoresis on Whatman no. 1 paper at 60 V cm^{-1} for 30 min; at pH 11·6, they employed the hanging strip technique at 5 V cm^{-1} for 17 h. In order to distinguish different amino acids, it is often sufficient to note whether they remain immobile, or become positively or negatively charged at the four pH values (Table 5.2).

Table 5.2 R_Fs of some non-protein amino acids

Amino acid	R_F ($\times$100) *in* solvent			*Ionic mobility*	*Colour reaction†*
	1	2	3		
γ-Hydroxyarginine	15	76	15	+ve in acid −ve at pH 11·5	scarlet with Sakaguchi
Homoserine	27	53	30	+ve at pH 1·9	violet with ninhydrin
Canavanine	15	59	20	+ve at pH 3·6 and 1·9	magenta with PCAF
β-Cyanoalanine	32	65	40	+ve at pH 1·9	green with ninhydrin
γ-Hydroxyornithine	10	60	19	−ve at pH 11·5	violet with ninhydrin
α, γ-Diaminobutyric acid	14	54	22	+ve at pH 3·6 and pH 1·9	brown-purple with ninhydrin

*Solvent 1= n-BuOH–HOAc–H$_2$O (12:3:5), 2 = PhOH–H$_2$O (4:1, w/v) plus NH$_4$OH (trace), 3 = n-BuOH–pyridine–H$_2$O (1:1:1). Support: Whatman no. 1 paper.
†Sakaguchi reagent = 0·1% 8-hydroxyquinoline in acetone, followed by 0·2% bromine in 0·5 M NaOH, PCAF (Fearon's reagent) = 1% sodium pentacyanoammonioferrate in copper-free distilled water, followed by phosphate buffer (pH 7).

(d) *Detection*

One of the easiest ways of recognizing many non-protein amino acids is by their atypical colour response with ninhydrin. Instead of giving the usual purple colour, they may go green, brown or deep red. This is often how they are first detected during two-dimensional chromatographic screening programmes. It is often useful, with certain types of amino acid, to employ more specific colour reagents. The Sakaguchi reagent, for example, is specific for arginine and its derivatives while Fearon's PCAF reagent (Table 5.2) is specific for the guanidine compounds canavanine and deaminocanavanine.

(e) *Authentic markers*

Standard kits of marker amino acid solutions, already prepared at the right concentration for PC or TLC, are available commercially. These kits include not only the common protein amino acids, but a number of non-protein amino acids as well.

5.2.4   Practical experiments

(a) *The amino acids of potato tubers*

When potato tubers are cut into thin slices, considerable changes occur in their metabolism. As a response to wounding, the respiration rate increases dramatically and there is also active synthesis of new cell wall protein. One of the major amino acids of this cell wall protein is hydroxyproline, and concomitant with the wound reaction, a significant accumulation of free hydroxyproline occurs in the slices.

Hydroxyproline, like proline, gives a distinctive yellow colour with ninhydrin so it is easy to observe this biochemical change on wounding, by two-dimensional chromatographic comparison of extracts of whole tuber tissue and of potato slices. The thin slices should be thoroughly washed in tap water and then kept for 5–7 days under slightly moist conditions (e.g. in Petri dishes on moist filter paper) in order for them to fully develop the wound response.

(b) *Procedure*

(1) Homogenize in a Waring blender about 30 g of freshly peeled potato tissue with 100 ml 80% ethanol. Filter and evaporate to dryness in a weighed flask. Determine the amount of solid present and redissolve the residue in sufficient distilled water to give 30 mg ml^{-1} water. Repeat the above extraction on about 30 g of potato slices, which have been kept for 5–7 days after slicing, as indicated above.

(2) Chromatograph the two extracts, separately, two-dimensionally on paper using phenol–water (3:1, w/v) in the first direction, *n*-butanol–acetic acid–water (4:1:1) in the second direction. Using a micropipette, apply 3 μl of each extract slowly, allowing the paper to dry between applications. Prepare a third chromatogram, running a mixture of amino acid markers for the sake of comparison. Note that chromatograms should be handled as little as possible with the fingers, since there are traces of ninhydrin-positive materials in human sweat.

(3) Evenly spray the three developed chromatograms with 0·2% ninhydrin in acetone, and heat the papers for 5 min at 90°C.

(4) Note the differences in amino acid patterns between the whole tuber and the slices, particularly the presence of hydroxyproline in the latter. From the chromatogram of the marker amino acids, and from the knowledge of their relative R_Fs, it should be possible to identify most of the other amino acids present. For further practical details of the simple chromatography of amino acids, see Witham *et al.* (1971).

(c) *A survey for canavanine*

Survey the seeds of as many different legumes as can be obtained commercially for the presence of the non-protein amino acid, canavanine. Include, if possible, in the survey, seed of the Jackbean, *Canavalia ensiformis*, the classical source of canavanine. A sample of canavanine itself can be obtained commercially, as the sulphate. Proceed as follows:

(1) Extract 0·1–0·3 g of powdered seed with 2 ml 0·1 M HCl at room temperature for 24 h.
(2) Chromatograph the extracts one-dimensionally in phenol–water (3:1) and in *n*-butanol–pyridine–acetic acid–water (4:1:1:2). Canavanine has R_F ($\times 100$) at 59 and 18 respectively.
(3) Carry out ionophoresis on 3 MM paper, at pH 7 phosphate buffer, at 5 V cm^{-1} for 3 h. Canavanine moves 7 mm towards the cathode.
(4) Spray papers and electrophoretograms with PCAF (see Table 5.2 for details); only canavanine gives a magenta colour. The distribution of canavanine in legume seeds is summarized by Bell *et al.* (1978).

5.3   AMINES

5.3.1   Chemistry and distribution

Plant amines can be considered simply as the products of decarboxylation of amino acids, formed by the reaction:

$$RCH(NH_2)CO_2H \rightarrow RCH_2NH_2 + CO_2.$$

This does not mean that they are inevitably synthesized *in vivo* by this means. In fact, biosynthesis by transamination of the corresponding aldehydes has been demonstrated as the main route of synthesis to aliphatic amines in some fifty plants. The most widespread plant amines are conveniently divided into three groups: aliphatic monoamines, aliphatic polyamines and aromatic amines.

Aliphatic amines are volatile compounds, ranging from simple compounds like methylamine, CH_3NH_2, to *n*-hexylamine, $CH_3(CH_2)_5NH_2$. They are widely distributed in higher plants and fungi and, when present in any concentration, have an unpleasant fish-like smell. They function in flowers (e.g. in the cow parsley, *Heracleum sphondylium*) as insect attractants by simulating the smell of carrion.

By contrast to the monoamines, diamines and other polyamines are less volatile, although they still possess offensive odours. Widespread polyamines include putrescine, $NH_2(CH_2)_4NH_2$, agmatine $NH_2(CH_2)_4NHC$ ($=NH)NH_2$, spermidine $NH_2(CH_2)_3NH(CH_2)_4NH_2$ and spermine $NH_2(CH_2)_3NH(CH_2)_4NH(CH_2)_3NH_2$. There are several others of more

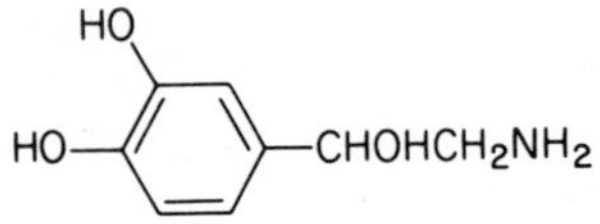

Noradrenaline

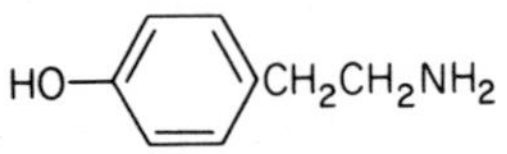

Tyramine

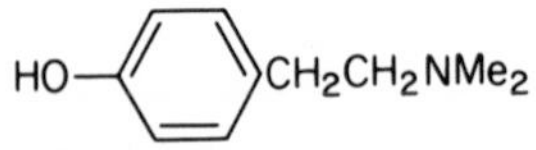

Hordenine

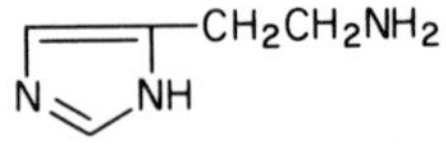

Histamine

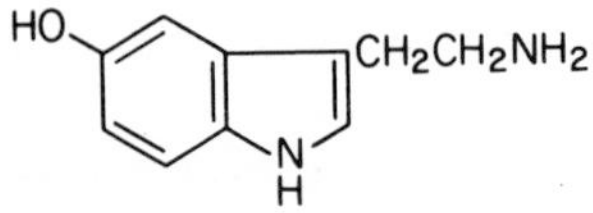

Serotonin

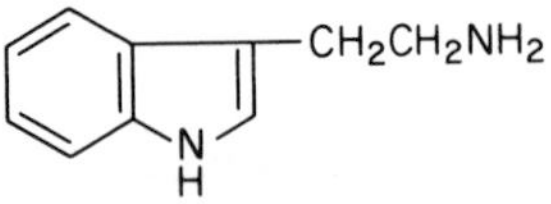

Tryptamine

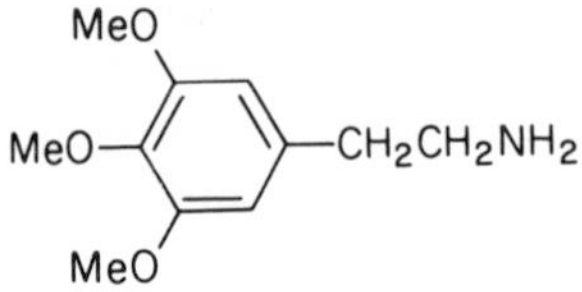

Mescaline

Fig. 5.4 Structures of aromatic amines of plants.

limited occurrence such as cadaverine $NH_2(CH_2)_5NH_2$ (Smith, 1981). Polyamines are of topical research interest because of their growth-stimulating activity in relation to their effect on ribosomal RNA.

The best known aromatic amine from plants is probably mescaline (see Fig. 5.4), the active principle of the flowering heads (peyote) of the cactus, *Lophophora williamsii*, and a powerful natural hallucinogenic compound. Indeed, many of the known aromatic amines are physiologically active and for this reason they are sometimes classified with the alkaloids (see Section 5.4). Three substances which are very important in animal physiology (e.g. in brain metabolism) are noradrenaline, histamine and serotonin; all three occur in plants, noradrenaline for example being present in the banana and the potato. Perhaps the most widespread aromatic amine is tyramine, which has been detected in fifteen of seventy-seven plant families surveyed.

All the amines so far mentioned are primary amines, i.e. have the general formula RNH_2. Secondary amines, general formula R_2NH and tertiary amines, R_3N, are known in plants but they are not very common. A typical secondary amine found in plants is dimethylamine; a typical tertiary amine is hordenine *N*-dimethyltyramine, which is the principal 'alkaloid' of barley.

(a) *Paper chromatography*

The simplest approach to the detection of volatile amines in plants is by one-dimensional PC in *n*-butanol–acetic acid–water (4:1:5, or other mixtures) (Kamienski, 1959). Primary amines can be detected by their giving a violet-red colour with the ninhydrin reagent. Secondary amines respond only feebly with ninhydrin and are detected with nitroprusside. A solution of 5 g sodium nitroprusside in 100 ml of 10% acetaldehyde is mixed just before spraying with an equal volume of 2% aqueous Na_2CO_3; secondary amines give blue colours. Tertiary amines can be detected on paper with Dragendorff's reagent (see p. 181).

A more specific reagent for primary amines is 2-acetoacetylphenol, which when sprayed as a 1% solution in ethanol, gives green-yellow fluorescent colours with primary amines (Baker *et al.*, 1952). The lack of specificity of some of the above reagents (e.g. ninhydrin, which also reacts with amino acids) can be overcome by a preliminary fractionation of the crude plant extract before chromatography. This can be done by either steam-distillation from an alkaline extract (see below) or by adsorption on a cation exchange resin and subsequent elution.

(b) *Gas liquid chromatography*

Because of their volatility, simple amines are readily separated by GLC and this is clearly the technique of choice for separating and identifying the individual components in complex mixtures of amines. Irvine and Saxby (1969), for example, used GLC on five columns (Table 5.3) to separate and characterize the thirty-six amines present in cured tobacco leaf. Their procedure involved steam-distillation from a 5 M alkaline leaf extract into 3 M HCl. This distillate was concentrated, the amines liberated with 5 M NaOH and taken into ether. The dry ether extract was passed through a Zeokarb 226 (SRC 45) ion exchange column, the amines being adsorbed and then con-

Table 5.3 Relative retention times of aliphatic amines on gas liquid chromatography

	Column support*				
Amine	MBMA	PPE	SE-30	DEGS	TXP
Methylamine	09	11	07	03	15
Ethylamine	12	13	12	04	41
Isopropylamine	12	11	23	27	33
n-Propylamine	22	23	26	61	53
n-Butylamine	41	44	43	81	69
n-Amylamine	63	74	63	100	85
n-Hexylamine	79	97	82	117	100

*MBMA = *m*-bis (*m*-phenoxyphenoxy)benzene plus Apiezon L (4:1); SE-30 = silicone oil; DEGS = diethyleneglycol succinate; PPE = polyphenyl ether OS-124; TXP = Trixylenyl phosphate.

Table 5.4 Thin layer chromatography and electrophoresis of aliphatic mines

	TLC R_F in			Electro-phoretic mobility (cm)	Ninhydrin colour
Amine	1	2	3*		
Monoamines					
Methylamine	—	63	24	10·5	pink
n-Propylamine	—	76	44	8·3	blue
iso-Butylamine	—	79	50	7·7	
iso-Amylamine	61	81	78	7·3	purple
Ethanolamine	95	59	50	8·0	
Polyamines					
Agmatine	22	43	07	7·3	
Putrescine	75	43	06	8·5	
Cadaverine	78	47	07	8·3	purple
Spermidine	63	23	02	7·5	
Spermine	50	13	00	6·6	

*Key: solvent = n-BuOH–MeCOEt–NH₄OH–H₂O (5:3:1:1); solvent 2 = methyl cellosolve–propionic acid–H₂O (70:15:15) satd. with NaCl; solvent 3 = n-BuOH–HOAc–water (4:1:1). Support: MN 300 cellulose. Time of development 2·5 h. From Smith (1970).

verted on the column to the trifluoroacetamides by reaction with trifluoro-acetic anhydride. The derivatives were washed off with ether and the resultant solution dried, concentrated and submitted to GLC. The separations achieved on different columns with seven of the simplest amines are indicated in Table 5.3.

(c) *Thin layer chromatography*

For separation and detection of plant polyamines, a combination of TLC and electrophoresis is recommended (Smith, 1970). TLC is carried out on MN 300 cellulose and electrophoresis in 0·5 M citrate buffer (pH 3·5) for 20 min at 25 V cm^{-1} on plates cooled in light petroleum. Mobilities and R_Fs for common aliphatic amines are collected in Table 5.4.

Aliphatic amines, like the amino acids, can be characterized as their DNP (dinitrophenyl) derivatives (see p. 181). The DNP derivatives separate on TLC using silica gel HF$_{254}$ and the systems pentane–acetic acid isoamyl ester–ammonia (70:29:1), benzene–ethyl acetate–petroleum (b.p. 40–60°C) (97:2:1) and pentane–benzene–triethylamine (9:9:2). The DNP derivatives are detected in the usual way as dark absorbing spots on a fluorescent background, in UV light of wavelength 254 nm (Ilert and Hartmann, 1972).

For quantitative determination of biologically important amines such as putrescine, spermine and spermidine, TLC separation on silica gel G, follow-ing by reaction with 1-dimethylaminonaphthalene-5-sulphonyl chloride (dansyl chloride) is recommended (Seiber, 1970; Smith, 1972). The dansyl derivatives are highly fluorescent and can be estimated by fluorimetry, after elution of the 'spots', or preferably by direct fluorimetric scanning of the TLC plate. The sensitivity is such that as little as 10^{-13} mol of amine can be detected.

5.3.3 Recommended techniques – aromatic amines

(a) *Paper chromatography and thin layer chromatography*

PC is still used to a considerable extent for separation of physiologically active aromatic amines, many of which are also 'phenolic' in nature. Fellows and Bell (1970), for example, in isolating these amines from legume seeds, employed PC using *n*-butanol–acetic acid–water (12:3:5), *iso*-propanol–1 M NH$_4$OH–water (20:1:2), *n*-butanol–pyridine–water (1:1:1) and *n*-propanol–1 M NH$_4$OH (5:1). The first two systems were also useful for two-dimensional separations. Some of the aromatic amines are rather labile in these solvents, and papers were run for 12–18 h in the dark. The substances were detected with Ehrlich's reagent (1% *p*-dimethylaminobenzaldehyde in acetone–10 M

HCl, 9:1). This gives pink colours with tryptamine derivatives. Ninhydrin can be used, but not all aromatic amines (e.g. hordenine) respond. Catecholamines will oxidize in air, or on treatment with ferricyanide, to give yellow-green fluorescent products. Phenolic amines can be detected by use of the phenolic spray, such as the Folin–Ciocalteu reagent (see Chapter 2, p. 42). Again, there is no completely specific reagent available for detecting aromatic amines as such.

TLC has also been used fairly extensively in the separation of aromatic amines (Stahl and Schorn, 1969). Two systems that can be used are light petroleum (b.p. 40–60°C)–acetone (7:3) and chloroform–ethyl acetate–acetic acid (6:3:1) on silica gel G plates (Booth and Boyland, 1964).

5.4 ALKALOIDS

5.4.1 Chemistry and distribution

The alkaloids, of which some 5500 are known, comprise the largest single class of secondary plant substances. There is no one definition of the term 'alkaloid' which is completely satisfactory, but alkaloids generally include 'those basic substances which contain one or more nitrogen atoms, usually in combination as part of a cyclic system'. Alkaloids are often toxic to man and many have dramatic physiological activities; hence their wide use in medicine. They are usually colourless, often optically active substances; most are crystalline but a few (e.g. nicotine) are liquids at room temperatures. A simple but by no means infallible test for alkaloids in fresh leaf or fruit material is the bitter taste they often impart to the tongue. The alkaloid quinine, for example, is one of the bitterest substances known and is significantly bitter at a molar concentration of 1×10^{-5}.

The most common precursors of alkaloids are amino acids, although the biosynthesis of most alkaloids is more complex than this statement suggests. Chemically, alkaloids are a very heterogeneous group, ranging from simple compounds like coniine, the major alkaloid of hemlock, *Conium maculatum*, to the pentacyclic structure of strychnine, the toxin of the *Strychnos* bark. Plant amines (e.g. mescaline) and purine and pyrimidine bases (e.g. caffeine) are sometimes loosely included in the general term alkaloid; these classes are, however, dealt with in separate sections in this book.

Many alkaloids are terpenoid in nature and some (e.g. solanine, the steroidal alkaloid of the potato, *Solanum tuberosum*) are best considered from the biosynthetic point of view as modified terpenoids. Others are mainly aromatic compounds (e.g. colchicine the tropolone alkaloid of autumn crocus bulbs), containing their basic group as a side-chain attachment. A considerable number of alkaloids are specific to one family or to a few related plants. Hence, the names of alkaloid types are often derived from the plant source, e.g. the

Table 5.5 Plant families with over fifty alkaloids present

Family	Typical source	Typical alkaloid(s)	Chemical type(s)
DICOTYLEDONS			
Apocynaceae	*Vinca rosea*	ajmalicine	benzyliso-quinoline
Compositae	*Senecio jacobaea*, ragwort	senecionine	pyrrolizidine
Leguminosae	*Cytisus laburnum*, broom	cytisine	quinolizidine
Loganiaceae	*Strychnos nux-vomica*	strychnine	
Menispermaceae	*Archangelisia flava*	berberine	protoberberine
Papaveraceae	*Papaver somniferum*, opium poppy	morphine, codeine, thebaine	morphine
Ranunculaceae	*Aconitum napellus*	aconite	diterpene
Rubiaceae	*Cinchona officinalis*, cinchona bark	quinine	quinoline
	Uragoga ipecacuanha roots	emetine	emetine-type
Rutaceae	*Skimmia japonica*	skimmianine	quinoline
Solanaceae	*Solanum tuberosum* potato	solanine, chaconine	steroidal
	Atropa belladonna, deadly nightshade	atropine	tropane
	Nicotiana tabacum, tobacco	nicotine	pyridine
MONOCOTYLEDONS			
Amaryllidaceae	*Narcissus pseudonarcissus*, daffodil	galanthamine, tazettine	
Liliaceae	*Colchicum autumnale*, autumn crocus	colchicine	tropolone

Atropa or tropane alkaloids and so on. To illustrate their natural occurrence, some of the more common alkaloids are mentioned (Table 5.5) in relation to their occurrence in particular families. This table shows only those families in which over fifty alkaloid structures have been detected (Hegnauer, 1967). These are the angiosperm families which are particularly rich in these bases, but it should be borne in mind that alkaloid distribution is very uneven and many families lack them altogether. Alkaloids are generally absent or infrequent in the gymnosperms, ferns, mosses and lower plants. The structures of some of the more common alkaloids are illustrated in Fig. 5.5.

The functions of alkaloids in plants are still largely obscure, although individual substances have been reported to be involved as growth regulators or as insect repellents or attractants. The theory that they act as a form of nitrogen storage in the plant is not now generally accepted.

The classic monograph on the plant alkaloids is by Henry (1949). A modern equivalent is the excellent introductory text of Cordell (1981). There is also an introductory text on their biochemistry by Robinson (1981). The other major

reference is the continuing series of volumes *The Alkaloids*, originally edited by
the late R. H. F. Manske and now by R. G. A. Rodrigo; the latest volume in the
series, No. 20, appeared in 1982.

5.4.2 Recommended techniques

(a) *Preliminary detection*

Because alkaloids are so heterogeneous chemically and there are so many of
them, they cannot be identified in plant extracts using a single chromato-

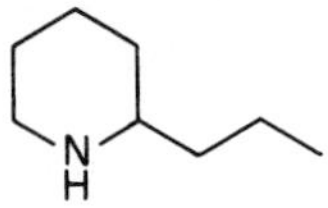

Coniine

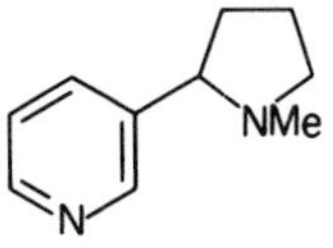

Nicotine

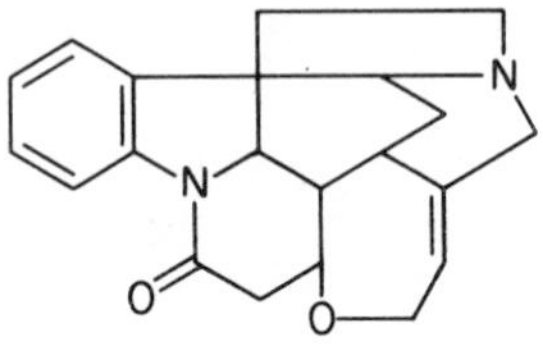

Strychnine

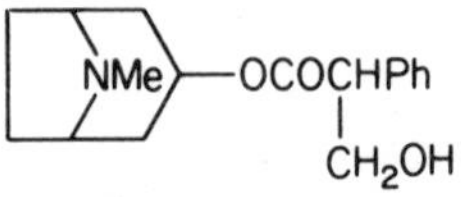

Atropine

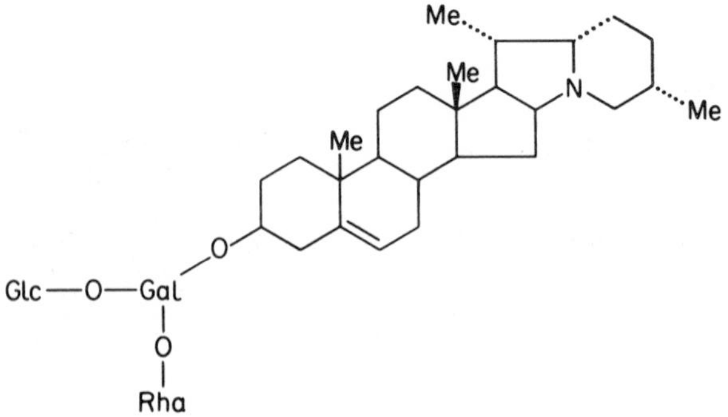

Solanine

Fig. 5.5 Structures of some common alkaloids.

graphic criterion. In general, it is difficult to identify an alkaloid from a new plant source without knowing approximately what type of alkaloid is likely to be found there. Also, because of the wide range of solubility and other properties of the alkaloids, any general screening procedure for alkaloids in plants may fail to detect particular compounds.

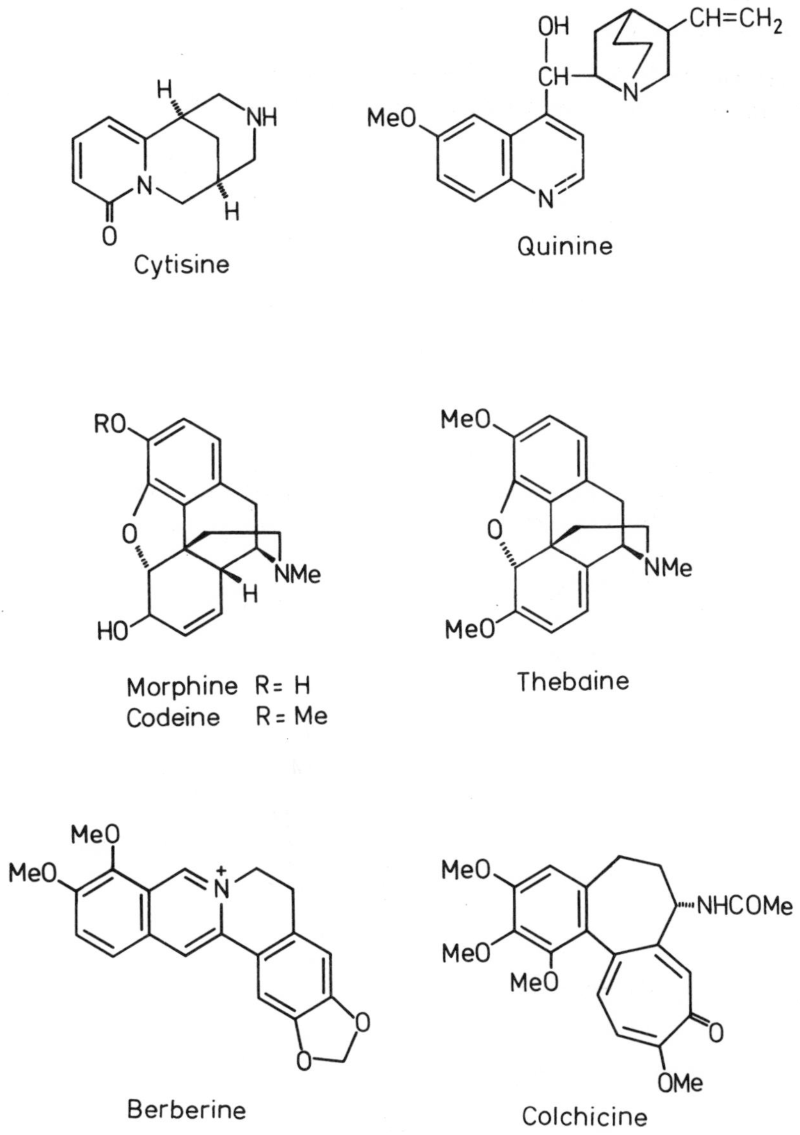

Cytisine

Quinine

Morphine R = H
Codeine   R = Me

Thebaine

Berberine

Colchicine

Fig. 5.5 *(continued)*.

Being bases, alkaloids are normally extracted from plants into a weakly acid (1 M HCl or 10% acetic acid) alcoholic solvent and are then precipitated with concentrated ammonia. Such a preliminary separation from other plant constituents may be repeated or further purification can be achieved by solvent extraction. Such relatively crude extracts can be tested for the presence of alkaloids, using several different 'alkaloid' reagents. Preferably, however, chromatography on paper and TLC in some generally applicable solvent systems should be carried out and then the papers and plates processed for alkaloid spots.

Alkaloid detection is especially important in forensic medicine, a particularly useful procedure being described by Clarke (1970). This procedure is mentioned here, with modification applicable to the survey of plants. For other attempts to provide a general system for alkaloid detection using TLC alone, see Hultin (1966) and Waldi *et al.* (1961).

(b) *Procedure*

Extract dried tissue with 10% acetic acid in ethanol, leaving to stand for at least 4 h. Concentrate the extract to one-quarter of the original volume and precipitate the alkaloid by dropwise addition of conc.NH_4OH. Collect by centrifugation, washing with 1% NH_4OH. Dissolve residue in a few drops of ethanol or chloroform.

Chromatograph an aliquot on sodium citrate-buffered paper in *n*-butanol–aqueous citric acid. Chromatograph another aliquot on silica gel G plates in methanol–conc.NH_4OH (200:3). Detect the presence of alkaloids on the paper and plate, first of all by any fluorescence in UV light and then by application of three spray reagents – Dragendorff, iodoplatinate and Marquis. R_Fs and colours of twelve of the commonest alkaloids are shown in Table 5.6.

Confirm the presence of a particular alkaloid by measuring the UV spectrum of a sample dissolved in 0·1 M H_2SO_4. Typical maxima values range from 250 to 303 nm (see Table 5.6). Alkaloids with aromatic rings in their structures may also absorb at longer wavelengths, e.g. colchicine λ_{max} 243 and 351 nm, berberine λ_{max} 265 and 343 nm. This test cannot be applied if more than one major alkaloid is present in the extract under examination.

(c) *Alkaloid reagents*

For the Dragendorff reagent, two stock solutions are prepared: (a) 0·6 g bismuth subnitrate in 2 ml conc.HCl and 10 ml water; (b) 6 g potassium iodide in 10 ml water. These stock solutions are mixed together with 7 ml conc.HCl and 15 ml water, and the whole diluted with 400 ml water. For spraying papers with the iodoplatinate reagent, 10 ml of 5% platinum chloride solution is mixed with 240 ml 2% potassium iodide and diluted to 500 ml with water. For

spraying plates, 10 ml of 5% platinum chloride, 5 ml conc. HCl and [
potassium iodide are mixed. The Marquis reagent can only be appl
TLC plates and consists of 1 ml formaldehyde in 10 ml conc. H_2SO_4.
corrosive acid).

Table 5.6 R_F data and colour properties of some well known alkaloids

Alkaloid*	R_F ($\times 100$)† on paper	TLC	Behaviour in UV light	Recommended reagent for detection‡	Spectral max (nm) in 0.1 M H_2SO_4
Cytisine	03	32	blue	Dragendorff	303
Nicotine	07	57	absorbs		260
Tomatine	08	62	invisible	Iodoplatinate	—
Morphine	14	34	absorbs		284
Solanine	15	52	invisible	Marquis	—
Codeine	16	35	absorbs		284
Berberine	25	07	fluorescent yellow		228
Strychnine	30	22			254
Thebaine	32	41	absorbs	Iodoplatinate	284
Atropine	37	18			258
Quinine	46	52	bright blue		250
Coniine	56	26	invisible		268

*Alkaloids are in order of R_F on paper. Data from Clarke (1970), who gives similar data for over 150 common alkaloids.
†Solvent on paper (previously buffered with 5% sodium dihydrogen citrate): n-BuOH–aqueous citric acid (870 ml : 4·8 g citric acid in 130 ml H_2O); solvent on silica gel: MeOH–NH₄OH (200:3).
‡Dragendorff: orange-brown spots on a yellow background; Marquis: yellow to purple spots; iodoplatinate: range of colours.

(d) *Further identification*

Once the type of alkaloid in a plant source has been determined, it then remains to isolate the base in some quantity and to compare it by a range of chromatographic and spectral methods (UV, IR, MS and NMR) with authentic samples. Classically, alkaloids were separated from other plant constituents as their salts and were often isolated as the crystalline hydrochloride or picrate. In modern laboratories, alkaloids are separated and isolated by any combination of a range of paper, TLC, column or GLC techniques. Column chromatography on silicic acid is commonly used, but the actual procedure to be adopted essentially depends on the type of alkaloid being studied. The more volatile alkaloids of tobacco or of broom, *Cytisus*, are probably best separated by GLC, whereas the higher molecular weight alkaloids of the opium poppy or of ergot are best examined by TLC. It is impossible in this brief review to cover all the many techniques that are available for studying the considerable number of different alkaloid groups now known. Some idea of the procedures that can be applied to just five groups is given in Table 5.7; the data are taken from Macek (1967) where further references can be obtained. The chapter by Santavy (1969) in Stahl's *Thin Layer Chromatography* should also be consulted.

Table 5.7 Chromatographic procedures for the separation of special alkaloid classes

Alkaloid class	Paper	TLC	GLC
Tobacco	as their salts in *t*-AmOH satd. with water or acetate buffer impregnated with buffer	on silica gel G in MeOH–CHCl$_3$ (3:17)	on 5·6% polyethylene glycol support (2 m× 5 mm column) temp. 170–200°C
Tropane	in EtOAc–25% HCO$_2$H (4:3); tank presatd. with aqueous phase for 14 h	on silica gel G mixed with 0·5 M KOH in 70% EtOH–25% NH$_4$OH (99:1)	on 1% dimethylpolysiloxane JXR in packed columns at temp. programme 12°C min^{-1} from 100–300°C
Opium*	in iso-BuOH–toluene (1:1) satd. with water on paper impregnated with Kolthoff buffer pH 3·5	on silica gel G impregnated with 0·5 M KOH in CHCl$_3$–EtOH (4:1)	on 2–3% silicone SE-30 on Chromosorb W at 204°C
Ergot†	in benzene–cyclohexane (1:1)/formamide plus ammonium formate: over-run solvent	on cellulose impregnated with 15% formamide in EtOAc–*n*-heptane–NHEt$_2$ (5:6:0·2)	
Rauwolfia‡	in heptane–MeCOEt (1:1)/formamide	on silica gel G in CHCl$_3$–NHEt$_2$ (9:1)	

*Easily detected on PC and TLC by intense blue fluorescence in UV.
†For detection of ergot alkaloids in higher plant tissue, see Taber *et al.* (1963).
‡Easily detected by green fluorescence in UV.

(e) *Quantitative determination*

In the past, a variety of procedures often based on the development of a colour have been devised for the analysis of individual alkaloids. One typical procedure is described in Section 5.4.3 for the steroidal alkaloid, solanine. In more recent work, reliance is often placed on either GLC for the more volatile alkaloids or HPLC. GLC is often coupled with MS for the simultaneous quantification and identification of alkaloids, for example for those of the pyrrolizidine series. New MS techniques have been discovered such as tandem mass spectroscopy which provide the direct identification of individual components in a mixture of related alkaloids and the methods are sensitive to below the microgram level (see Phillipson, 1982).

Analytical HPLC has been applied to most classes of alkaloid, and procedures for separating a few of the better known classes are listed in Table 5.8. These are all based on isocratic separation with UV detection, but gradient elution and other methods of detection have also been used.

Table 5.8 HPLC procedures for the quantitative analysis of some common alkaloids

Alkaloid class (and typical individual)	Type and size of column	Solvent system	Wavelength of UV detection
Quinolizidine (e.g. cytisine)	LiChrosorb SI 100 (50×0·3 cm)	15% MeOH in Et_2O: 2·5% NH_4OH (50:1)	220 and 310
Morphine-type (e.g. morphine)	5 μm porous silica gel (30×0·39 cm)	Hexane–$CHCl_3$–EtOH–Et_2NH (60:6:8:0·1)	285
Indole (e.g. harman)	7 μm Merck RP–8 (25×0·2 cm)	MeOH–H_2O–HCO_2H (166:34:1) buffered with Et_3N to pH 8·5	330
Emetine-type (e.g. emetine)	Merckosorb SI 60 (20×0·2 cm)	$CHCl_3$–MeOH (17:3)	254
Steroidal (e.g. solanidine)	Zorbax SIL (50×0·46 cm)	Hexane–MeOH–Me_2CO (18:1:1)	213
Protoberberine (e.g. magnoflorine)	Toya Soda LS 170 (60×0·40 cm)	H_2O–MeCN–Et_3N–HOAc (75:25:0·8:0·3)	254

(f) *Screening plants for alkaloids*

Many different procedures have been reported for detecting the presence or absence of alkaloids in plant tissues. Such procedures are part of the economic exploitation of higher plant species for their useful drugs. Indeed, more plants have probably been surveyed for alkaloids than for any other class of secondary constituent.

A typical procedure, used by Hultin and Torssell (1965) for screening 200 Swedish plant species, involves a preliminary extraction of 4 g dried tissue of each sample with methanol. The aqueous solution of the acid-soluble portion of this methanol fraction is basified with conc.NH_4OH and then extracted with chloroform and chloroform–ethanol. These concentrates are then tested for alkaloids with six reagents and alkaloid is only recorded present if all six reagents react positively.

Lüning (1967) in surveying orchids, recommends a survey of fresh tissue, since these plants are difficult to dry and during drying the alkaloids are likely to be destroyed or resinified. In order to use fresh tissue, screening can be done in the field, using one of the several portable test kits that have been devised for such purposes (e.g. Lawler and Slaytor, 1969).

(g) *Authentic markers*

Samples of practically all the well known plant alkaloids can be obtained commercially.

5.4.3   Practical experiments

(a) *Detection of alkaloids in plant tissues*

Experience in the TLC and PC of alkaloids can be gained simply by extracting a selected range of plant materials using the general procedure outlined above and chromatographing the concentrated extracts on paper and thin layer plates against markers of authentic alkaloids. For example, the tobacco present in a cigarette, on extraction with hot alcohol, will yield sufficient nicotine for detection on TLC plates with Dragendorff's reagent. Other suitable plant sources include potato berries or sprouts (for solanine), unripe tomatoes (for tomatine), *Berberis* root (for berberine), broom or laburnum seeds (for cytisine), *Atropa belladonna* berries (for atropine) and hemlock (for coniine). Because of impurities in crude extracts, R_F values of these alkaloids may not always correspond precisely with the marker compounds.

(b) *Isolation and determination of solanine in potato tissue*

Solanine, a steroidal glycoside, is the major alkaloid of the potato plant. Its toxicity is relatively low and the trace amounts present in the cultivated potato tuber are insufficient to have any physiological effects. However, solanine concentration can be exceptionally high in tubers growing near the soil surface which have undergone 'greening'. In rare instances, deaths have resulted from the ingestion of such greened potatoes, particularly in farm animals. Because of the toxicity of solanine at high concentrations, methods of isolating and determining solanine content in the potato are of some importance. From the practical point of view, it is much easier to isolate solanine from the sprout, berry or flower of the potato plant, since these parts contain much higher (often lethal) concentrations of the alkaloid (sprouts contain about 0·04% fresh weight). However, tests can still be carried out on the tuber (solanine content *ca.* 0·001%) if these other tissues are not available.

Isolation
(1) Extract the tissue by maceration with 5% acetic acid (15–20 parts) and filter the extract to remove cellular debris. Warm to 70°C and add conc.NH_4OH dropwise until the pH is 10.
(2) Centrifuge and discard the supernatant. Wash the precipitate with 1% NH_4OH and recentrifuge.

(3) Collect, dry and weigh the crude solanine so obtained. This may be purified by dissolving in boiling methanol (in which it is sparingly soluble), filtering and concentrating until the alkaloid starts crystallizing out.

Determination
(1) Dissolve a weighed amount of the crude alkaloid in 96% ethanol–20% H_2SO_4 (1:1) so that the concentration of alkaloid is between 0·2 and 3·0 mmol l^{-1} (mol. wt. of solanine is 868·04).
(2) The alkaloid solution (1 ml) is mixed without special cooling with 5 ml 60% H_2SO_4 and, after 5 min, 5 ml of 0·5% solution of formaldehyde in 60% H_2SO_4 is added. Leave to stand for 180 min and then measure the absorbance at 565–570 nm.
(3) The actual amount of alkaloid in the crude isolate can then be determined by reference to absorbance measurements on pure alkaloid solutions treated in the same way with formaldehyde in 1 M H_2SO_4.

The purity of the solanine can also be checked by PC in ethyl acetate–pyridine–water (3:1:3) or TLC on silica gel G in acetic acid–ethanol (1:3) (R_F 46).

(c) *Berberine from Berberis*

The yellow alkaloid berberine, a quaternary ammonium salt, shows anti-microbial and cytotoxic activity. It occurs in most parts of the *Berberis* (or barberry) shrub, its presence being obvious from the yellow coloration it imparts to the tissues. It is easily isolated by paper electrophoresis and quantified by spectrophotometry.

Procedure
(1) Separately extract weighed amounts (5–10 g) of different *Berberis* tissues (wood, stem, leaf, flower) with 100 ml hot ethanol–acetic acid (9:1). Concentrate the extracts on a rotary evaporator and make each up to a known volume (2–5 ml).
(2) Streak aliquots (e.g. 0·4 ml) of the extracts along the middle of a Whatman 3 MM paper (25×13 cm) and electrophorese in acetate/formate buffer pH 2·2 at 20 V cm^{-1} for 1·5 h, using a low-voltage apparatus.
(3) The yellow cationic bands, which move about 3·5 cm from the origin are cut out and eluted separately with a known volume (5–10 ml) of ethanol–acetic acid (9:1). Measure the absorbance at 350 nm and from these measurements, it is possible to calculate the relative amounts of berberine in wood, stem and leaf.
(4) Identify the isolate as berberine by measuring the complete spectrum in

EtOH–HOAc; there is a max. at 267, shoulder at 340, max. at 350 and weak max. at 435 nm. Finally, co-chromatograph with an authentic sample of berberine in several systems, e.g. on silica gel it has R_F 14 in methanol–NH_4OH (200:3); on cellulose plates, it has R_F 40 in 15% aqueous acetic acid and R_F 67 in *n*-butanol–acetic acid–water (4:1:5).

5.5   CYANOGENIC GLYCOSIDES

5.5.1   Chemistry and distribution

The fact that bitter almonds from *Prunus amygdalus* trees are capable of producing the toxic gas, prussic acid or hydrogen cyanide (HCN) has been known from time immemorial. It has also been known for many years, since about 1800, that release of HCN is related to the presence in intact plants of substances called cyanogenic glycosides, which give HCN on either enzymic or non-enzymic hydrolysis. The ease of detection of HCN, by its smell of 'bitter almonds' or more reliably (and safely) with picrate paper, has meant that many surveys of plants for cyanogenesis have been carried out. By contrast, chemical studies of the cyanogenic glycosides have been relatively restricted and at present only some thirty such compounds have been fully characterized.

HCN is released from cyanogenic glycosides according to the scheme:

(R$_1$ and R$_2$ are alkyl, aromatic or other substituents)

The first step is normally enzymic and cyanogenic plants contain a specific hydrolase (slightly different from the common β-glucosidase of plants) for carrying this out. The second step, the liberation of HCN and a substituted ketone from the unstable cyanohydrin intermediate, occurs spontaneously.

The most common cyanogenic glycosides are linamarin and lotaustralin, compounds usually found together in plants such as flax, *Linum usitatissimum*, clover, *Trifolium repens* and birdsfoot trefoil, *Lotus corniculatus*. The toxic principle of bitter almonds is amygdalin, which differs from all other known cyanogenic glycosides in having gentiobiose as the sugar unit. Prunasin, which has the same aglycone (mandelonitrile) as amygdalin but the sugar is glucose, also occurs in *Prunus* seeds. Another aromatic cyanogenic glycoside is dhurrin (the glucoside of *p*-hydroxymandelonitrile) which occurs in *Sorghum*. Structures of these five compounds are shown in Fig. 5.6.

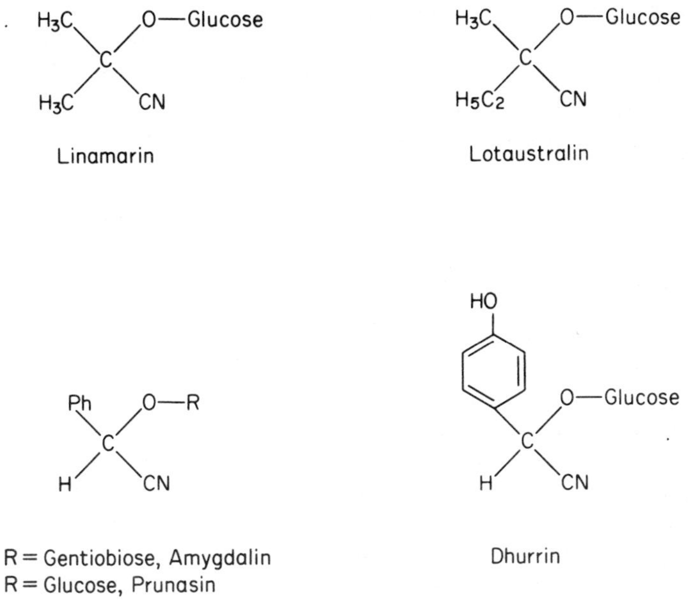

Linamarin

Lotaustralin

R = Gentiobiose, Amygdalin
R = Glucose, Prunasin

Dhurrin

Fig. 5.6 Structures of cyanogenic glycosides.

Cyanogenesis has been detected in at least 1000 species representing approximately 100 families and 500 genera. It is a character of some chemotaxonomic interest (Hegnauer, 1977) and also of genetic interest, since it varies within populations of plant species such as *Trifolium repens* and *Lotus corniculatus* (Daday, 1954). Biosynthetically, cyanogenic glycosides are formed from the structurally related amino acids *via* aldoximes and nitriles by a five-step pathway (Conn and Butler, 1969). A major function of cyanogenic glycosides in clover is to protect young seedlings from being eaten by slugs and snails, but the interaction between plant and animal in the evolution of cyanogenesis is a complex one (Jones, 1972).

5.5.2 Recommended techniques

(a) *Detection of hydrogen cyanide*

Picrate papers are prepared by dipping rectangular pieces of filter paper in saturated (0·05 M) aqueous picric acid previously neutralized with NaHCO₃ and filtered. Dried papers will keep indefinitely. Two to three leaves (or similar amounts of other tissues) of plant to be tested are placed in a test tube, a drop of·water and two drops of toluene added and the material is briefly crushed with a glass rod. The tube is then firmly corked, with a moistened picrate paper suspended inside from the cork and left to incubate at 40°C for 2 h. A colour change from yellow to reddish-brown indicates the enzymic release of HCN from the plant. If the reaction is negative, the tube should be left at room temperature for a further 24–48 h and then re-examined for any non-enzymic release of HCN (Pusey, 1963). Intensity of the colour change is related to the amount of cyanogen present and it is possible to observe colour ratings as a measure of concentration of linamarin and lotaustralin in *Lotus* plants (Grant and Sidhu, 1967). Positive records have been made on herbarium plant tissue, but in general, the test is preferably carried out on fresh plants.

Picrate paper is not entirely specific to cyanogens since it will respond falsely to volatile isothiocyanates released from wild *Brassica* species and it is also lacking in sensitivity. For these reasons, another test paper is often used in conjunction with picrate, based on the work of Feigl and Anger (1966). Filter paper strips are prepared by dipping them for 2 min in a 1:1 mixture of two freshly prepared solutions: (1) 1% (w/v) 4,4-tetramethyldiamine diphenylamine in chloroform; and (2) 1% (w/v) copper ethyl acetoacetate in chloroform. The dried papers can be stored in a glass jar until used. The Feigl–Anger papers turn from a faint blue-green to a bright blue in the presence of HCN; they will detect as little as 1 μg HCN.

(b) *Chromatography of cyanogenic glycosides*

Evidence that a particular cyanogenic glycoside occurs in a plant can only be obtained by more detailed chromatographic examination of a concentrated alcoholic extract. Linamarin and lotaustralin can be separated on Whatman no. 4 paper in butanone–acetone–water (15:5:3) and have R_Fs ($\times 100$) of 52 and 63 respectively. They can be detected by spraying with ammoniacal $AgNO_3$ or, more specifically, by spraying with linamarase and detecting the HCN produced on an adjacent paper soaked in alkaline picrate (Wood, 1966).

Linamarin and lotaustralin can also be separated by TLC on silica gel G in chloroform–methanol (5:1). They are visualized by spraying with 2% α-naphthol in ethanol, then with conc. H_2SO_4 and then the plate is heated. GLC can also be used on the trimethylsilyl ethers; they have been separated on columns of 10–20% SE-20 on silanized Chromosorb W (80–100 mesh) at 210°C (Bissett *et al.*, 1969).

Aromatic cyanogenic glycosides such as dhurrin and other mandelonitrile derivatives are usually chromatographed on paper in solvents such as *n*-butanol–ethanol–water (40:11:14) and *n*-butanol–acetic acid–water (12:3:5) (Sharples and Stoker, 1969). They have absorption in the short UV and can be detected on paper with a 254 nm UV lamp. Dhurrin can be monitored during purification by its UV absorption (λ_{max} at 230 nm shifting to 255 nm in the presence of alkali).

(c) *Quantitative analysis*

The official method of HCN analysis used by food chemists is based on the hydrolysis of cyanogenic glycoside by endogenous enzymes in a closed system, steam distillation of the HCN released into a trap, and determination of the HCN (Horwitz, 1965). Conn (1979) has developed a micro version, which can be used on 250 mg samples of fresh plant material, assuming a concentration of 0·1% fresh weight. The HCN released by enzymic hydrolysis is allowed to diffuse into a trap of NaOH in a closed vessel and the NaCN formed is then determined colorimetrically. Alternatively, the intact cyanogenic glycosides of a plant, after mild extraction, can be quantitated by GLC of the trimethylsilyl ethers (Seigler, 1977).

(d) *Isolation*

Cyanogenic glycosides can be isolated and purified by general procedures used for other plant glycosides, but it is important to de-activate glycosidases present in the plant tissue during the isolation process. Linamarin and lotaustralin can be isolated as a pure mixture of glycosides from clover by chromatography on cellulose columns in *n*-butanol–water (9:1); 3·8 kg of fresh clover will yield 3·8 g pure glycoside (Maher and Hughes, 1971). A procedure for obtaining pure prunasin and vicianin from ferns is described by Kofod and Eyjolfsson (1969).

5.6 INDOLES

5.6.1 Chemistry and distribution

By far the most important plant indole is indole 3-acetic acid (IAA), the naturally occurring growth regulator of universal occurrence in the plant kingdom. Its detection, estimation and identification is fundamental to the study of plant physiology and many papers have been published on methods of IAA detection.

A major problem in the study of plant auxins is the very low concentration normally present in plants; the amount of IAA in vegetative tissue normally ranges between 0·5 and 15 μg kg^{-1}. Another is the fact that IAA is present in equilibrium with bound forms and any estimation must include measurement of the concentrations of such compounds. Bound forms include the simple glucose or inositol ester, the corresponding nitrile, aldehyde or lactic acid and more complex forms such as glucobrassicin (see Fig. 5.7).

A third problem of isolation is the lability of IAA, both *in vivo* (it undergoes oxidation to inactive products by the IAA oxidase enzyme complex) and *in vitro* (it is slowly destroyed in basic conditions); necessary precautions must be

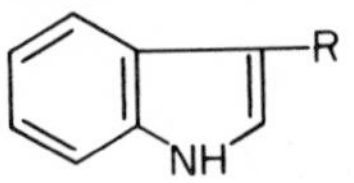

R = CH$_2$CO$_2$H   Indole 3−acetic acid (IAA)

R = CH$_2$CHO   Indole 3−acetaldehyde

R = CH$_2$CN   Indole 3−acetonitrile

R = CH$_2$CHOHCO$_2$H   Indole 3−lactic acid

R = OGlc   Indican

R = CH$_2$C(SGlc): NOSO$_3$H   Glucobrassicin

Fig. 5.7 Formulae of auxin and derivatives.

taken to avoid such loss. IAA is synthesized in plants from the aromatic amino acid tryptophan, by any one of five biosynthetic routes (Wightman and Cohen, 1968; MacMillan, 1980). Any method of estimation depends for its success in separating IAA from the precursor tryptophan, from intermediates in the pathway and also from any other indoles that may be present in a particular plant. These include tryptamine and serotonin (already discussed in Section 5.3 under amines) and indican, the indigo precursor present in wood, *Isatis tinctoria*.

In spite of the importance of auxin in growth, relatively little is known of the distribution of IAA derivatives in plants. Comprehensive studies of native plant indoles have recently been carried out in woad, in corn *Zea mays* (Stowe *et al.*, 1968) and in barley and tomato shoots (Schneider *et al.*, 1972). Further similar studies, using modern analytical methods (see below), on a wider range of plant tissues are badly needed.

5.6.2   Recommended techniques

(a) *Extraction*

Fresh tissue is macerated in pre-chilled methanol at 4°C and left for 24 h. This is filtered and the filtrate is evaporated at 45°C to a small volume and refiltered through celite. This is then fractionated by extraction into ether at different pHs (Schneider *et al.*, 1972); into a neutral fraction (indole 3-acetaldehyde) at pH7, an acid fraction (mainly IAA) at pH 3·0 and a basic fraction (mainly tryptamine) at pH 11. Any tryptophan present will remain in the aqueous residue. These fractions are then concentrated and compared with standards by chromatography, spectral measurements and by bioassay.

(b) *Chromatography*

R_F values in selected PC and TLC solvents of auxin and its derivatives are shown in Table 5.9. The recommended spray DMAC is made up by dissolving 0·1 g *p*-dimethylaminocinnamaldehyde in 10 ml conc.HCl and diluting to 200 ml with acetone just before use. The paper is dipped in the reagent, then dried and heated at 65°C for 2·5 min. For TLC plates, the reagent is sprayed on as a 0·25% solution of the aldehyde in ethanol–conc.HCl (1:1) and the colours will develop overnight at room temperature. An alternative and popular reagent for PC is the Salkowski reagent, which is 0·001 M $FeCl_3$ in 5% perchloric acid. Although only one TLC system is mentioned in the table, a range of other solvents and supports have been used (Kaldewey, 1969).

Table 5.9 R_F values, colours and spectra of indole 3-acetic acid (IAA) and its derivatives

| | R_F (×100) *in** | | | | *Colour with* | *Spectral max.* (nm) |
Indole	IAW	BAW	NaCl	CEF	DMAC	*in methanol*
Indole 3-acetic acid	46	92	64	90	blue-purple	274,282,290
Indole 3-lactic acid	44	87	63	65	blue-purple	275,282,289
Indole 3-acetaldehyde	95	—	36	86	green-blue†	244,260,292
Tryptamine	83	69	49	10	blue-purple‡	272,279,288‖
5-Hydroxytryptamine	64	50	36	05	blue‡	276,296‖
Tryptophan	32	54	54	05	purple‡	275,281,290

*Key: IAW = iso-PrOH–NH_4OH–H_2O (8:1:1); BAW = *n*-BuOH–HOAc–H_2O (12:3:5); NaCl = 8% NaCl in H_2O; CEF = $CHCl_3$–EtOAc–HCO_2H (5:4:1) on silica gel G plates (other three solvents on paper) (data from Schneider *et al.*, 1972).
†Can be distinguished from IAA by its reaction with 2,4-dinitrophenylhydrazine.
‡These give brown or red colours with ninhydrin.
‖Spectra measured in H_2O (not MeOH).

(c) *Identification*

UV spectral measurements in methanol (Table 5.9) are useful for purposes of identification and IR measurements (as KBr discs) can also provide useful confirmation of identity. UV spectral measurements in conc.H_2SO_4 solution at 30°C and 70°C can be used for distinguishing differently substituted indoles in μg quantities (Mollan *et al.*, 1973). The most characteristic property of IAA is its fluorescence spectrum, which varies according to the pH of the solution it is measured in. At pH 5, for example, IAA has a fluorescence peak at 365 nm and an activation peak at 285 nm (Burnett and Audus, 1964).

(d) *Bioassay*

The standard procedure is to study the effect on growth of *Avena* coleoptile sections. Coleoptiles, cut from seedlings grown in red light, are first washed with basal medium (2% sucrose in 0·01 M KH_2PO_4 buffer pH 4·5) to remove endogenous auxin and are then treated for 10 h with basal medium containing different strengths of the isolated IAA material. Increased lengths of sections over control sections can be related to auxin concentration, a standard curve being prepared from measurements made from synthetic auxin solutions (Witham *et al.*, 1971). A similar procedure using wheat coleoptiles can also be used (Fawcett *et al.*, 1960).

(e) *Quantitative estimation*

A procedure for the analysis of IAA in plants has been described by McDougall and Hillman (1978). This involves elaborate purification by solvent fractionation, column chromatography and TLC, followed by GLC–MS, in which the amount of IAA is determined by single ion monitoring. More recently, Ek *et al.* (1983) have measured IAA production in mycorrhizal fungi by GLC–MS directly on dry ether extracts of culture medium or mycelium, which were silylated in pyridine before the analysis. Loss of material during the estimation could be measured by adding deuterated IAA to the original extracts as an internal standard.

5.7 PURINES, PYRIMIDINES AND CYTOKININS

5.7.1 Chemistry and distribution

The purines and pyrimidines present in plants can be considered under four headings. First of all, there are the well known bases of nucleic acids: the purines, adenine and guanine; and the pyrimidines, cytosine, uracil (only in RNA) and thymine (only in DNA). These are common to all living tissues. These five bases, besides occurring in the structures of the nucleic acids, also occur at least in trace amounts, bound in low-molecular-weight form in plants. Each base can occur with the appropriate sugar ribose or deoxyribose attached (as nucleosides) or with sugar and phosphate (as nucleotides). Certain nucleotides such as adenosine triphosphate (ATP) have very important functions in primary metabolism. Others such as uridine diphosphate glucose (UDPG) are involved in carbohydrate metabolism. The phytochemical detection of such sugar derivatives will be mentioned in a later chapter on carbohydrates (see Chapter 6, p. 230).

Secondly, there are a number of unusual bases found in plants which are closely related in structure to the nucleic acid bases. These are present either bound in nucleic acid, as is 5-methylcytosine, in the DNA of wheat germ, or in low-molecular-weight form. Examples of the latter are the two pyrimidine

Pyrimidines

Cytosine

Uracil, R = H
Thymine, R = Me

5-Methylcytosine

Vicine, R = NH$_2$
Convicine, R = OH

H_2N—CH$_2$—CH—CO$_2$H with NH$_2$

Lathyrine

Tautomerism

OH^- H^+

Uracil tautomers

Purines

Adenine, R$_1$ = NH$_2$, R$_2$ = H
Guanine, R$_1$ = OH, R$_2$ = NH$_2$

Theobromine, R = H
Caffeine, R = Me

Kinetin, R = furfuryl
Zeatin,
R = CH$_2$—CH=C(Me)CH$_2$OH
Dihydrozeatin,
R = CH$_2$CH$_2$CH(Me)CH$_2$OH

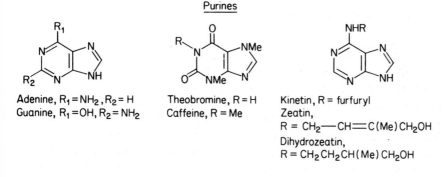

Nucleoside

Ribose

Guanosine

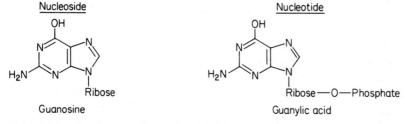

Nucleotide

Ribose—O—Phosphate

Guanylic acid

Fig. 5.8 Formulae of plant purines and pyrimidines.

glycosides, vicine and convicine (see Fig. 5.8), which are found in certain legume seeds of the genera *Vicia* and *Pisum*. Another example is the non-protein amino acid lathyrine, which has a pyrimidyl ring in its structure and which also occurs in legumes, in *Lathyrus* seed (see Section 5.1).

A third group of bases of importance in plants are the methylated purines such as theobromine and caffeine, which are highly valued as stimulants. They occur in relatively high concentration in tea, coffee and cocoa but are also found as trace constituents in a number of other plants.

The fourth and final group are also purines, this time substituted in the 6-position, and are known collectively as cytokinins. They comprise an important group of plant growth regulators and are primarily responsible for initiation of cell division during growth. The first cytokinin to be described was kinetin (6-furfurylaminopurine), actually a breakdown product of an animal nucleic acid preparation discovered during plant tissue culture studies (Miller *et al.*, 1956). Subsequently, naturally occurring cytokinins were detected in a number of plants and presumably they are of universal occurrence in the plant kingdom. The first natural cytokinin to be discovered was zeatin from *Zea mays* (Letham, 1963), but a number of other closely related substances, including dihydrozeatin, N^6-methylaminopurine and N^6-(Δ^2-isopentenyl)adenine, have since been implicated as cytokinins in other plants (for review, see MacMillan, 1980).

Purines and pyrimidines are colourless crystalline relatively stable compounds. They are generally only slightly soluble in water, being more soluble in either weak acid or alkaline solution. Distribution studies have been handicapped by the fact that there is no specific colour reagent (like ninhydrin for the amino acids) which allows for their easy detection in micro amounts in plant extracts. Reliance has to be placed during their chromatography on their appearance as dark absorbing spots in short UV light and this test is by no means entirely specific for these bases.

Once isolated, purines and pyrimidines, however, can be satisfactorily characterized by their UV spectra, particularly since these undergo specific shifts according to the pH of the solution. This is because purines and pyrimidines can exist in more than one tautomeric form; in basic solution, the hydroxyl groups will ionize and this form will be stabilized whereas in acid solution, the amino groups will ionize and the keto tautomer will be stabilized (see Fig. 5.7).

5.7.2 Recommended techniques

(a) *Paper chromatography*

This procedure was widely used until recently for separation of the common purines and pyrimidines and is still convenient for separating simple mixtures

(Markham, 1955). Indeed, the common bases can be separated one-dimensionally on Whatman no. 1 paper in water, adjusted to pH 10 with 1 M NH$_4$OH. R_Fs ($\times$100) are adenine 37, guanine 40, uracil 76, thymine 74 and cytosine 70 (ascending PC at 22–23°C). These five bases can also be separated in butanol saturated with water (R_Fs 38, 15, 31, 52 and 22 respectively). The bases are detected on paper using short UV (wavelength 253 nm) light as dark absorbing spots on a fluorescent background and the limit of detection is 0·5 to 1·0 μg ml^{-1}. For quantitative work the spots are detected by UV printing techniques on photographic plates and the spots then cut out, eluted with 0·1 M HCl and the concentration determined from the absorbance at the UV max. (between 245 and 265 nm).

Other purines and pyrimidines can be satisfactorily separated on paper using simple solvents. Caffeine and theobromine, for example, have R_Fs ($\times$100) of 65 and 33 in *n*-butanol–conc.HCl (9:1) and vicine and convicine have R_Fs 44 and 52 in *n*-butanol–pyridine–water, 18 and 24 in isopropanol–1 M NH$_4$OH–water (7:1:2). There are no generally applicable chromogenic reagents, although individual bases can sometimes be detected by special procedures (Lederer and Lederer, 1957).

(b) *Thin layer chromatography*

This has the advantage of sharpness of resolution and greater sensitivity and should always be used when complex mixtures of bases are being separated. Cellulose MN 300 is the preferred support, although silica gel and celite–starch mixtures have also been used (Mangold, 1969). Typical R_Fs for eight bases and four nucleosides in two of the most useful solvents are given in Table 5.10. Ion exchange cellulose layers (e.g. ECTEOLA and EPI–cellulose) are important for separation of nucleotides (Randerath and Randerath, 1967). Sephadex and DEAE–Sephadex are also applicable. For complex mixtures, thin layer electrophoresis on cellulose in 0·05 M formate buffer pH 3·4 (1500 V, 25 mA) followed by TLC in methanol–conc.HCl–water (64:17:18) can be employed.

(c) *High performance liquid chromatography*

This is a potentially important analytical technique, since purines and pyrimidines are readily detected at low concentrations by UV monitoring. The bases can be separated on an ion exchange column (e.g. Zipax SCX) using as mobile phase 0·01 M HNO$_3$ and 0·05 M NH$_4$NO$_3$. At a flow rate of 1·5 cm min^{-1}, the bases come off the column in sequence at 1·0 min (uracil), 1·5 min (cytidine), 2·2 min (guanine), 4·2 min (cytosine) and 5·0 min (adenine). Reversed-phase separation is also possible on ODS-treated silica columns, using gradient elution with 0·05 M KH$_2$PO$_4$ and varying amounts of methanol (Mischke and Wickstrom, 1980).

Table 5.10 Thin layer chromatography and ultraviolet spectra of purines and pyrimidines

Purine or pyrimidine	R_F ($\times 100$) TLC *on cellulose* MN 300 Solvent 1	Solvent 2*	UV spectral max. (nm) *in* 0·1 M HCl
Common bases			
Adenine	30	49	262
Guanine	20	25	249
Uracil	68	47	260†
Cytosine	48	43	276
Thymine	75	62	265
Unusual bases			
5-Hydroxymethylcytosine	46	34	279
Xanthine	26	21	—
Orotic acid	61	21	—
Nucleosides			
Adenosine	33	44	259†
Guanosine	31	22	252†
Uridine	67	37	262†
Cytidine	49	36	262

*Solvent 1 = MeOH–conc.HCl–H_2O (7:2:1); solvent 2 = *n*-BuOH–MeOH–H_2O–conc.NH_4OH (60:20:20:1).
†Measured in water as solvent.

(d) *Identification*

As already mentioned, purines and pyrimidines are identified by measuring their UV spectra in solutions of different pHs. There are sufficient differences in their spectra and shifts with pH to enable one to identify an unknown from such spectral measurements. Spectral maxima in water or 0·1 M HCl for a few bases are given in Table 5.10; further details of their spectral properties can be found in Beaven *et al.* (1955) and Venkstern and Baer (1966).

(e) *Cytokinins*

Procedures for purification and identification have been reviewed by Horgan (1978). After extraction of plant tissue with alcohol at 0°C for 12 h, the cytokinin extract is cleaned up by (a) passing through a cellulose phosphate column, washing with dilute acetic acid before eluting with dilute ammonia; and (b) solvent partition into ethyl acetate or *n*-butanol. After column chromatography on Sephadex LH20, final purification is achieved by PC, TLC or HPLC. PC has the advantage over TLC that recovery from paper strips is more complete than from silica gel. Identification of the separated cytokinins is based on co-chromatography with authentic materials in several systems, coupled with UV spectral comparisons at acid, neutral and basic pHs and MS. The value of the MS fragmentation pattern for identifying zeatin has already been illustrated in Fig. 1.6. GLC–MS has also been used, although there are some practical problems in its application to particular compounds (Horgan, 1981).

On paper, cytokinins have been separated in solvents such as *n*-butanol–water–conc.NH₄OH (172:18:10) and *n*-butanol–acetic acid–water (12:3:5), the former solvent separating zeatin from dihydrozeatin, two compounds which run close to each other in many systems. Cytokinins can be detected as blue spots after spraying the paper with the bromophenol blue/silver nitrate reagent. TLC separation on silica gel in chloroform–methanol (9:1) is useful for separating *cis*- and *trans*-isomers, e.g. *cis*- and *trans*-zeatin. Reversed-phase HPLC will resolve mixtures of closely related cytokinin glucosides if an ODS Hypersil column is used with a linear gradient of 5–20% acetonitrile in water over 30 min, and with UV detection at 265 nm (Horgan, 1981). HPLC is at present being developed for other cytokinin separations and it may well provide an ideal system for quantitative estimation. Other procedures such as bioassay and radioimmunoassay (Weiler, 1983) are also available.

5.7.3 Practical experiments

The simplest way of gaining experience in handling the common purines and pyrimidines is to hydrolyse nucleic acid samples, preferably isolated from plant tissues (see Chapter 7, p. 247), and to detect the bases produced.

(a) *Procedure*

(1) RNA is hydrolysed with 1 M HCl for 1 h at 100°C; this gives free purines, adenine and guanine, but the pyrimidine glycosidic links are more resistant to acid hydrolysis and the nucleosides cytidylic and uridylic acids are formed.
(2) DNA is hydrolysed with 72% perchloric acid at 100°C for 2 h and all four bases (adenine, guanine, cytosine and thymine) are obtained in the free state.
(3) Aliquots of these acid hydrolysates can be chromatographed directly in iso-propanol–conc.HCl–water (17:4:4) or similar solvent. Practical details PC of the five bases, for the beginner, are given in Witham *et al.* (1971).

(b) *Extraction of caffeine from tea leaves*

The extraction of caffeine from tea leaves provides a good practical example of an isolation of a purine from plants (Ikan, 1969).

(1) Finely powdered tea leaves (100 g) are extracted with 400 ml ethanol in a Soxhlet for 3 h.
(2) The caffeine is then adsorbed on to magnesium oxide (50 g), acidified with 50 ml 10% H_2SO_4 and desorbed and extracted into chloroform. Caffeine is thus obtained as soft silky needles, m.p. 235°C which can be recrystallized from a small volume of hot water. The UV maximum of caffeine in water is at 278 nm.
(3) A simple colour test for caffeine consists of dissolving a sample in three drops of conc.HNO_3, evaporating to dryness and adding two drops conc.NH_4OH; a purple colour is formed.

5.8 CHLOROPHYLLS

5.8.1 Chemistry and distribution

The chlorophylls are the essential catalysts of photosynthesis and occur universally as green pigments in all photosynthetic plant tissues. They occur in the chloroplasts in relatively large amounts, often bound loosely to protein but are readily extracted into lipid solvents such as acetone or ether.

Chemically, they each contain a porphyrin (tetrapyrrole) nucleus with a chelated magnesium atom in the centre and a long-chain hydrocarbon (phytyl) side chain attached through a carboxylic acid group. There are at least five chlorophylls in plants, all with the same basic structure but which show variations in the nature of the aliphatic side chains attached to the porphyrin nucleus (see Fig. 5.9). Thus, the structure of chlorophyll *b* only differs from that of *a* in having an aldehyde group instead of a methyl substituent attached to the top right-hand pyrrole ring (Fig. 5.9). Chlorophylls *a* and *b* occur in higher plants, ferns and mosses; chlorophylls *c* to *e* are only found in algae, while yet other chlorophylls are confined specifically to certain bacteria (Jackson, 1976; Vernon and Seely, 1966).

Chlorophylls are relatively labile and during isolation it is necessary to protect them from degradation. For example, plants contain an active chlorophyllase enzyme which removes the phytol side chain, giving rise to chlorophyllides. The central magnesium atom is also relatively easily lost during handling, with the formation of protochlorophylls.

Other porphyrin pigments occur in plants, but usually in much smaller amounts. The cytochromes, for example, essential units in the respiratory chain in plants and animals, are iron–porphyrin–protein complexes which are distinguished by spectroscopic methods. Again, some algae contain

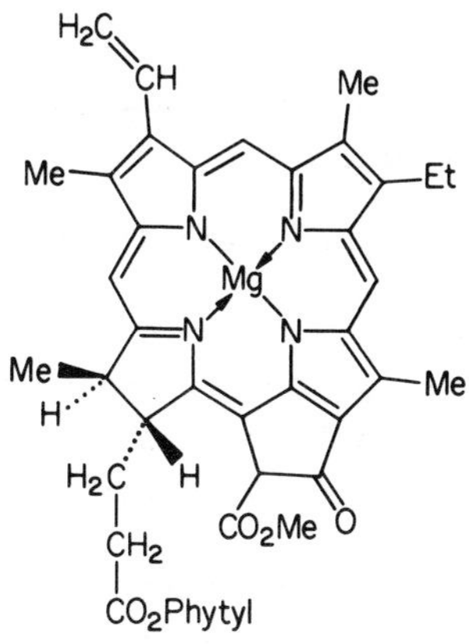

Chlorophyll *a* 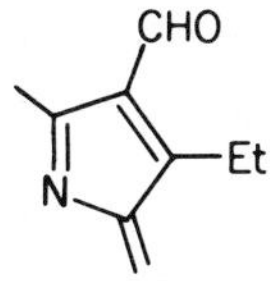   Chlorophyll *b*

Fig. 5.9 The structures of chlorophylls *a* and *b*.

relatively large amounts of red and blue pigments, called phycobilins (O'Carra and O'Eocha, 1976) which are pyrrole-based and proteinaceous, although the four pyrrole nuclei are linked in an open-chain structure (unlike the porphyrins) and there is no chelating metal. Finally, phytochrome, the universal photoreceptor pigment of plants, has a structure very similar to that of the phycobilins.

5.8.2 Recommended techniques

(a) *Chlorophyll estimation*

The determination of total chlorophyll content is frequently required in plant analysis. It is always better to extract fresh tissue and make measurements immediately, although extracts can be stored in the dark in acetone containing traces of Na_2CO_3 at -20 to $-30°C$ without appreciable loss. As a general precaution, it is advantageous to work in dim light to avoid pigment losses. The fresh tissue is ground in a mortar or macerator in the presence of excess acetone or methanol until all the colour is released from the tissue. $CaCO_3$ is added to prevent pheophytin formation and the extract is filtered on a Buchner funnel, the brei being washed with fresh acetone until colourless. The extract and washings are then made up to a known volume and stored in the refrigerator.

Measurement of chlorophylls *a* and *b* can then be made by direct determination of the absorbance at different wavelengths, using a standard spectrophotometer.

Assuming an 80% acetone extract, the absorbance should be measured at 663 and 646 nm in 1 cm cells. The concentrations can then be calculated from the following formulae.

Total chlorophyll (mg l^{-1}) = $17.3 A_{646} + 7.18 A_{663}$
Chlorophyll a (mg l^{-1}) = $12.21 A_{663} - 2.81 A_{646}$
Chlorophyll b (mg l^{-1}) = $20.13 A_{646} - 5.03 A_{663}$

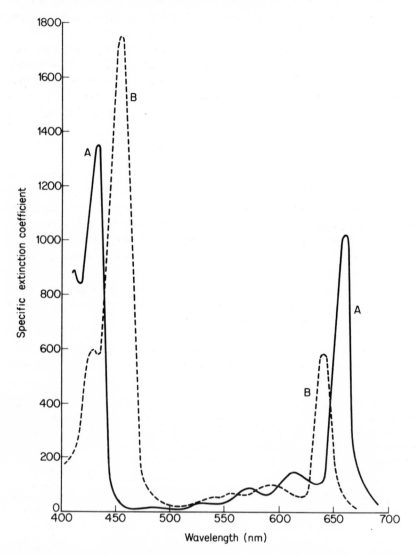

Fig. 5.10 Absorption spectra of chlorophylls a and b. Curve A, chlorophyll a; curve B, chlorophyll b.

The quantitative absorption curves of chlorophylls *a* and *b* in 80% ace. intersect at 652 nm, the specific absorption at this point being 36·0. The tota. amount of chlorophyll can therefore also be obtained by measuring the absorbance at this point and calculating the concentration in mg l^{-1} from 1000 $A_{652}/36$ or 27·8 A_{652}. This provides a useful check on measurements made at the wavelength maxima.

For absorption coefficients in other solvents and for other details of spectro-photometric estimations, see Holden (1976) and Lichtenthaler and Wellburn (1983).

Another method of estimating chlorophyll content is to separate *a* from *b* by TLC (see below), followed by elution and spectral measurement. A simpler procedure is to use HPLC, since the results are automatically quantified following separation. One system, which works at the picomole level, is isocratic separation on a porous silica gel column (Nucleosil 50·5) eluted with iso-octane–ethanol (9:1) (Stransky, 1978). Other procedures for the HPLC of porphyrin pigments are reviewed by Dolphin (1983).

(b) *Thin layer chromatography*

Although PC has been much used for chlorophylls, the procedures are rather complicated (Holden, 1976) and TLC is probably now the method of choice. TLC can be carried out on cellulose MN 300 in the dark using petroleum (b.p. 60–80°C)–acetone–*n*-propanol (90:10:0·45) in a 30 min run (Bacon, 1965). The plates should then be examined both in daylight (green colours) and in UV light (red fluorescences). Breakdown of chlorophylls during chromat-ography is inevitable and since each pigment gives rise to three artifacts, eight spots will be observed on the plate (Table 5.11).

Table 5.11 R_F values and colours of chlorophylls *a* and *b* on thin layer chromatography

Pigment	R_F	Colour in daylight	Nature of artifact
Pheophytin a	93	grey ⎫	magnesium-free
Pheophytin b	80	yellow-brown ⎭	chlorophylls
Chlorophyll a	60	blue-green	
Chlorophyll b	35	yellow-green	
Pheophorbide a	18	grey ⎫	magnesium-free
Pheophorbide b	07	yellow-brown ⎭	chlorophyllides
Chlorophyllide a	03	blue-green ⎫	chlorophylls without
Chlorophyllide b	02	yellow-green ⎭	phytyl side chains

TLC on MN 300 cellulose in petroleum (b.p. 60–80°C)–acetone–*n*-propanol (90:10:0·45) for 30 min run on 20 × 20 cm plate. Separation can be also achieved on a microscope slide (5 min run).

For quantitative work, it is essential to remove the water from an acetone pigment extract; otherwise, some chlorophyll remains on the baseline. This can be done either by repeated evaporation at low temperature adding fresh acetone each time or by adding ammonium acetate ($1 \text{ g } 5 \text{ ml}^{-1}$ extract) and taking the pigment into ether.

(c) *Spectral identification*

The chlorophylls are readily differentiated by their relatively complex UV and visible absorption curves. They show major peaks at about 400 nm, a number of minor peaks, between 500 and 600 nm, and a major peak above 625 nm (see Fig. 5.10). For chlorophylls *a* and *b*, the long wave peaks are at 663 and 645 nm, respectively, for *c* and *d* they are at 630 and 688 nm. Bacterial chlorophylls *a* and *b* have maximal absorbances at 780 and 750 nm, while *Chlorobium* chlorophylls absorb at 650 and 660 nm. The exact positions of the visible peaks depend on the solvent used and there are spectral changes as a result of pigment degradation due to loss of the magnesium atom or the phytyl side chain. Finally, it should be pointed out that the *in vivo* spectra of the chlorophylls do not match those measured *in vitro*; chlorophyll *a* shows three peaks at 673, 683 and 694 nm *in vivo* and *b* absorbs at 650 nm.

REFERENCES

General references

Beevers, L. (1976) *Nitrogen Metabolism in Plants*, Edward Arnold, London.

Brenner, M., Niederweiser, A. and Pataki, G. (1969) Amino acids and derivatives. in *Thin Layer Chromatography* (ed. E. Stahl), George Allen and Unwin, London, pp. 730–86.

Clarke, E.G.C. (1970) The forensic chemistry of alkaloids. in *The Alkaloids* (ed. H.F. Manske) (Vol. XII), Academic Press, New York, pp. 514–90.

Conn, E.E. and Butler, G.W. (1969) Biosynthesis of cyanogenic glycosides and other simple nitrogen compounds. in *Perspectives in Phytochemistry* (eds J.B. Harborne and T. Swain), Academic Press, London, pp. 47–74.

Cordell, G.A. (1981) *Introduction to Alkaloids*, John Wiley, New York.

Fowden, L. (1981) Non-protein amino acids. in *Biochemistry of Plants (Secondary Plant Products*, Vol. 7) (ed. E.E. Conn), Academic Press, New York, pp. 215–48.

Hegnauer, R. (1967) Comparative phytochemistry of alkaloids. in *Comparative Phytochemistry* (ed. T. Swain), Academic Press, London, pp. 211–30.

Hegnauer, R. (1977) Cyanogenic glycosides as systematic markers in Tracheophyta. *Plant System. Evol.*, Suppl. 1, 191–210.

Henry, T.A. (1949) *The Plant Alkaloids*, Churchill, London.

Holden, M. (1976) Chlorophylls. in *Chemistry and Biochemistry of Plant Pigments*, 2nd edn. (ed. T.W. Goodwin) (Vol. 2), Academic Press, London, pp. 1–37.

Horgan, R. (1978) Analytical procedures for cytokinins. in *Isolation of Plant Growth Substances* (ed. J.R. Hillman), Cambridge University Press, Cambridge, pp. 97–114.

Horgan, R. (1981) Modern methods for plant hormone analysis. *Prog. Phytochem.*, **7**, 137–70.

Ikan, R. (1969) *Natural Products: A Laboratory Guide*, Academic Press, London, pp. 178–260.

Jackson, A.H. (1976) Structure, properties and distribution of chlorophylls. in *Chemistry and Biochemistry of Plant Pigments*, 2nd edn (ed. T.W. Goodwin), Academic Press, London, pp. 1–63.

Jones, D.A. (1972) Cyanogenic glycosides and their function. in *Phytochemical Ecology* (ed. J.B. Harborne), Academic Press, London, pp. 103–24.

Kaldewey, H. (1969) Simple indole derivatives and plant growth regulators. in *Thin Layer Chromatography* (ed. E. Stahl), George Allen and Unwin, London, pp. 471–93.

MacMillan, J. (ed.) (1980) *Hormonal Regulation of Development, Molecular Aspects of Plant Hormones*, Springer-Verlag, Berlin.

Miflin, B.J. (ed.) (1981) *Biochemistry of Plants (Amino Acids and Derivatives*, Vol. 5), Academic Press, New York.

Randerath, K. and Randerath, E. (1967) Thin layer methods for nucleic acid derivatives. in *Methods in Enzymology* (Vol. XII A), Academic Press, New York, pp. 323–47.

Robinson, T. (1981) *The Biochemistry of Alkaloids*, 2nd edn, Springer-Verlag, Berlin.

Rodrigo, R.G.A. (ed.) (1982) *The Alkaloids (Chemistry and Physiology*, Vol. 20), Academic Press, New York.

Rosenthal, G.A. (1982) *Plant non-protein Amino and Imino Acids*, Academic Press, New York.

Seiber, N. (1970) in *Methods of Biochemical Analysis* (ed. D. Glick), John Wiley and Sons, New York, pp. 259–337.

Smith, T.A. (1981) Amines. in *Biochemistry of Plants (Secondary Plant Products*, Vol. 7) (ed. E.E. Conn), Academic Press, New York, pp. 249–68.

Venkstern, T.V. and Baer, A.A. (1966) *Absorption Spectra of Minor Bases, their Nucleosides, Nucleotides and of Selected Oligoribonucleotides*, Plenum Press, New York.

Vernon, L.P. and Seeley, G.R. (1966) *The Chlorophylls*, Academic Press, New York.

Supplementary references

Bacon, M.F. (1965) *J. Chromatog.*, **17**, 322.

Baker, W., Harborne, J.B. and Ollis, W.D. (1952) *J. Chem. Soc.*, 3215.

Beaven, G.H., Holiday, E.R. and Johnson, E.A. (1955) in *The Nucleic Acids* (ed. E. Chargaff and J.N. Davidson) (Vol. 1), Academic Press, New York, pp. 493–533.

Bell, E.A., Lackey, J.A. and Polhill, R.M. (1978) *Biochem. System. Ecol.*, **6**, 201.

Bell, E.A. and Tirimanna, A.S.L. (1965) *Biochem. J.*, **97**, 104.

Bieleski, R.L. and Turner, N.A. (1966) *Analyt. Biochem.*, **17**, 278.

Bissett, F.H., Clapp, R.C., Coburn, R.A., Ettlinger, M.G. and Long, L. (1969) *Phytochemistry*, **8**, 2235.

Blackburn, S. (1965) *Meth. Biochem. Anal.*, **13**, 1.

Booth, J. and Boyland, E. (1964) *Biochem. J.*, **91**, 364.

Burnett, D. and Audus, L.J. (1964) *Phytochemistry*, **3**, 395.

Conn, E.E. (1979) in *Herbivores, their Interaction with Secondary Plant Metabolites* (eds G.A. Rosenthal and D.H. Janzen), Academic Press, New York, pp. 387–412.

Daday, H. (1954) *Heredity*, **8**, 61, 377.

Dolphin, D. (1983) in *Chromatography, Fundamentals and Applications, Part B* (ed. E. Heftmann), Elsevier, Amsterdam, pp. 377–406.

Dunnill, P.M. and Fowden, L. (1965) *Phytochemistry*, **4**, 933.

Ek, M., Ljungquist, P.O. and Stenstrom, E. (1983) *New Phytologist*, **94**, 401.

Fawcett, C.H., Wain, R.L. and Wightman, F. (1960) *Proc. R. Soc. Ser. B*, **152**, 231.

Feigl, F. and Anger, V.A. (1966) *Analyst*, **91**, 282.

Fellows, L.E. and Bell, E.A. (1970) *Phytochemistry*, **9**, 2389.

Grant, W.F. and Sidhu, B.S. (1967) *Can. J. Bot.*, **45**, 639.

Hanes, C., Harris, C.K. and Moscarella, M.A. (1961) *Can. J. Biochem. Physiol.*, **39**, 163, 439.

Horwitz, W. (ed.) (1965) *Official Methods of Analysis of the Association of Official Agricultural Chemists*, 10th edn, AOAC, Washington DC, U.S.A., p. 341.

Hultin, E. (1966) *Acta Chem. Scand.*, **20**, 1588.

Hultin, E. and Torsell, K. (1965) *Phytochemistry*, **4**, 425.

Ilert, H.I. and Hartmann, T. (1972) *J. Chromatog.*, **71**, 119.

Irvine, W.J. and Saxby, M.J. (1969) *Phytochemistry*, **8**, 473.

Kamienski, E.S. van (1959) in *Papier Chromatographie in der Botanik* (ed. H.F. Linskens), Springer-Verlag, Berlin.

Kipps, A. (1972) Ph.D. Thesis, University of Durham.

Kofod, H. and Eyjolfsson, R. (1969) *Phytochemistry*, **8**, 1509.

Kuster, T. and Niederwieser, A. (1983) in *Chromatography, Fundamentals and Applications, Part B* (ed. E. Heftmann), Elsevier, Amsterdam, pp. 1–51.

Lawler, L.J. and Slaytor, M. (1969) *Phytochemistry*, **8**, 1959.

Lederer, E. and Lederer, M. (1957) *Chromatography, A Review of Principles and Applications*, Elsevier, Amsterdam.

Letham, D.S. (1963) *Life Sci.*, **8**, 569.

Lichtenthaler, H.K. and Wellburn, A.R. (1983) *Biochem. Soc. Trans.*, **11**, 591.

Lüning, B. (1967) *Phytochemistry*, **6**, 852.

McDougall, J. and Hillman, J.R. (1978) in *Isolation of Plant Growth Substances* (ed. J.R. Hillman), Cambridge University Press, Cambridge, pp. 1–26.

Macek, K. (1967) in *Chromatography* (ed. E. Heftmann), Reinhold, New York, pp. 606–26.

Maher, E.P. and Hughes, M.A. (1971) *Phytochemistry*, **10**, 3005.

Mangold, H.K. (1969) in *Thin Layer Chromatography* (ed. E. Stahl), George Allen and Unwin, London, pp. 786–804.

Markham, R. (1955) in *Modern Methods of Plant Analysis* (eds K. Paech and M.V. Tracey) (Vol. IV), Springer Verlag, Berlin, pp. 246–304.

Miller, C.O., Skoog, F., Okumura, F.S., Von Saltza, M.H. and Strong, F.M. (1956) *J. Am. Chem. Soc.*, **78**, 1375.

Mischke, C.F. and Wickstrom, E. (1980) *Analyt. Biochem.*, **105**, 181.

Mollan, R.C., Harmey, M.A. and Donnelly, D.M.X. (1973) *Phytochemistry*, **12**, 447.

Moore, S. and Stein, W.H. (1956) *Adv. Protein Chem.*, **11**, 191.

O'Carra, P. and O'Eocha, C. (1976) in *Chemistry and Biochemistry of Plant Pigments*, 2nd edn (ed. T.W. Goodwin) (Vol. 1), Academic Press, London, pp. 328–77.

Phillipson, J.D. (1982) *Phytochemistry*, **21**, 2441.

Pusey, J.G. (1963) in *Teaching Genetics* (eds C.D. Darlington and D.A. Bradshaw), Oliver and Boyd, Edinburgh, pp. 99–104.

Santavy, F. (1969) in *Thin Layer Chromatography* (ed. E. Stahl), George Allen and Unwin, London, pp. 421–70.

Schneider, E.A., Gibson, R.A. and Wightman, F. (1972) *J. Exp. Bot.*, **23**, 152.

Seigler, D.S. (1977) *Prog. Phytochem.*, **4**, 83.

Sharples, D. and Stoker, J.R. (1969) *Phytochemistry*, **8**, 597.

Smith, T.A. (1970) *Phytochemistry*, **9**, 1479.

Smith, T.A. (1972) Private communication.

Stahl, E. and Schorn, P.J. (1969) in *Thin Layer Chromatography* (ed. E. Stahl), George Allen and Unwin, London, pp. 494–505.

Stowe, B., Vendrell, M. and Epstein, E. (1968) in *Biochemistry and Physiology of Plant Growth Substances* (eds F. Wightman and G. Setterfield), Runge Press, Ottawa, pp. 173–82.

Stransky, H. (1978) *Z. Naturforsch.*, **33c**, 836.

Taber, W.A., Vining, L.C. and Heacock, R.A. (1963) *Phytochemistry*, **2**, 65.

Waldi, D., Schnackerz, K. and Munter, F. (1961) *J. Chromat.*, **6**, 61.

Weiler, E.W. (1983) *Biochem. Soc. Trans.*, **11**, 485.

Wightman, F. and Cohen, D. (1968) in *Biochemistry and Physiology of Plant Growth Substances* (eds F. Wightman and G. Setterfield), Runge Press, Ottawa, pp. 273–88.

Witham, F.H., Blaydes, D.F. and Devlin, R.M. (1971) *Experiments in Plant Physiology*, Van Nostrand, New York, pp. 44–6.

Wood, T. (1966) *J. Sci. Food Agr.*, **17**, 85.

CHAPTER SIX

Sugars and their Derivatives

6.1 Introduction
6.2 Monosaccharides
6.3 Oligosaccharides
6.4 Sugar alcohols and cyclitols

6.1 INTRODUCTION

Carbohydrates or sugars occupy a central position in plant metabolism so that methods for their detection and estimation are very important to the plant scientist. Not only are sugars the first complex organic compounds formed in the plant as a result of photosynthesis, but also they provide a major source of respiratory energy. They provide a means of storing energy (as starch) and transport of energy (as sucrose) and also the building blocks of the cell wall (cellulose). In addition, many other classes of plant constituent, e.g. the nucleic acids and the plant glycosides, contain sugars as essential features of their structures. Finally, sugars play a number of ecological roles, in plant–animal interactions (flower nectars are mainly sugar), in protection from wounding and infection and in the detoxification of foreign substances.

Sugars are conveniently classified into three groups, on the basis of molecular size: the simple monosaccharides (e.g. glucose, fructose) and their derivatives; the oligosaccharides, formed by condensation of two or more monosaccharide units (e.g. sucrose); and the polysaccharides which consist of long chains of monosaccharide units, joined head to tail, either as straight chains or with branching. The first two groups are dealt with in this chapter, discussion of polysaccharides being reserved for Chapter 7.

In their chemistry, low-molecular-weight sugars have many properties in common. They are optically active, aliphatic polyhydroxy compounds, which are usually very water-soluble. They are often difficult to crystallize, even when pure, and are frequently isolated as derivatives (e.g. as osazones by reaction with phenylhydrazine). The sugars are relatively labile, and they easily undergo isomerization (enzymic or otherwise), and/or ring opening during extraction and the concentration of such extracts. Therefore, care must be taken to avoid extremes of heat or pH when isolating them.

Sugars are colourless substances and when present in micro amounts have to be detected by reaction with a suitable chromogenic reagent. Reducing sugars such as glucose, classically detected by a yellow-red precipitate with Fehling's solution, are easily detected on chromatograms by using one of a range of phenolic or amine reagents (e.g. resorcinol–H_2SO_4 or aniline hydrogen phthalate). Non-reducing sugars are less responsive to these reagents and are usually detected by their rapid oxidation with periodate or lead tetracetate. A general reagent for all sugars is alkaline $AgNO_3$, but this is not entirely specific for sugars since it also reacts with certain other plant substances, such as phenols.

6.2 MONOSACCHARIDES

6.2.1 Chemistry and distribution

The major free sugars in plants are the monosaccharides, glucose and fructose (and the disaccharide sucrose), together with traces of xylose, rhamnose and galactose. Other sugars present in trace amounts are the sugar phosphates, which are involved in metabolism; these are very easily hydrolysed during manipulation to the parent sugars and special procedures (see later) are required to detect them. The bulk of carbohydrate occurs in plants in bound form, as oligo- or polysaccharide, or attached to a range of different aglycones, as plant glycosides. Since the free sugar pool is relatively uniform in higher plants, analysis of monosaccharides is therefore most frequently concerned with the identification of sugars in hydrolysates of plant glycosides, oligo-saccharides or polysaccharides.

Five sugars are commonly found as components of glycosides and poly-saccharides and most plant analyses are concerned with their separation and identification. Two are hexoses, glucose and galactose, two are pentoses, xylose and arabinose, and one is a methylpentose, rhamnose. Of fairly common occurrence are the uronic acids, glucuronic and galacturonic acids and a third hexose, mannose, is not uncommon in polysaccharides. The pentose sugars ribose and deoxyribose, components of RNA and DNA respectively should be mentioned here, but these sugars are rarely encountered in plants in any other association. The only common keto sugar is fructose, not often present in plant glycosides, but a frequent component of oligosaccharides (e.g. sucrose) and of the polysaccharides known as fructans. A range of rarer sugars occur in plant glycosides, one example being the five-carbon branched sugar apiose, present as the flavone glycoside apiin in parsley seed.

The structures of most of the monosaccharides mentioned above are given in Fig. 6.1. It is to be remembered that each sugar can exist as more than one optically active isomer; however, in plants, only one form is normally encountered. Thus, glucose is usually the β-D-isomer, rhamnose the α-L form and so on. Chromatography does not normally distinguish between optical enantiomers. Again, each sugar can theoretically exist in both a pyrano-(six-membered) and furano-(five-membered ring) form, although one or other is usually favoured. Thus, glucose normally takes up the pyrano-configuration, whereas fructose is usually the furano-form (see Fig. 6.1).

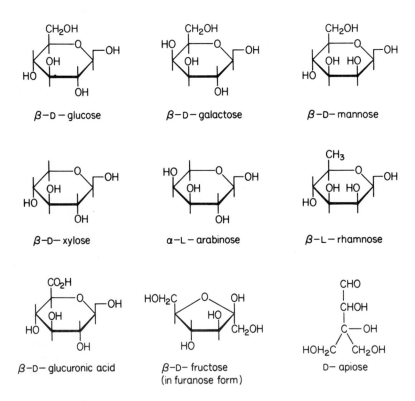

Fig. 6.1 Formulae of common monosaccharides.

More detailed accounts of the simple plant sugars can be found in most organic chemistry or biochemistry textbooks. For accounts of their comparative distribution in the plant kingdom, see Bell (1962) and Loewus and Tanner (1982).

6.2.2 Recommended techniques

(a) *Preparation of sample*

Analysis of the free sugars in leaf or other plant tissue is usually accomplished by extraction of fresh tissue with 95% ethanol (or methanol). This is followed by concentration of the extract, which removes all the alcohol, and the freeing of the concentrated aqueous extracts of any precipitate, by filtration through celite or by centrifugation. It may be useful to include at some stage a washing of the extract with light petroleum to remove lipids, but this is not always necessary. The clear aqueous extract so obtained can be spotted directly on to chromatograms.

Sugar analysis on the hydrolysate of a polysaccharide or a plant glycoside requires prior removal of mineral acid used for the hydrolysis and, in the case of plant glycosides, removal by extraction into ether or ethyl acetate of the aglycone moiety. After hydrolysis with 1 M H_2SO_4, the acid may be removed as $BaSO_4$ by adding $BaCO_3$ solution. Small amounts of 1 M HCl can be removed *in vacuo* at the water pump, but the temperature must be kept below 40°C. Larger amounts of HCl can be conveniently removed from an aqueous

Table 6.1 R_Fs colours and solvents for paper chromatography of sugars

Sugar	R_F ($\times$ 100) *in**				Colour with aniline hydrogen phthalate†
	BAW	BEW	BTPW	PhOH	
Glucose	12	16	24	34	brown
Galactose	12	16	21	38	brown
Allose	15	16	25	44	brown
Arabinose	18	22	29	52	red
Xylose	20	26	35	43	red
Rhamnose	32	37	47	60	yellow-brown
Glucuronic acid	{ 16	21	03	13	orange-yellow
	{ 30	33	52	55	orange-yellow
Galacturonic acid	15	19	03	15	orange-yellow
Ribose	27	36	45	62	red
Fructose	19	21	29	54	yellow
Apiose	22	24	54	59	pink
Mannose	17	23	29	44	brown

*Solvent key: BAW = BuOH–HOAc–H_2O (4:1:5, top)
    BEW = *n*-BuOH–EtOH–H_2O (4:1:2·2)
    BTPW = *n*-BuOH–toluene–pyridine–H_2O (5:1:3:3)
    PhOH = PhOH satd. with H_2O.
†Made up by dissolving aniline (9·2 ml) with phthalic acid (16 g) in *n*-BuOH (490 ml), Et_2O (490 ml) and H_2O (20 ml).

hydrolysate by repeated washing with a 10% solution of di-*n*-octylmethyl-amine in chloroform. (Note: chloroform is heavier than water and makes up the lower layer.) Ion exchange resins are also widely used to remove acid under these conditions.

(b) *Paper chromatography*

Sugar extracts are run one-dimensionally by descent in four different solvents (Table 6.1) on Whatman no. 1 paper, with a standard mixture containing glucose, galactose, arabinose, xylose and rhamnose. The standard mixture should be made up in 10% isopropanol (this prevents bacterial contamination), each sugar being present at a concentration of 0·5% by weight; about 3–5 μl of this solution should be applied to the paper. The plant extracts should be applied at several different concentrations (e.g. one, two and five applications) if the amount of sugar present in them is unknown.

Good separations of the common sugars can be achieved with 18–24 h runs, but for best results, it is necessary to allow the solvent to run off the paper and leave for 36 h. The dried papers are then dipped in aniline hydrogen phthalate, made up in ether–butanol, and redried. Finally, the papers are heated at 105°C for 5 min in order to develop the distinctive colours (Table 6.1). The paper should be routinely examined in UV light, since the sugar spots fluoresce and this provides a more sensitive means of detection when sugar concentrations are low.

The only two common sugars which are at all difficult to distinguish on chromatograms are glucose and galactose. However, these sugars do separate well if the paper is developed for at least 24 h. The best solvents are BTPW and phenol, and it is to be noted that the relative positions of the two sugars in BTPW are reversed in phenol. Glucose and fructose are readily separated, but not if sucrose is also present with them (as in plant nectars, etc.) and special solvents are necessary (see below). Glucuronic and galacturonic acids have similar mobilities, but glucuronic acid is easily distinguished, since it lactonizes during chromatography to glucuronolactone and a second spot of approximate equal intensity but with a higher R_F is apparent on chromatograms (Table 6.1).

Recently, the monosaccharide allose, an isomer of glucose, has been found in glycosidic combination in plants, e.g. as a flavone glycoside in *Veronica fili-formis*. Allose runs close to glucose in three solvents (Table 6.1), but is clearly separated in phenol (Chari *et al.*, 1981). The close similarity of allose to glucose in most sugar solvents points to the importance of co-chromatography in as many solvents as possible to be sure of correct identification. In the case of the *Veronica* glycoside, identity of allose as one of the combined sugars was confirmed by ^{13}C-NMR spectroscopy.

(c) *Confirmation of identity*

This can be done on a microscale by measuring the spectral absorption of the coloured product formed between a particular sugar and either resorcinol–1 M H_2SO_4 or aniline hydrogen phthalate. With the former reagent, rhamnose gives three peaks in the visible region at 412, 488 and 545 nm (see Fig. 6.2), whereas glucose has peaks at 489 and 555 nm, with an inflection at 430 nm, and galactose has peaks at 422 and 495 nm with an inflection at 550 nm (Harborne, 1960).

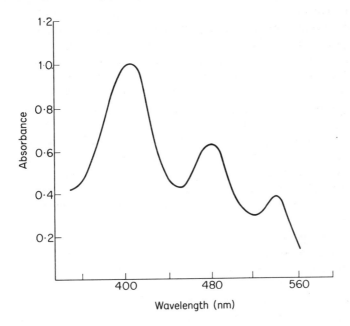

Fig. 6.2 Visible spectrum of coloured product from rhamnose and resorcinol–H_2SO_4.

Glucose and galactose can also be identified enzymically on a microscale by use of glucose oxidase and galactose oxidase respectively. The enzyme, together with a trace of catalase, is added to the sugar solution, or can be sprayed on to a chromatogram. Within a short time, the sugar is specifically oxidized to a product which no longer reacts with aniline hydrogen phthalate.

(d) *Quantitative analysis*

It is often necessary to measure the amounts of different monosaccharides present in plant extracts or in hydrolysates of plant glycosides. One standard procedure of the many available for this (Harborne, 1960) is as follows: Place a given volume of the sugar solution, in triplicate, using a micropipette along the start line of a chromatogram. Allow the paper to dry and develop in any of the four standard sugar solvents (Table 6.1). Develop the paper by dipping in aniline hydrogen phthalate reagent, drying and heating for 5 min at 105°C. Cut out the coloured spots and elute each with 3 ml methanol containing 1% stannous chloride. Measure the absorbance at the visible maximum in a spectrophotometer (for glucose at 397 nm, for rhamnose at 375 nm) and take the average of the three readings.

It is essential to have run control chromatograms, containing varying known amounts of authentic sugars, under identical conditions in order to calculate the concentration of each sugar in the original extract. In addition, absorbances should be measured against eluates of blank spots of the same size as the coloured spots, which are cut out at the same time from the control chromatograms.

(e) *Authentic markers*

Almost all the sugars mentioned above can be obtained from commercial suppliers. Apiose can be readily obtained from parsley seed as follows. Extraction of the powdered seed with hot alcohol and acid hydrolysis of the residue obtained after removal of alcohol from the extract will furnish a mixture of sugars, one of which is apiose, clearly distinguished by its high mobility (Table 6.1). It can be obtained on a preparative scale by chromatographing the sugar mixture as a streak on Whatman no. 3 paper in one of the four sugar solvents.

6.2.3   Alternative techniques

(a) *Thin layer chromatography*

Most of the separations achieved on paper can be obtained by TLC on microcrystalline cellulose using the same solvents as in Table 6.1. Silica gel, a popular TLC absorbent, is often modified by pretreatment with a suitable buffer (phosphate or borate) before it is used for sugar separations. A system that will separate the common sugars on plain silica gel is the mixture *n*-butanol–acetic acid–ether–water (9:6:3:1). The development time is 4 h. The sugars may be detected by spraying with aniline hydrogen phthalate. An alternative spray is 0·2% naphthoresorcinol in *n*-butanol containing 10% phosphoric acid. On heating the plate for 10 min at 100°C, ketoses give pink, pentoses green and hexoses blue colours. Typical R_F values ($\times 100$) in this system are: sucrose 09, glucose 22, fructose 27, xylose 40, ribose 47 and rhamnose 55.

(b) *Gas liquid chromatography*

GLC of sugars is a less convenient technique than PC but it has the advantage of being more sensitive. The level of detection by GLC is 0·1 μg. compared to 5 μg on paper. However, sugars have to be separated on GLC as their trimethylsilyl ethers, and it is necessary to take aqueous solutions containing sugars to dryness and to conduct all subsequent operations in the absence of moisture. The residue (assuming it contains microgram amounts of sugar) is treated with hexamethyldisilazane (0·2 ml), trimethylchlorosilane (0·1 ml) in dry pyridine (0·5 ml). Excess reagents are removed in high *vacuo* and the sugar ethers are dissolved in dry heptane for injection into the gas chromatograph. Chromatography can be carried out on a silanized Chromosorb W column coated with 3% SE 52 at a column temperature of 180°C and inlet pressure of 15 p.s.i. Typical retention times under these conditions are 4 and 5·5 min for α- and β-rhamnose, 12 and 18 min for α- and β-glucose; it must be noted that all sugars give two peaks, due to anomeric formation on the GLC column. Derivatization prior to GLC can be simplified by using the commercial reagent Trisil Z, which is available in 1 ml ampoules. After reaction with the sugar sample, a portion of the solution can be injected directly into the GLC column.

The GLC of sugars was originally described by Sweeley *et al.* (1963). Its use for identification of sugars of flavonoid glycosides is referred to by Mabry *et al.* (1970). Recent reviews of the GLC of sugars are by Holligan (1971), Holligan and Drew (1971) and Churms (1983).

(c) *High performance liquid chromatography*

HPLC is largely a complementary technique to GLC for sugar estimations. At present, it lacks the sensitivity of GLC (detection limit 10 μg compared with 0·1 μg for GLC) but it is otherwise excellent for rapidly resolving mixtures of closely related sugars. It is incidentally probably a better technique to use than PC for separating and quantifying homologous series of oligosaccharides (see below). A disadvantage of HPLC is that detection is perforce limited to refractive index measurements, which lack sensitivity and do not allow for gradient elution of the solvents through the column. If however the sugars are derivatized (e.g. as the benzoate esters) before separation, then UV detection can be employed, but one of the advantages over GLC is then lost.

A variety of columns and eluents have been developed for the HPLC of carbohydrates (Churms, 1983; Folkes and Taylor, 1982). One popular method for un-derivatized sugars is separation on a bonded-phase column (Waters Bondapak, Partisil-10 PAC or Spherisorb S5NH$_2$) and elution with acetonitrile—water, the proportions of which depend on the nature of the compounds to be separated. An HPLC technique for determining sucrose, glucose and fructose contents in plant tissues has been developed by McBee and Maness (1983). Dried plant tissue is extracted with water and cleaned up on a protective ion exchange precolumn of Aminex HPX–85H. Separation is then achieved on an Aminex HPX–87 column eluted with water at 85°C.

(d) *Paper electrophoresis*

This is an important technique for the separation of sugars, but it is more useful for oligosaccharides and monosaccharide derivatives (see later) than for the parent monosaccharides. The other techniques described above normally provide sufficient armoury for distinguishing the common plant sugars. It could, however, be used as a screening technique for distinguishing carbohydrates with acidic or basic substituents (glucuronic acid, sugar phosphates, amino sugars) from neutral monosaccharides by carrying out electrophoresis in the absence of complexing reagents (i.e. in acetate buffer, pH 5 for instance). Thus neutral sugars would be immobile, since they only migrate in the presence of molybdate, borate or germanate salts, with which they form charged complexes. A standard procedure for electrophoresis of common sugars is to use 0·05 M sodium borate buffer (pH 9–10) and voltages of 25–30 V cm^{-1} for 90 min. Under these conditions, glucose and xylose have the same mobility (1·0) with galactose at 0·93, arabinose at 0·96 and rhamnose at 0·52. For further details, see reviews by Foster (1957) and Churms (1983).

(e) *Sugar phosphates*

Special procedures are required for the detection of sugar phosphates in plant extracts (Selvendran and Isherwood, 1967; Isherwood and Selvendran, 1970). Plants are first extracted with trichloracetic acid and the crude extract fractionated on an ion exchange (Dowex 1) column. TLC is then carried out on cellulose plates in iso-butyric acid–1 M NH$_4$OH–0·1 M EDTA (100:60:1·6) or ethanol–1 M ammonium acetate (3:7). PC can also be used, but only with solvents which do not contain strong acids.

6.2.4   Practical experiments

(a) *Example of sugar detection: analysis of floral nectars*

The nectars of angiosperm flowers are well known to have an important ecological role in attracting insects, humming birds and other animals to the flower for the purposes of pollination (Faegri and van der Pijl, 1979). Floral nectars consist of nearly pure sugar solutions, the main sugars being glucose, fructose and sucrose. Trace amounts of such oligosaccharides as raffinose, maltose and melibiose may also be present. Nectars have been extensively surveyed by Percival (1961) using PC and she found that plants fell into one of three groups according to sugar content: sucrose predominant (e.g. *Berberis*); sucrose, glucose and fructose in approximately equal amounts (e.g. *Abutilon*); and fructose and glucose predominant (e.g. many Cruciferae). A simple practical for students involves their collecting nectars from a range of garden plants, separating the sugars by PC and determining the nectar type from the proportions of the sugars present. A suitably diluted sample of honey might be included on the chromatogram, since this is of course derived from flower nectars by the bees. As a result of invertase action, the concentration of sucrose compared to glucose and fructose in the honey is low.

(b) *Procedure*

(1) Collect nectars from various flowers available locally using capillary tubes. In many cases, careful removal of the corolla will reveal a drop of nectar present on the sepal.
(2) Make three dilutions of the nectar by adding an equal amount, twice the amount and four times the amount of distilled water.
(3) Apply the nectar solutions along the start line of a paper chromatogram, leaving 2·5 cm between samples and applying a marker solution of glucose, fructose and sucrose at each end.
(4) Develop the chromatogram by descent in *n*-butanol–acetone–water (4:5:1) for 18 h.
(5) Dip the dried paper in aniline hydrogen phthalate in *n*-butanol–ether; dry again and heat in the oven at 100°C for 10 min.
(6) Note the R_F and concentration of the sugars in each nectar and determine to which nectar group each flower sampled belongs.

(c) *Further study*

PC of nectars could be repeated and results confirmed in other solvents, e.g. ethyl acetate–acetic acid–formic acid–water (18:3:1:4) or ethyl acetate–pyridine–water (8:2:1). GLC offers a more sensitive approach to detection of nectar sugars. For procedures, see Baskin and Bliss (1969) who surveyed extrafloral nectars of the Orchidaceae, Bowden (1970) who analysed the nectaries of *Andropogon gayanus,* and Rix and Rast (1975) who examined the nectars of *Fritillaria* species.

6.3   OLIGOSACCHARIDES

6.3.1   Chemistry and distribution

Most of the common plant oligosaccharides contain from two (e.g. sucrose, maltose) to five (e.g. verbascose) monosaccharide units. Even when there are only two units, these can be joined together by ether links in a number of different ways (i.e. through different hydroxyls and by α- or β-links) so that one of the main problems in oligosaccharide identification is distinguishing different isomers. In the case of disaccharides containing glucose, eight isomeric structures are possible and all are known. They can, however, be distinguished from each other by suitable chromatographic procedures (Table 6.2).

The number of oligosaccharides which accumulate as such in plants are relatively few. Sucrose (2-α-glucosylfructose) is the only one which is of universal occurrence. Fairly common are α,α-trehalose (α-glucosyl-α-glucose), raffinose(6^G-α-galactosylsucrose), stachyose (6^G-α-digalactosylsucrose) and verbascose (6^G-α-trigalactosylsucrose). Raffinose and stachyose

Table 6.2 R_G and M_G values of oligosaccharides

Sugar	R_G in			M_G in
	BAW	BEW	BTPW*	*Borate*†
Disaccharides containing glucose				
Sophorose ($\beta 1 \rightarrow 2$)	58	71	61	25
Laminaribiose ($\beta 1 \rightarrow 3$)	69	74	71	58
Cellobiose ($\beta 1 \rightarrow 4$)	50	58	51	16
Gentiobiose ($\beta 1 \rightarrow 6$)	42	50	38	63
Maltose ($\alpha 1 \rightarrow 4$)‡	54	67	60	29
Kojibiose ($\alpha 1 \rightarrow 2$)	—	—	—	30
Nigerose ($\alpha 1 \rightarrow 3$)	—	—	—	58
Isomaltose ($\alpha 1 \rightarrow 6$)	45	58	45	—
Disaccharides containing different sugars				
Sambubiose	65	76	60	25
Lathyrose	64	65	54	—
Rutinose	74	78	77	51
Neohesperidose	87	86	91	16
Robinobiose	70	73	64	31
Melibiose	35	55	34	—
Trisaccharides				
Umbelliferose	64	70	62	—
2^G-Glucosylrutinose	46	53	42	—
2^G-Xylosylrutinose	55	61	57	—
Tetrasaccharide				
Stachyose	11	25	09	—

*For solvent key, see Table 6.1. R_G = R_F relative to glucose, M_G = mobility relative to glucose.

†Borate buffer (pH 10) at 15 V cm^{-1}.

‡Isomaltose ($\alpha 1 \rightarrow 6$) is best distinguished, by NaBH$_4$ reduction to isomaltol, followed by electrophoresis.

are present, for example, in many legume seeds. One of the more uncommon oligosaccharides is umbelliferose (2^G-α-galactosylsucrose), which is mainly restricted in its distribution to members of the Umbelliferae.

A large number of the known plant oligosaccharides are not present in the free state, but occur combined to other organic molecules as plant glycosides. Among flavonoid pigments, for example (see Chapter 2), the following oligo-saccharides are frequently found: rutinose (6-α-rhamnosylglucose), neohes-peridose (2-α-rhamnosylglucose), sophorose (2-β-glucosylglucose), sambu-biose (2-β-xylosylglucose) and robinobiose (2-α-rhamnosylgalactose). Other groups of plant glycoside containing many different oligosaccharides are the saponins and steroidal alkaloids. Structures of a few oligosaccharides are illustrated in Fig. 6.3.

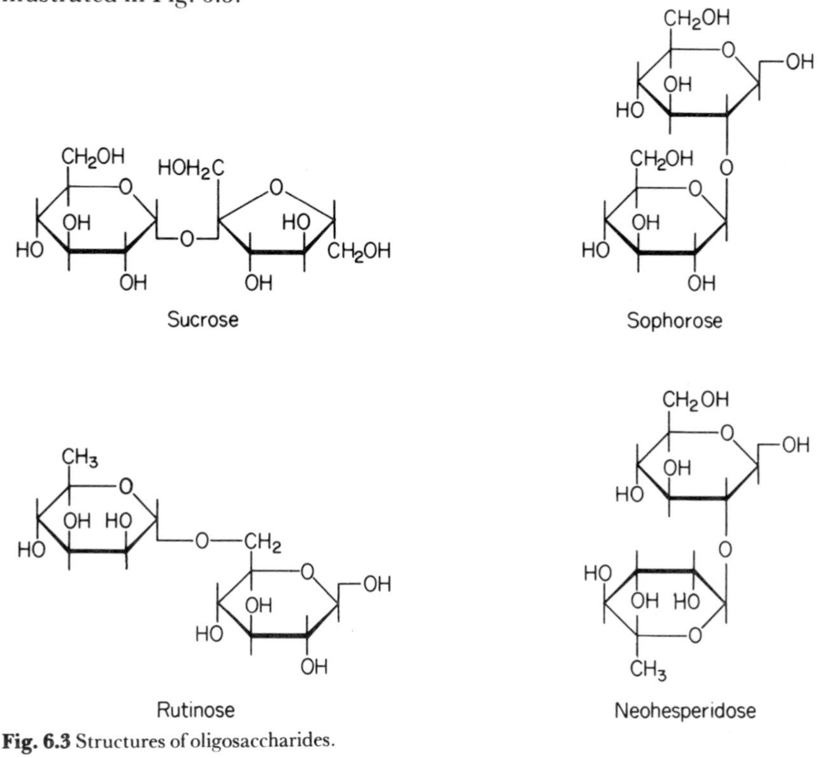

Fig. 6.3 Structures of oligosaccharides.

Oligosaccharides resemble monosaccharides in their chemical properties and methods used for their separation are largely similar.

6.3.2 Recommended techniques

The separation of oligosaccharides by PC is more difficult than that of mono-saccharides, since they are not very mobile in the usual sugar solvents. In order to overcome this difficulty, it is usual practice to allow the chromato-gram to develop for a much longer period (from 48 to 96 h), allowing the solvent at the same time to drip off the end of the paper. Since it is not possible under these conditions to measure R_F values, mobilities are related to the movement of glucose, and given as R_G values (Table 6.2). In order to be certain of separating closely similar oligosaccharides (e.g. rutinose and neo-hesperidose), it is probably essential to carry out one-dimensional paper electrophoresis in conjunction with PC. Electrophoresis in borate buffer (pH 10) at low voltages (e.g. 15 V cm^{-1}) for 2–4 h is quite satisfactory for this purpose (Table 6.2). HPLC is also useful for separating oligosaccharides (see Section 6.2.3).

Data for a range of oligosaccharides are collected in Table 6.2 and it is clearly possible to identify known oligosaccharides on a microscale by co-chromatography and co-electrophoresis with authentic materials using these systems. Such identifications should be confirmed by colour tests and other procedures indicated below.

(a) *Confirmation of identity*

Oligosaccharides can be further characterized by carrying out various colour tests using more specific spray reagents on chromatograms after development (Bailey and Pridham, 1962). For example, triphenyltetrazolium chloride (4% in methanol plus equal vol 1 M NaOH) gives a red colour with reducing oligosaccharides which have $\beta 1$ to 4 or $\beta 1$ to 6 linkages (e.g. with gentiobiose, but not with sophorose). Again, orcinol–1 M HCl reacts specifically with oligosaccharides containing ketose units (e.g. sucrose, raffinose) with the development of a red colour. Oligosaccharides may also be tested for hydrolysis with β-glucosidase and maltase, to see if they contain β- or α-linkages, respectively. Other standard techniques which can be applied, if sufficient of an unknown sugar is available, are measurements of the rate of periodate oxidation and of the optical rotation.

(b) *Authentic markers*

A reasonable range of pure oligosaccharides can be obtained from commercial suppliers. Included are gentiobiose, maltose, sucrose, laminaribiose and tre-halose. Certain others can readily be obtained by acid hydrolysis of commer-cially available plant glycosides (e.g. rutinose from rutin, see below) or from polysaccharides (e.g. nigerose from nigeran).

6.3.3   Practical experiment

(a) *Example of oligosaccharide separation: rutinose and neohesperidose*

The water-soluble bitter principles of citrus fruits are certain flavanone glycosides present in the peel (Horowitz, 1964). Not all flavanone glycosides are bitter; the major flavonoid of the tangerine orange, hesperidin (hesperitin 7-rutinoside), for example, is tasteless. The requirement for bitterness is the combination of flavanone with a particular disaccharide which is isomeric with rutinose (6-α-rhamnosylglucose); this is neohesperidose (2-α-rhamnosylglucose). The chemical structures of these two sugars are given in Fig. 6.2. Thus, two bitter glycosides are naringin (naringenin 7-neohesperidoside) and neohesperidin (hesperitin 7-neohesperidoside), both of which are present in the Seville orange and have bitterness, on a molar basis, one-fifth and one-fiftieth that of quinine.

Bitterness can be reversed by chemical alteration to the flavanone part of the molecule. By ring opening with alkali, and reduction of the isolated double bond with sodium-amalgam, the corresponding dihydrochalcone glycosides are formed and these are now intensely sweet. The compound formed from naringin has the same intensity of sweetness as saccharin and has been proposed as a replacement for saccharin in foods and fruit juices. Again, the sugar component is essential for sweetness and dihydrochalcone glycosides with rutinose instead of neohesperidose are tasteless.

The distinction between rutinose and neohesperidose is thus vital in terms of the organoleptic properties of these natural plant flavanones. Simple chromatographic experiments will clearly separate these two isomeric disaccharides. Samples may be obtained readily by hydrogen peroxide oxidation of flavonoid glycosides, which are available commercially. Rutinose may be obtained from rutin (quercetin 3-rutinoside) and neohesperidose from naringin or neohesperidin.

(b) *Procedure*

(1) Dissolve separately 2–3 mg of rutin and naringin in 1 ml of 0·1 M NH_4OH and add to each solution two drops of 100 volume hydrogen peroxide. Leave at room temperature for 4 h. During the oxidation, much of the colour of the flavonoids in alkaline solution is discharged due to oxidation of the flavonoid nucleus.

(2) These two solutions can then be spotted on to the start line of two chromatograms, at two or three different dilutions, and chromatograms developed for 48 h (use a glucose marker on each chromatogram) in BAW and BTPW. After drying, dipping in aniline hydrogen phthalate and

developing, the positions of the two isomeric disaccharides can be noted and R_Fs measured.

(3) An electrophoretogram on Whatman no. 3 paper in borate buffer (pH 10) for 4 h at 10 V cm^{-1} should also be run. This very clearly separates the two sugars (Table 6.2).

(c) *Additional experiments*

(1) Naringin is intensely bitter in aqueous solution and can still be tasted in 10^{-4} to 10^{-5} M solutions. Prepare a series of naringin solutions of decreasing concentration and determine at which strength its bitterness can no longer be detected. Note that the molecular weight of naringin is 580, but that a dried sample m.p. 171°C still contains two molecules of water of crystallization. It has a limited solubility in water, but 1 g will dissolve in 1000 ml.

(2) Naringin can be isolated very easily by extracting grapefruit peel with hot water (Ikan, 1969). Heat 1 part of chopped peel with 6–8 parts of water at 90°C for 15 min and then filter hot through celite. Concentrate the aqueous filtrate to about one-ninth of its volume *in vacuo* and allow to crystallize in the refrigerator. This yields the octahydrate, m.p. 83°C. Recrystallization from isopropanol (100 ml 8·6 g naringin) will give the dihydrate, m.p. 171°C.

6.4 SUGAR ALCOHOLS AND CYCLITOLS

6.4.1 Chemistry and distribution

The aldehyde group of the common hexoses and pentoses is readily reduced to an alcohol, by sodium borohydride or similar reducing reagents, and such a reduction is sometimes useful during identification procedures. Reduction at the anomeric carbon atom, however, alters the possibilities for isomerism and the same sugar alcohol may be formed from several different reducing sugars. Sorbitol, for example, is obtainable from glucose, gulose and fructose. Such sugar alcohols, produced in this way in the laboratory by the use of sodium borohydride, also occur fairly widely in plants. They have similar solubility and R_F properties to the common monosaccharides but do not react with some of the common sugar spray reagents and may thus be overlooked during routine plant surveys.

Glycerol (CH_2OH–$CHOH$–CH_2OH) is undoubtedly the best known sugar alcohol, but this is a building block of plant lipids and its identification is dealt with elsewhere in this volume (see Chapter 4). Of the other sugar alcohols, mannitol (by reduction of mannose) is very common in algae, fungi and lichens as well as in higher plants (Lewis and Smith, 1967a). Two which are relatively frequent are sorbitol (from glucose) and dulcitol (from galactose). Sorbitol is widely distributed in the Rosaceae (e.g. in rose hips) and dulcitol in the Celastraceae. The major function of sugar alcohols is in the storage of energy, but mannitol may also be involved in the mechanism of translocation in phloem in higher plants. Other possible functions include osmo-regulation and protection of plants from desiccation and frost damage. Structures of the three common sugar alcohols are shown in Fig. 6.4.

```
   CH₂OH          CH₂OH          CH₂OH
    |              |              |
  HOCH           CHOH           CHOH
    |              |              |
  HOCH           HOCH           HOCH
    |              |              |
   CHOH           CHOH           HOCH
    |              |              |
   CHOH           CHOH           CHOH
    |              |              |
   CH₂OH          CH₂OH          CH₂OH

  Mannitol        Sorbitol       Dulcitol
```

Fig. 6.4 Structure of common sugar alcohols.

A group of plant alcohols related to the alicyclic sugar alcohols are the carbocyclic inositols. These are alcohols based on cyclohexane which usually have six hydroxyl groups, one or more of which may be methylated (see Fig. 6.5). Optical isomerism is again a structural feature; for example, four hexahydroxy inositols are known, *myo*-inositol (or *meso*-inositol), L-inositol, D-inositol and scyllitol (or *scyllo*-inositol). One cyclitol is of universal occurrence, namely *myo*-inositol, best known as a lipid constituent in the bound form of phytic acid. Other common cyclitols are pinitol (3-methyl ether of D-inositol) present in at least thirteen angiosperm families and quebrachitol (2-methyl ether of L-inositol) present in over eleven angiosperm families. Rarer inositols are sequoyitol (5-methyl ether of *myo*-inositol), which is only found in gymnosperms, and mytilitol (2-*C*-methylscyllitol), found among plants only in the red algae. The natural distribution of cyclitols is reviewed by Plouvier (1963), but see also Loewus and Tanner (1982).

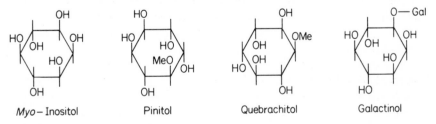

| *Myo* – Inositol | Pinitol | Quebrachitol | Galactinol |

Fig. 6.5 Structure of common cyclitols.

One inositol derivative of current interest in carbohydrate metabolism is galactinol (1-*O*-α-D-galactosyl-*myo*-inositol), first isolated from sugar beet, but since found in a number of other plants (see Fig. 6.5). It is an intermediate for the transfer of galactose units in the synthesis of such storage oligosaccharides as verbascose and stachyose.

6.4.2   Recommended techniques – sugar alcohols

(a) *Sample preparation*

Extract plant tissue with 80% ethanol under reflux, using three changes of solvent. Remove the alcohol under reduced pressure and take up the residue in water, filtering or centrifuging to remove any solid. De-proteinize if necessary, finally de-ionizing the extract with cation and anion exchange resins (Lewis and Smith, 1967b).

(b) *Paper chromatography*

Standard sugar solvents should be avoided since the alcohols tend to have similar mobilities (hexitols all have R_F *ca.* 21 in BTPW) and do not separate from reducing sugars. One general solvent to be used is *n*-propanol–ethyl acetate–water (7:1:2). The alcohols can be detected with alkaline $AgNO_3$ or with bromocresol purple (Bradfield and Flood, 1950). The $AgNO_3$ is prepared as a saturated solution in water, which is mixed with acetone in the proportions (1:200) and used as a dip. After drying, the chromatogram is then dipped into a solution of 0·5% NaOH in ethanol. Brown spots appear on a white background within 10 min at room temperature; the background can be kept white by finally washing the paper in 2 M NH_4OH and drying.

A second solvent recommended for sugar alcohols (Lewis and Smith, 1967b) is the mixture: methyl ethyl ketone–acetic acid–water saturated with boric acid (9:1:1). In order to detect the spots, it is necessary first to break down the borate complexes formed between the alcohols and the boric acid in the solvent. This can be done by dipping the dried chromatogram in a solution of hydrofluoric acid (1–4 vols of 40% acid) in acetone (40 vols). The paper should then be dried for 5–10 min at room temperature, when the sugar alcohols can be revealed by spraying with ammoniacal $AgNO_3$. They appear as brown to black spots on a white background. In the above solvents, the sugar alcohols are more mobile than most of the common sugars. If the R_F of mannitol is taken as 100, the mobilities of the simple sugars are: sucrose 12, glucose and galactose 39, fructose 65, arabinose 79 and xylose 106. Mobilities of the sugar alcohols themselves are: dulcitol 96, mannitol 100, sorbitol 116, arabitol 152 and xylitol 156.

(c) *Paper electrophoresis*

It is advisable to confirm paper chromatographic detection of sugar alcohols by low-voltage electrophoresis. This can be carried out in basic lead acetate (5·8%) solution, when the alcohols have the following mobilities, relative to ribose at 100: dulcitol 32, sorbitol 47 and mannitol 23. After electrophoresis, the alcohols are detected as yellow-green spots on a pink background, after spraying with chromium trioxide–$KMnO_4$–1 M H_2SO_4 mixture (Frahn and Mills, 1959). In making up this spray reagent, it is important to use a 2·5% stock solution of $KMnO_4$ (not 0·5% as recommended by Frahn and Mills). An alternative medium for paper electrophoresis is sodium borate buffer (0·05 M, pH 10). Relative mobilities are: glucose 100, dulcitol 97, sorbitol 83 and mannitol 91. For detection, it is necessary to pretreat the paper with hydrofluoric acid (see above).

6.4.3 Recommended technique – cyclitols

(a) *Sample preparation*

Dried plant material (1 g) (or 10 g fresh) is homogenized in a blender with 30 ml benzene and centrifuged. The residue is then homogenized with 50 ml boiling water for 10 min shaken with 100 mg charcoal, 100 mg celite, 10 ml each of cation and anion exchange resins for 10 min and centrifuged at 18 000 rev/min. The supernatant is concentrated *in vacuo* to 1–2 ml, transferred to a 5 ml narrow flask and dried. The syrup is dissolved in 100 μl water by warming over a micro-bunsen. 10 μl are used for two-dimensional TLC on microcrystalline cellulose in acetone–water (17:3) followed by *n*-butanol–pyridine–water (10:3:3). If there is a great excess of glucose present compared to the cyclitols, it can be preferentially oxidized with catalase and glucose oxidase to D-gluconic acid, which solidifies and can be removed by centrifugation.

(b) *Paper chromatography*

Cyclitols are commonly separated and identified by a combination of PC and electrophoresis (Table 6.3) (Kindl and Hoffmann-Ostenhof, 1966). Besides the two solvents given in Table 6.3 for chromatography, *n*-propanol–ethyl acetate–water (7:1:2) may be used on Whatman no. 3 paper with overnight development. Typical R_Fs in this system are: *myo*-inositol 33, (+)-inositol 47, sequoyitol 74 and pinitol 100. Cyclitols can be detected on chromatograms using the same reagents as for sugar alcohols; alkaline $AgNO_3$ is probably the most widely used.

Galactinol is less mobile than *myo*-inositol in all solvents: comparative R_F values are, for example, 25 and 31 in *n*-butanol–ethyl acetate–acetic acid–water (8:6:5:8) and 18 and 24 in ethyl acetate–pyridine–water (10:6:5).

6.4.4 Alternative techniques

TLC can be used in place of PC for polyol and cyclitol separations. For example, mannitol, sorbitol and dulcitol can be separated on silica gel in n-butanol–acetic acid–ether–water (9:6:3:1) (Hay *et al.*, 1963). Cyclitols, on the other hand, can be chromatographed on microcrystalline cellulose, using the same solvents as for paper. A suitable solvent on silica gel plates is n-butanol–pyridine–water (10:3:1). Detection on these plates may be achieved by spraying with a fresh solution of 1% lead tetracetate in ethanol. On heating for 2–3 min at 100°C, the cyclitols appear as white spots on a maroon background. Both polyols and cyclitols can be separated by GLC as their trimethylsilyl ethers, using similar techniques as for the monosaccharides.

HPLC can also be employed for separating polyols and cyclitols and, in general, similar procedures work as for the reducing sugars (see Section 6.2.3). In analysing the free sugars and polyols of crude lichen extracts, Gordy *et al.* (1978) found it advantageous to use two columns, the Waters μBondapak for carbohydrates and Aminex Qu-150S, the latter being more efficient for resolving the polyols. If complex mixtures of polyols are encountered, separation should be tried of the 4-nitrobenzoyl esters on a 5 μm silica column eluted with n-hexane–chloroform–acetonitrile (5:2:1) (Schwarzenbach, 1977).

One other technique, used for surveying plants for sugar alcohols, deserves mention for its sheer elegance. This method, pioneered by O. Kandler and his associates (see e.g. Sellmair *et al.*, 1977) involves exposing plant leaves to $^{14}CO_2$ and then detecting the incorporation of photo-assimilation in the free sugar pool by two-dimensional PC followed by radio-autography. By this

Table 6.3 R_G and M_G values of cyclitols

Cyclitol	R_G ($\times 100$) *in**		M_G ($\times 100$) *in* 0·05 M†	
	AW	BPW	Germanate	Borate
Scyllitol	36	29	20	5
myo-Inositol	39	30	70	53
D-Inositol	58	48	100	63
epi-Inositol	44	—	180	73
Methyl ethers				
Sequoyitol	77	66	50	18
Pinitol	106	98	100	66
Mytilitol	—	—	20	20
Quebrachitol	98	92	—	—

*AW = acetone–water (17:3); BPW = n-butanol–pyridine–water (10:3:3).
†90 min at 25–30 V cm⁻¹.

means, it is possible to detect sugars in leaf extracts below the level that direct PC would reveal them. A survey of 550 plant species for hamamelitol, using this method, showed it to be present only in species of the genus *Primula*.

REFERENCES

General references

Bailey, R.W. and Pridham, J.B. (1962) The separation and identification of oligosaccharides. *Chromat. Rev.*, **4**, 114–36.

Bell, D.J. (1962) Carbohydrates. in *Comparative Biochemistry* (eds M. Florkin and H.S. Mason) (Vol. III), Academic Press, New York, pp. 288–355.

Churms, S.C. (1983) Carbohydrates. in *Chromatography, Fundamentals and Applications* (ed. E. Heftmann). Elsevier, Amsterdam, pp. 223–86.

Folkes, D.J. and Taylor, P.W. (1982) Determination of carbohydrates. in *HPLC in Food Analysis* (ed. R. Macrae), Academic Press, London, pp. 149–66.

Foster, A.B. (1957) Zone electrophoresis of carbohydrates. *Adv. Carbohyd. Chem.*, **12**, 81.

Holligan, P.M. (1971) A review of techniques used for the gas liquid chromatography of carbohydrates. *New Phytol.*, **70**, 239–69.

Holligan, P.M. and Drew, E.A. (1971) Quantitative analysis by gas liquid chromatography of soluble sugars and polyols from plant tissues. *New Phytol.*, **70**, 271–97.

Lewis, D.H. and Smith, D.C. (1967a) Sugar alcohols in fungi and green plants. I. Distribution. *New Phytol.*, **66**, 143–84.

Lewis, D.H. and Smith, D.C. (1967b) Sugar alcohols in fungi and green plants. II. Methods of detection and estimation. *New Phytol.*, **66**, 185–204.

Loewus, F.A. and Tanner, W. (1982) *Plant Carbohydrates I, Intracellular Carbohydrates*, Springer-Verlag, Berlin.

Mabry, T.J., Markham, K.R. and Thomas, M.B. (1970) *The Systematic Identification of Flavonoids*, Springer-Verlag, Berlin.

Plouvier, V. (1963) The distribution of aliphatic polyols and cyclitols. in *Chemical Plant Taxonomy* (ed. T. Swain), Academic Press, London, pp. 313–36.

Supplementary references

Baskin, S.I. and Bliss. C.A. (1969) *Phytochemistry*, **8**, 1139.

Bowden, B.N. (1970) *Phytochemistry*, **9**, 2315.

Bradfield, A.E. and Flood, A.E. (1950) *Nature* (Lond.), **166**, 264.

Chari, V.M., Barkmeijer, R.J.G., Harborne, J.B. and Osterdahl, B.G. (1981) *Phytochemistry*, **20**, 1977.

Faegri, K. and Pijl, L. van der (1979) *The Principles of Pollination Ecology*, 3rd edn, Pergamon, Oxford.

Frahn, J.L. and Mills, J.A. (1959) *Aust. J. Chem.*, **12**, 65.

Gordy, V., Baust, J.G. and Hendrix, D.L. (1978) *Bryologist*, **81**, 532.

Harborne, J.B. (1960) *Biochem. J.*, **74**, 202.

Hay, G.W., Lewis, B.A. and Smith, F. (1963) *J. Chromatog.*, **11**, 479.

Horowitz, R.M. (1964) in *Biochemistry of Phenolic Compounds* (ed. J.B. Harborne), Academic Press, London, pp. 545–72.

Ikan, R. (1969) *Natural Products, a Laboratory Guide*, Academic Press, London, p. 10.

Isherwood, F.A. and Selvendran, R.R. (1970) *Phytochemistry*, **9**, 2265.

Kindl, H. and Hoffmann-Ostenhof, O. (1966) *Phytochemistry*, **5**, 1091.

McBee, G.G. and Maness, N.O. (1983) *J. Chromatog.*, **264**, 474.

Percival, M.S. (1961) *New Phytol.*, **60**, 235.

Rix, E.M. and Rast, D. (1975) *Biochem. System. Ecol.*, **2**, 207.

Schwarzenbach, R. (1977) *J. Chromatog.*, **140**, 304.

Sellmair, J., Beck, E., Kandler, O. and Kress, A. (1977) *Phytochemistry*, **16**, 1201.

Selvendran, R.R. and Isherwood, F.A. (1967) *Biochem. J.*, **105**, 723.

Sweeley, C.C., Bentley, R., Makito, M. and Wells, M.M. (1963) *J. Am. Chem.*, **85**, 2497.

CHAPTER SEVEN

Macromolecules

7.1 Introduction
7.2 Nucleic acids
7.3 Proteins
7.4 Polysaccharides

7.1 INTRODUCTION

The macromolecules of plants are distinguished from all other constituents by their high molecular weight. This may vary from 10000 to over 1000000, whereas in other plant metabolites the molecular weight is rarely above 1000. Chemically, macromolecules consist of long chains of small structural units or 'building blocks', linked covalently in a number of different ways. Chemical characterization in the first instance therefore depends on identifying these smaller units. Proteins, for example, are long chains of amino acids (up to twenty different ones) joined together through peptide (–CO–NH–) links. Polysaccharides are similarly derived from the union of simple sugar units, such as glucose, joined through ether (–O–) links. The nucleic acids, by contrast, are more complex and have three types of structural unit: purine and pyrimidine bases, pentose sugars and phosphate groups. The three main classes of macromolecules found in plants are thus proteins, polysaccharides and nucleic acids. However, mixed polymers are also known, such as the glycoproteins, which contain both sugars and amino acids in covalent linkage.

While the above polymers all have an ordered structure, there are compounds peculiar to plants which are random polymers, formed by oxidative, non-enzymic polymerization of simple phenolic units. There are three main groups of these: the lignins, derived from phenylpropanoid units; the condensed tannins, derived from flavonoid units; and the plant melanins, derived from the polymerization of catechol. Methods for their analysis have already been described in Chapter 2.

Most of the macromolecules of plants function in a similar way to those of animal organisms. Thus, DNA and RNA are involved in the storage and transcription of genetic information, proteins act as catalysts in enzymic reactions, polysaccharides provide a storage form of energy and so on. Some of the macromolecules of plants, however, have a function specifically related to plant growth. For example, the synthesis of lignin, a polymer which, together with cellulose, forms a complex matrix in the cell wall, provides the plant with the ability to develop a rigid stem or trunk.

Methods of separating and identifying plant macromolecules differ in a number of ways from those used with low-molecular-weight constituents. For example, it is often difficult to solubilize the polymers and special procedures may be needed. In a simple case, macromolecular constituents may be dissolved by homogenizing plant tissue with salt solution and then precipitated by changing the pH of the extract. Because the concentration of some macromolecules (particularly nucleic acids) is low in plant tissues generally, it may be advantageous to disrupt the plant cells and separate the different organelles by centrifugation, before isolation. The mitochondrial fraction, thus separated, will be rich in enzymic protein, while the nuclear fraction should be rich in nucleic acid.

Because of the structural complexity of macromolecules in the living cell, any method of isolation will almost certainly cause some degradation or artifact formation, unless special precautions are taken. Enzymic protein is particularly susceptible to degradation and loss of activity during isolation procedures and it is important to avoid extremes of pH and of temperature. Again, nucleic acids are difficult to obtain pure because of their close association with histone proteins in the nucleus.

The study of macromolecules, whether from plant or animal tissues, is fundamental to an understanding of molecular biology and much has been written elsewhere, in general terms, of methods required for their characterization. There are also many detailed texts devoted to the methodology of isolating proteins and nucleic acids, mainly from animal but also from plant tissue. In the space available here, it will not be possible to cover all aspects of this vast subject. This account will be concerned mainly with the problems particularly associated with the isolation of macromolecules from plant, as opposed to animal, tissues. Comparative aspects of plant polymers will also be discussed in some detail.

7.2   NUCLEIC ACIDS

7.2.1   Chemistry and distribution

The basic chemistry of the nucleic acids, DNA (deoxyribonucleic acid) and RNA (ribonucleic acid), is so well known that only the main facts need to be re-iterated here. Both DNA and RNA are high-molecular weight polymers containing phosphate, four nitrogen bases and a pentose sugar, either deoxyribose (in DNA) or ribose (in RNA). DNA contains two pyrimidine bases, cytosine and thymine and two purines adenine and guanine. RNA has the same base complement except that the pyrimidine uracil replaces the thymine of DNA. A few other bases are also occasionally found, mainly as trace constituents, in nucleic acids. For example, plant DNAs tend to have a significant proportion (up to 30%) of 5-methylcytosine in place of cytosine. Formulae of the nucleic acid bases are given in Fig. 5.8 (p. 209).

The nitrogen bases in the nucleic acids are joined via sugar linkages to long chains of alternating sugar and phosphate. The three-dimensional structure of DNA, based on the original proposal of Watson and Crick, consists of a two-stranded α-helix with each strand consisting of a long chain of poly-nucleotides and the strands joined through the bases by hydrogen bonding. The two strands are complementary in their base sequence, since adenine in one chain is always paired with thymine on the other (and *vice versa*) and similarly, guanine is paired with cytosine (see Fig. 7.1).

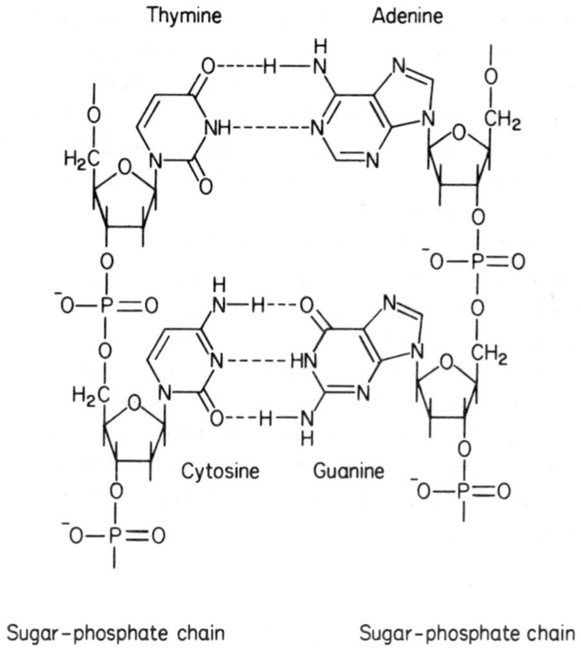

Fig. 7.1 The basic structure of deoxyribonucleic acid.

DNA as it occurs in the chromosomal material of the nucleus is closely associated with a particular type of protein, called histone, and this has to be removed before the properties of DNA can be studied in full. Besides being present in the nucleus, DNA also occurs in chloroplasts and mitochondria of plant cells and in both these cases, the base composition is different from that of the corresponding nuclear DNA.

Less is known of the three-dimensional structure of RNA, but it is assumed to be somewhat similar to that of DNA. There are three types of RNA: transfer (or soluble), messenger and ribosomal, all with different functions in the transfer and transcription of genetic information within the cell. The molecular weight (mol. wt.) of RNA is generally lower than that of DNA. Molecular weight variation in the nucleic acids ranges from 20 000 to 10 000 000.

The sequence of bases in DNA is now known to 'code' for protein synthesis in such a way that different triplets of bases determine, in turn, the amino acid sequence of that protein. It follows that since the sequence of bases in the chain of DNA is the source of genetic information, this sequence must include at least parts which are different in different organisms. A study of such differences can thus provide a measure of the ultimate taxonomic distance between organisms. Direct comparison of the DNA of different plants is, however, confounded by the fact that the nuclear DNA is highly complex, consisting as it does of some single copy or unique DNA together with considerable amounts of repetitive DNA, identical sequences of varying length which occur in varying numbers. It has been estimated that only 1% of the nuclear DNA of the higher plant cell is concerned in coding for protein synthesis. Because of the presence of repetitive DNA, it may be advantageous when comparing sequences of different organisms to separate single copy from repetitive DNA at an early stage. By contrast with nuclear DNA, chloroplastic and mitochondrial DNAs are simple in organization, consisting of single circular threads, as in bacteria, and direct comparison of sequences is immediately possible.

There are three ways of measuring differences in DNA between organisms. The first and simplest is to compare base composition. The second is to determine the ability of single-stranded DNA from one organism to 'complement' with single-stranded DNA of another. The third is to compare actual base sequences or partial sequences.

The first method of comparing DNAs requires a simple measurement of the amounts of the different bases formed on hydrolysis. Since cytosine (C) and guanine (G) always complement each other, they are present in equal amounts; adenine (A) complements with thymine (T) and these bases also occur in equivalent concentrations. The amounts of the two pairs (G+C) and (A+T), however, vary in different DNAs. These variations are expressed either as a percentage of the total base content, i.e. as (G+C)%, or as the fraction

$$(G+C)/(A+T).$$

The (G+C)% composition of DNA varies between wide limits in the bacteria, from 30% in *Clostridium*, 50% in *Escherichia coli* to 80% in *Streptomyces*, and the measurement is a useful one in taxonomic studies (De Ley, 1964). There is also a wide range in (G+C)% among the fungi and correlations with existing taxonomic classification have been noted (Storck and Alexopoulos, 1970). In land plants, the variation is much less. In angiosperms, the range is between 36 and 43% (G+C) with rather higher average values (48–49%) in the family of grasses, the Gramineae (Biswas and Sarkar, 1970). The DNAs of ferns and allied forms have similar (G+C) compositions (37–41%), with those of *Selaginella* showing slightly higher values (45–50%) (Green, 1971).

Base compositions can also be measured among RNAs, the only major difference being the replacement of thymine by uracil. Comparisons have been made of (G+C/A+U) ratios among the large and small subunits of ribosomal RNA of a range of organisms and significant differences have been noted at the level of phylum and family in the plant kingdom (Loening, 1973).

The second method of comparison requires the isolation of pure DNA from the different organisms and the separation of single copy from repetitive sequences. These DNAs are then sheared to break them into smaller pieces and then converted to single-stranded DNA by warming them in solution; the degree of complementation occurring between single-stranded DNAs of two species can then be determined by mixing them and allowing them to recombine, the amount of recombination being measured by changes in UV absorbance. Such DNA–DNA re-association experiments have been conducted using single-copy DNA by Belford *et al.* (1971) and using repetitive DNA by Flavell *et al.* (1977).

The third method of comparison is by sequence analysis. Methods for sequencing nucleic acids have been reviewed (Marcus, 1981; Hindley and Studen, 1983). In the case of higher plants, relatively few DNA sequences are as yet available. Some progress has been made with mitochondrial and chloroplastic DNA and the DNA coding for the large subunit of RUBP carboxylase for maize, *Zea mays*, has been sequenced (McIntosh *et al.*, 1980). A simpler approach than full sequencing is to fragment chloroplast DNA, using different restriction endonucleases, enzymes which recognize and cleave the DNA duplex at specific base pair sequences from 2 to 8 base pairs long. The fragments so produced can be separated by gel electrophoresis. The chloroplast DNA from related plants will produce different sets of fragments, according to the number of base pair substitutions which separate them. This procedure has been successfully applied to DNA comparisons among species of *Lycopersicon* and *Atriplex* (Palmer *et al.*, 1983).

7.2.2 Recommended techniques

(a) *General outline*

Isolation of nucleic acids from plant tissues normally presents a number of problems not associated with similar separations from microbial or animal systems. The methodology of extraction and purification has been extensively explored in recent years (Ingle, 1963; Holdgate and Goodwin, 1965; Cherry and Chroboczek, 1966; Hall and Davies, 1979) and several alternative procedures are now available. The method chosen may always have to be modified to remove interfering contaminants present in the particular plant under study. Isolation requires a series of steps, which can be carried out in slightly different sequences, and these are discussed in the following paragraphs.

Choice of tissue is important, particularly when DNA is being studied, since it occurs in such low amounts in leaf tissue. The DNA content of pollen, root tips or seed is generally much higher than the leaf. For RNA isolation, leaf tissue is satisfactory since there is usually about 0·01% dry weight present.

A major problem is separation from other macromolecules, especially protein and polysaccharide, and also from RNA, if DNA is being isolated. The enzymes which break down nucleic acids, namely DNase and RNase, can be deactivated by plunging the fresh tissue into boiling ethanol.

Protein is removed by denaturing or coagulating it with trichloracetic or perchloric acid. Three successive extractions with 10% trichloracetic acid (TCA) at 0°C are generally sufficient to dissolve away protein and any non-nucleic acid phosphorus compounds from plant material. Protein may also be eliminated by high speed homogenization in the presence of chloroform–ethanol (8:1).

Lipid is best removed by washing at this stage with ethanol saturated with sodium acetate, which also gets rid of traces of TCA left over from the protein extraction. Further washings with chloroform and chloroform–ethanol will dissolve any remaining lipid.

Nucleic acids are released from the residue left from previous extractions by either using phenol–buffer or a neutral salt solution (e.g. 0·1 M NaCl).

A dialysis may be necessary at some stage to remove any small molecules present as contaminants. Other procedures may occasionally be required to remove such compounds. Chlorogenic acid, for example, which is present in large amount in coffee beans, forms a green complex with magnesium ion and may interfere with nucleic acid isolation (De *et al.*, 1972). Its presence in nucleic acid fractions can be prevented by adding 0·05% EDTA to the extraction medium.

The methods described below refer to the isolation of total DNA. When studying the DNA of individual organelles, fractionation of the cell constituents by centrifugation is a necessary preliminary to isolation. Because of differences in molecular weight, it may be possible to separate mitochondrial and chloroplastic from nuclear DNA at a later stage from a total cell extract. Centrifugation at high speeds for long periods in 8 M caesium chloride is used to separate different DNAs according to their molecular weights or buoyant densities.

(b) *Measurement of DNA and RNA in leaf tissues*

The following procedure is taken from Detchou and Possingham (1972). Fresh tissue (100 mg) is homogenized in 1 ml 5% $HClO_4$ at 0°C for 1 min at 350 rev min^{-1}. After 15 min the homogenate is centrifuged at 2000 g for 10 min, the supernatant then being discarded. The residue is re-extracted twice with 5% $HClO_4$ at 0°C to remove the last traces of soluble phosphate. The residue is then extracted (×4) with 1 ml ethanol–ether–chloroform (2:2:1) at room temperature to remove phospholipid. The final residue is dried and then hydrolysed in 0·5 ml 5% TCA at 90°C for 30 min.

The total nucleic acid is obtained by measuring the absorbancy of the hydrolysate at 268·5 nm against a suitable blank. Total content is calculated using an appropriate conversion factor (Logan *et al.*, 1952). The DNA content is determined by reaction of the hydrolysate with diphenylamine (Burton, 1956; see also below). The RNA is then calculated as the difference between the DNA and the total nucleic acid content.

(c) *Extraction of RNA from seedling tissue*

This procedure is that of Kirby (1965) as modified by Koller and Smith (1972). The tissue is homogenized in 4 ml 0·01 M tris-HCl buffer, pH 7·4, containing 50 mmol NaCl and 1% *p*-aminosalicylic acid. An equal volume of water-saturated phenol, containing 0·1% 8-hydroxyquinoline and 14% *m*-cresol, is added and the whole centrifuged.

The phenol layer is removed and 1 ml 1·3 M NaCl and 4 ml phenol–cresol added and the mixture is recentrifuged. The tris layer is removed and re-extracted with 5 ml phenol–cresol. The final tris layer is made 0·02 M with respect of sodium acetate and the RNA is precipitated with 3 vols of ethanol at 0°C. The precipitate is washed with 80% ethanol, centrifuged and dissolved in 2 ml 0·15 M sodium acetate containing 0·25% sodium lauryl sulphate. The pure RNA is finally precipitated from this solution by adding 6 ml ethanol at −20°C and standing for 1 h.

A procedure for the isolation of RNA from banana pulp is described by Wade *et al.* (1972).

(d) *Isolation of DNA from leaf tissue*

This procedure is from Bolton and Britten (1965); see also Bolton and Bendich (1967). Fresh seedlings (10 g), from which the cotyledons have been removed, are minced and ground in a mortar for 1 min. 10 ml of a solution containing 1% sodium dodecyl sulphate, 0·1 M EDTA–sodium salt and 0·3 M NaCl are added and the grinding continued for 1 min. The thick paste is shaken for 30 s with 20 ml chloroform containing 1% octanol and centrifuged. The upper aqueous layer, which contains the DNA, is heated to 70°C for 5 min, cooled quickly to room temperature and is adjusted to contain 1 M $NaClO_4$. It is again shaken with chloroform–octanol and centrifuged. The aqueous layer is placed in a beaker and two vols 95% ethanol are layered over the extract and the DNA can be recovered by winding on to a glass rod. If it will not wind efficiently, the DNA can be collected by centrifugation. The yield is 2–3 mg in the case of pea seedlings.

The DNA can be further purified by dissolving it in 0·15 M NaCl–0·015 M Na citrate and adding in sequence α-amylase (to remove starch), ribonuclease (to remove RNA) and protease (to remove residual protein). After a clean-up by solvent partition, the DNA is finally recovered as a pellet by centrifugation (Smith and Flavell, 1974).

(e) *Other techniques*

More elaborate separation techniques than those described above have been applied to the purification of nucleic acids, to the separation of RNA from DNA and to the fractionation of the different molecular weight components within the total DNA or RNA material. Polyacrylamide gel electrophoresis has been applied (Loening, 1967), the gels being scanned at 265 nm with a chromoscan recording densitometer in order to detect the separated bands. Fractionation of nucleic acids is commonly carried out on columns of methylated albumin Kieselguhr, ECTEOLA–cellulose, DEAE–cellulose or DEAE–Sephadex (Cohn, 1967; Grossman and Moldave, 1968). For other methods of purifying DNA, see the recent review by Cowling (1983). The hydrolysis of nucleic acids and identification of their pyrimidine and purine bases is given in Chapter 5, p. 213.

(f) *Authentic markers*

Samples of DNA and RNA, which can be used as standards in testing out the efficiency of purification procedures, are available commercially.

7.2.3   **Practical experiment**

(a) *Nucleic acid from cauliflower heads*

A simple procedure for the extraction and detection of nucleic acids present in cauliflower florets is described as a class experiment by Witham *et al.* (1971). It is suitable for obtaining experience in isolation techniques and can be carried out in 2–3 h. The procedure, with slight modification, is as follows.

(b) *Procedure*

(1) Cauliflower (100 g) is homogenized in a blender with an equal vol of 95% ethanol for 4 min. The homogenate is filtered on a suction pump. The residue is re-extracted with a further 100 ml ethanol and refiltered.

(2) The residue is then homogenized in a blender with 100 ml acetone for 4 min. After filtering, the residue is dried between filter papers.

(3) The dried residue is suspended in 100 ml ice-cold 5% $HClO_4$, filtered and the filtrate discarded. The residue is then washed, first with 100 ml ethanol and then 100 ml acetone and finally redried.

(4) The dried residue is extracted with 100 ml 10% NaCl, previously adjusted to pH 7·5 with $NaHCO_3$, by heating at 100°C for 15 min. The cooled solution is filtered and the filtrate should contain most of the RNA.

(5) The residue from the NaCl extraction is then treated with 100 ml 0·5 M $HClO_4$ at 70°C for 20 min. The residue is then discarded and the solution should contain the DNA.

(6) DNA can be tested for in the final extract by reaction with diphenylamine. The reagent is freshly prepared by adding 2·75 ml conc.H_2SO_4 to 1 g diphenylamine dissolved in 100 ml acetic acid. DNA solution (1 ml) (containing approx. 5 mg 100 ml^{-1}) is heated with 2 ml diphenylamine reagent at 100°C for 10 min and a blue colour should develop.

(7) RNA can be tested for in the NaCl extract by reaction with orcinol. The reagent is freshly prepared by dissolving 1 g orcinol in a solution of 0·5 ml $FeCl_3$ (10%) diluted to 100 ml with conc.HCl. RNA solution (1 ml) and 1 ml reagent are heated at 100°C for 10 min and a deep green colour should develop.

7.3 PROTEINS

7.3.1 Chemistry and distribution

The proteins in plants, as in other organisms, are high-molecular-weight polymers of amino acids. The amino acids are arranged in a given linear order, as determined by the triplet base code of the DNA in the nucleus, and each protein has a specific amino acid sequence (see Fig. 7.2). In the simplest cases, a protein may consist of a single chain of polypeptide. Several identical chains may however, form a higher-molecular-weight hydrogen-bonded aggregate and it is sometimes difficult to determine the exact form in which the protein exists *in vivo*.

Because of the stereochemistry of the peptide bond, a polypeptide does not usually have a random form but the chain is often coiled in the form of an α-helix and this, in turn, can fold on itself and adopt a particular three-dimensional shape. This is described as the secondary and tertiary structure of

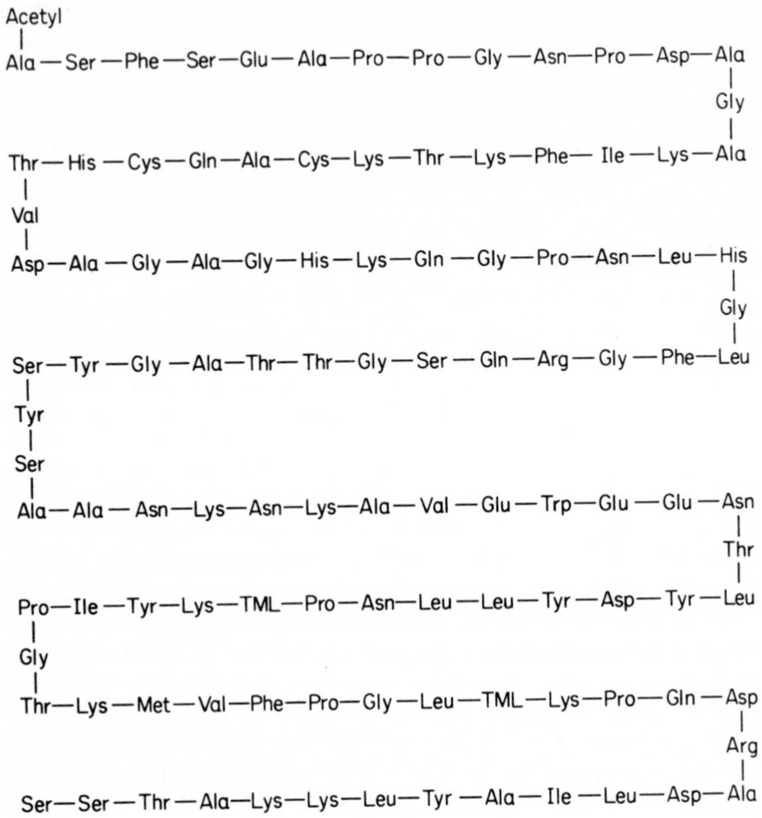

Acetyl
|
Ala — Ser — Phe — Ser — Glu — Ala — Pro — Pro — Gly — Asn — Pro — Asp — Ala
                                                                          |
                                                                        Gly
                                                                          |
Thr — His — Cys — Gln — Ala — Cys — Lys — Thr — Lys — Phe — Ile — Lys — Ala
  |
Val
  |
Asp — Ala — Gly — Ala — Gly — His — Lys — Gln — Gly — Pro — Asn — Leu — His
                                                                          |
                                                                        Gly
                                                                          |
Ser — Tyr — Gly — Ala — Thr — Thr — Gly — Ser — Gln — Arg — Gly — Phe — Leu
  |
Tyr
  |
Ser
  |
Ala — Ala — Asn — Lys — Asn — Lys — Ala — Val — Glu — Trp — Glu — Glu — Asn
                                                                          |
                                                                        Thr
                                                                          |
Pro — Ile — Tyr — Lys — TML — Pro — Asn — Leu — Leu — Tyr — Asp — Tyr — Leu
  |
Gly
  |
Thr — Lys — Met — Val — Phe — Pro — Gly — Leu — TML — Lys — Pro — Gln — Asp
                                                                          |
                                                                        Arg
                                                                          |
Ser — Ser — Thr — Ala — Lys — Lys — Leu — Tyr — Ala — Ile — Leu — Asp — Ala

For amino acid abbreviations, see Fig. 5.1, p. 178, TML = $\epsilon - N$-trimethyl-lysine

Fig. 7.2 Amino acid sequence of the protein of cytochrome *c* from wheat germ.

the protein. Many proteins are, in fact, roughly rounded in shape and hence are called globular proteins.

Proteins which contain other structural elements besides amino acids are classified as conjugated proteins. The nature of the linkage between the polypeptide chain and the other structural moiety is not always precisely known. A simple conjugated protein is one which contains a metal such as iron, copper or molybdenum (perhaps 3 or 4 atoms per molecule), complexed in its structure. Other conjugated proteins (Table 7.1) contain lipid, phosphate, carbohydrate or nucleic acid. Chromoproteins have a chromophoric group attached and are coloured. The respiratory enzyme, cytochrome *c*, for example, is yellow and consists of a polypeptide chain (see Fig. 7.2) linked to a haem (iron–porphyrin) chromophore.

Table 7.1 The main classes of plant proteins

Class name	Properties	Example(s)
SIMPLE PROTEINS		
Albumins	soluble in water and dilute salt, coagulated by heat	proteins of legume seeds, e.g. vicilin, legumin,
Globulins	insoluble in water but soluble in salt	arachin
Glutelins	soluble in dilute acid and alkali	proteins of cereal seeds,
Prolamines	soluble in 70–80% ethanol	e.g. gliadin
Histones	soluble in water but insoluble in dilute NH_4OH; rich in lysine and arginine	proteins in the cell nucleus
CONJUGATED PROTEINS		
Chromoproteins	coloured; contain a chromophoric attachment	phytochrome, cytochromes
Lipoproteins	contain lipid	proteins of the chloroplast
Metalloproteins	complexed with metal such as Fe, Co, Zn, Cu, Mo	nitrate reductase, phenoloxidases
Glycoproteins	contain carbohydrate	barley albumin

Plant proteins were at one time specifically classified (Osborne, 1924) on the basis of solubility properties and this is still a useful criterion for distinguishing different types of storage protein (Table 7.2). The classes are separated according to their solubility or lack of solubility in water, aqueous acid and alkali, and 70% alcohol. Seed proteins are still conveniently fractionated on the basis of differential solubility in aqueous solvents, although complete separation of two protein classes is rarely achieved in a single operation.

An alternative system of classification is according to molecular weight. This can vary from 12000 in cytochrome *c*, and 60000 in alcohol dehydrogenase to 500000 in urease. Molecular weights of over 1000000 are also

Table 7.2 Purification of the enzyme RuBP carboxylase from comfrey leaves

Procedure	Protein* (mg)	μmol $^{14}CO_2$ fixed min^{-1} mg protein^{-1}	Degree of purification
1. Buffer extraction	2455	0·05	1
2. 40% $(NH_4)_2SO_4$ ppte	107	0·54	11
3. Pooled sucrose gradient	60	0·97	20
4. DEAE–cellulose chromatography	26	1·3	26

*From 100 g fresh weight deribbed leaves of comfrey, *Symphytum officinale*.

known. Proteins separate according to molecular weight when they are subjected to gel filtration on a column of Sephadex, a cross-linked dextran. The higher the molecular weight is, the faster is the movement of the protein down the column. Separation on Sephadex is so precise that the technique can be used as a means of determining the molecular weight of a new protein (see p. 258). Separation of proteins by gel electrophoresis is also partly determined by their molecular size, although the charge properties of the protein are equally important in determining their mobility on the gel. The net charge on a protein depends on the number of basic and acidic amino acids present in the polypeptide chain. Most plant proteins have more acidic than basic amino acids and thus have a net negative charge and move towards the anode.

A third method of protein classification is on the basis of function. Many proteins are also enzymes, catalysing particular steps in either primary or secondary metabolism, and can clearly be distinguished from other proteins by their enzymic properties. All enzymes are, of course, not necessarily functional and the purpose of some of the enzymes most easily detected in plant tissues, the peroxidases and esterases, is still not entirely clear. On the other hand, non-enzymic protein is definitely known to be used either for storage purposes, particularly in seeds and roots, or as part of the internal structure of the cell, in the membrane or in the cell wall.

Plant proteins have a few special features which distinguish them from proteins in animals. They are, in general, relatively low by comparison in their content of sulphur amino acids (methionine and cysteine). They also occasionally contain amino acids other than the so called twenty protein amino acids. Plant (but not animal) cytochrome c may have the compound trimethyl-lysine present as one of its amino acid constituents. Again, hydroxyproline is found as an amino acid constituent of the cell wall protein in many plant tissues. From the nutritional viewpoint, some cereal proteins and notably the prolamins are undesirable because of their relatively low content of the basic amino acid lysine. This problem has been rectified by breeding cereal varieties, the so-called 'high lysine' strains, in which there is a better balance of protein types and in which prolamin synthesis is partly suppressed. Another feature of certain plant proteins is their deleterious effect on mammalian systems; two examples are the toxic abrin from *Abrus precatorius*, which agglutinates the red blood cells, and a protein present in soya bean, which inhibits the activity of the digestive enzyme trypsin.

Proteins occur throughout the plant in all types of tissue and even a simple organ like the leaf may contain several thousand, mainly enzymic, proteins. The problem of separation and isolation of a particular individual protein can therefore be a formidable one. It is often preferable to work with plant tissue which is rich in protein, and this is the reason why many protein studies have been based on the seeds of legume plants such as beans and peas. For isolating enzymes, it is wise to employ very young seedling tissue, since this tends to have a particularly active enzyme complement.

Although the amino acid composition of the total leaf protein is practically identical whatever the plant source, plants do vary in their protein make-up. Particular enzymes involved in secondary metabolism may be completely absent from many plants. The enzyme myrosinase, for example, which hydrolyses glucosinolates to yield mustard oils, is only found in those plants which synthesize glucosinolates (see p. 169). Other enzymes involved in intermediary metabolism may show variation in their primary amino acid sequence, particularly since it is known that only part of that sequence is essential for enzymic activity. Variation in primary sequence has been extensively explored in the case of three electron-carrier proteins in plants, namely cytochrome *c*, ferredoxin and plastocyanin. Partial or complete sequence data are also available for some enzymes (e.g. ribulose 1,5-bis-phosphate carboxylase), for protease inhibitors and for storage proteins (see Boulter and Parthier, 1982).

There are many general reference books on proteins, which may deal *inter alia* with those specifically present in plants. Books on plant proteins include those of Boulter and Parthier (1982), Harborne and van Sumere (1975) and Norton (1978). There is also a book specifically on seed proteins (Daussant *et al.*, 1983). Plant enzymes are discussed in detail in Volumes V and VI of *Modern Methods of Plant Analysis* (Linskens and Tracey, 1963, 1964) and methods for their extraction are reviewed by Rhodes (1977). An excellent general book on protein methodology is that of Bailey (1967). Methods of protein sequence analysis are reviewed by Croft (1980).

7.3.2   Recommended techniques

(a) *General considerations*

In isolating proteins from plant tissues, there are four particular hazards which must be dealt with: the highly acidic cell sap and the presence of phenolase, tannins and proteolytic enzymes. The effects of the acid cell sap (pH 2-4), which could denature a protein, can be avoided by macerating the tissue in the presence of a neutral (pH 6-8) buffer. The phenolases liberated on maceration react with phenols in the presence of oxygen to give quinones, which then react irreversibly with protein to form brown pigments. The deleterious effect of phenolase can be counteracted by adding a reducing compound, ascorbic acid or methyl mercaptan, or by carrying out the protein extraction in a nitrogen atmosphere. The third hazard, natural tannin, which complexes readily with protein, can be removed by absorbing it into polyvinyl-pyrrolidone (PVP, Polyclar-AT), which can be added to the buffer during preliminary extraction. Finally, the effects of plant proteolytic enzymes can only be avoided by carrying out the isolation as quickly as possible and using low temperatures (0-4°C).

An alternative to buffer extraction, which is used particularly for isolating enzymes from plants, is to prepare an acetone powder. This is done by macerating the tissue in an excess of very cold acetone at $-20°C$. This has the advantage that, with leaf tissue, the acetone removes chlorophyll and most of the low-molecular-weight constituents. The dried powder produced can be used directly for enzyme assay or else extracted with buffer and the enzyme further purified.

After the preliminary extraction, the protein, enzymic or otherwise, is purified by a series of fractionations, the purpose of which is to remove other proteins and also other high-molecular-weight components. For example, the protein might be purified by ammonium sulphate fractionation, followed by ion exchange chromatography on DEAE–cellulose and finally by gel filtration on Sephadex. The actual method used will depend on the plant source and the type of protein being isolated. For purposes of illustration, two examples of isolation and purification procedures will be given here, one for enzymic protein and the other for storage protein.

Once a protein has been obtained pure, it is usual to determine its amino acid composition, nitrogen content and molecular weight. It may also be characterized by a number of other analytical procedures. Only a few of these procedures can be mentioned here, attention being given to protein estimation, molecular weight determination and gel electrophoresis. For detailed accounts of other procedures, see Bailey (1967).

(b) *Separation of RuBP carboxylase from angiosperm leaves*

A good illustration of current procedures needed to purify plant enzymes is that of the isolation of ribulose 1,5-bisphosphate carboxylase (RuBP carboxylase), the key enzyme of photosynthesis, from leaves of comfrey (Simpson *et al.*, 1983). Compare also the method of Reger *et al.* (1983) for purifying the same enzyme from maize leaves. The properties of this enzyme from different plants are summarized in the review of Akazawa (1979). In brief, the isolation from comfrey involves (1) buffer extraction, (2) ammonium sulphate fractionation, (3) centrifugation in a sucrose gradient and (4) column chromatography on DEAE–cellulose. The degree of purification achieved at each stage is indicated in Table 7.2.

(1) 50 g samples of fresh leaves are homogenized for 30 s in 600 ml buffer A, containing 50 mM Tris–HCl, 50 mM $NaHCO_3$, 10 mM $MgCl_2$, 5 mM $Na_2S_2O_3$, 1·0 mM EDTA and 0·1 M PMSF in the presence of 48 g insoluble PVP.

(2) After filtration and centrifugation, the supernatant is saturated to 40% with $(NH_4)_2SO_4$ (244 g l^{-1}) and then centrifuged at 1300 g for 30 min. The pellets are redissolved in 20 ml buffer A and recentrifuged at 13 000 g for 30 min.

(3) 3 ml of the supernatant are layered on top of each of six Polyallomer tubes containing 34 ml of a linear $0 \cdot 2 - 0 \cdot 8$ M sucrose gradient prepared in buffer B (same as A, except that 2 mmol dithiothreitol replaces $Na_2S_2O_3$). This is spun for 20 h at 270 000 rev min^{-1} and 1 ml fractions are collected from the bottom and the enzyme located by its absorbance at 280 nm.

(4) The appropriate fractions are placed on a $3 \times 1 \cdot 5$ cm DEAE–cellulose column equilibrated beforehand in buffer B and then the enzyme is eluted from the column with buffer B made $0 \cdot 25$ M with NaCl. Enzyme activity is monitored at each stage (see Table 7.2) by incubation in the presence of RuBP and radioactive $NaHCO_3$ and measurement of the $^{14}CO_2$ fixed.

For general hints on extracting and purifying other plant enzymes, see the review by Rhodes (1977).

(c) *Isolation of seed proteins*

Legumin and vicilin, the two main globular proteins of legume seeds, may be isolated as follows: The seeds, e.g. peas, *Pisum sativum*, are ground to a fine powder and extracted overnight with 1 M NaCl buffered to pH 7 with sodium phosphate. After centrifuging, the supernatant is saturated with ammonium sulphate to 70% saturation and the precipitate collected and dried. This crude protein is then fractionated by extraction with acetate buffer pH 4·5 containing $0 \cdot 2$ M NaCl. After 24 h the heavier component (legumin, mol. wt. 330 000) is collected by centrifugation, dissolved in buffer pH 7 and dialysed against water, when it separates out. The supernatant is also dialysed against water and the protein vicilin (mol. wt. 186 000) eventually precipitates. The acetate buffer extraction and centrifugation must be repeated two or three times to complete the purification of the two globulins.

When isolating proteins from seeds which have a high oil content, a preliminary defatting is necessary. For example, groundnuts, *Arachis hypogaea* require maceration in light petroleum (200 ml 50 g nuts^{-1}). The protein is then extracted with $0 \cdot 1\%$ NaOH, the solution filtered and the protein precipitated by acidification to pH 5. The protein globulin is filtered, washed and dried. It can be fractionated into two components, arachin and conarachin by extraction with 10% NaCl. The solid is removed and the supernatant diluted with water (4 vols) to precipitate the arachin. Heating the aqueous filtrate coagulates the conarachin, which can then be collected and dried (Ikan, 1969).

General procedures for these and other plant seeds can be found in Paech and Tracey (1955b). The isolation of albumins and globulins from crucifer seed is described by Vaughan *et al.* (1966). The use of ascorbic acid in the extraction medium for isolating legume proteins is discussed by Wright and Boulter (1973) and McLeester *et al.* (1973). A final word of caution: PVP should be added to the extraction medium whenever isolating protein from seeds with a significant tannin content.

(d) *Estimation of protein*

The most widely used method for determining the protein content of any plant preparation is that of Lowry *et al.* (1951). This is a colour reaction between the Folin–Ciocalteu reagent and the peptide bond. A solution of the sample (0·1–0·2 ml) (which should contain between 10 and 100 μg protein) is mixed with 1 ml of a reagent (made by adding 1 ml 0·5% $CuSO_4.5H_2O$ in 1% sodium citrate to 50 ml 2% Na_2CO_3 in 0·1 M NaOH) and the solution is left to stand for 10 min at room temperature. 0·10 ml of the Folin–Ciocalteu reagent (commercial reagent diluted with water to give a 1 M acid solution) is pipetted in, mixed and the absorbance measured at 750 nm after 0·5–2 h. A standard curve relating absorbance at 750 nm to protein content should be prepared within the range of the protein concentration in the natural samples, using a pure protein as standard. Free glycine must be absent from the test solution, but the estimation is not affected by the presence of inorganic salts commonly used in eluting proteins from a chromatography column. The method can therefore be employed for monitoring protein separations.

Other methods are also used occasionally for protein estimation. For example, proteins show general UV absorption at 215 nm and, in the absence of any other UV-absorbing molecules, such measurements can be related to protein concentration. There is also UV absorption between 250 and 300 nm due to aromatic amino acids and UV spectral measurements in alkaline solution can be used for determining the amount of tyrosine, (λ_{max} 294 nm) and tryptophan (λ_{max} 280 nm) in a protein sample. Protein nitrogen is routinely determined by the micro-Kjeldahl method which requires 0·5 to 1 mg samples and involves digestion with H_2SO_4 in the presence of K_2SO_4–$CuSO_4$, and titration of the ammonia liberated (Bailey, 1967).

(e) *Molecular weight determination by gel filtration*

The estimation of molecular weight of proteins by Sephadex gel filtration was first described by Andrews (1964) and this method has been widely used ever since. Essentially, it consists of taking some Sephadex powder, grade G-75 or G-100, allowing it to swell up in water for 2 days and then packing it into a narrow column (2·5 × 50 cm). The column, after settling and prewashing with buffer, is equilibrated with 0·05 M tris-hydrochloric buffer, pH 7·5 containing 0·1 M KCl and a standard flow rate (*ca.* 30 ml h^{-1}) establishing by applying a suitable pressure at the top of the column. The protein sample in solution is then introduced at the top of the column. After flowing through, protein is detected in the column effluent by its colour (e.g. cytochrome *c* has λ_{max} 408 nm), by its UV absorption at 215 nm or by its enzymic properties. The rate of flow of the protein in the gel is determined by its molecular weight and the volume required to elute it from the column is inversely related to the

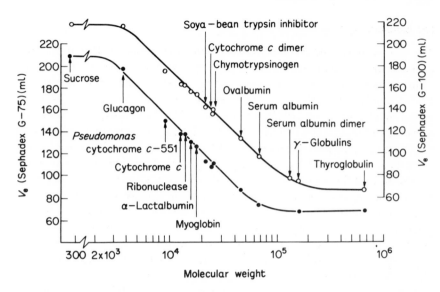

Fig. 7.3 Plot of elution volumes (V_e against log MW for proteins on Sephadex G-75 and G-100 (from Andrews, 1964).

logarithm of the molecular weight (see Fig. 7.3). A standard straight-line relationship must be obtained for each particular column, by running through several proteins of different known molecular weights. Among standard proteins which are suitable and commercially available, are cytochrome *c* from horse heart (mol. wt. 12 400), trypsin inhibitor from soya bean (mol. wt. 21 500), chymotrypsinogen (mol. wt. 25 000), ovalbumin (mol. wt. 45 000), bovine serum albumin (mol. wt. 67 000) and human γ-globulin (mol. wt. 160 000).

The molecular weight can alternatively be determined, less accurately, by separation of the protein on thin layers of Sephadex G-100, where mobility is, again, inversely related to molecular weight (for practical details, see Ikan, 1969).

(f) *Polyacrylamide gel electrophoresis*

Besides being used for purifying proteins, gel electrophoresis is a valuable technique for directly comparing the proteins or enzyme complement of different organisms. Such comparisons are preferably made on direct plant extracts so that all the major proteins (or enzymes) are included in the survey. It is generally better to use mature seed tissue, since the protein content is not subject to the environmental or physiological pressures that affect leaf protein. The results are obtained as a series of bands of different intensity and mobility (see Fig. 7.4) and are useful for relating protein patterns to plant systematics

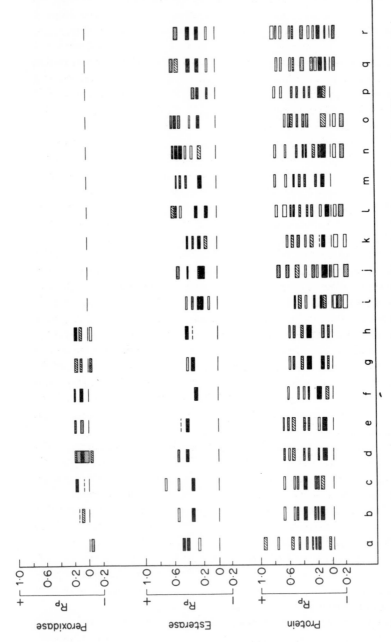

Fig. 7.4 Interpretative diagram illustrating the gel electrophoretic distributions of protein, esterase and peroxidase components in seed extracts of Umbelliferae species.

(Boulter *et al.*, 1967; Fairbrothers, 1968; Crowden *et al.*, 1969). The distance the different bands move relative to the buffer front are described as R_P values.

The samples are prepared by grinding whole seeds (about 25–100 mg) to a fine powder with an equal weight of sand and grinding with 6–8 drops of extracting buffer to produce a thick slurry. After standing for 1 h, the slurry is centrifuged and the supernatant (40 μl) used directly for electrophoresis. Protein content of these solutions should be 2–3% and the concentration should be adjusted accordingly. The extracting buffer should contain hydroxymethylamine methane (35 μM), citric acid (2·5 μM), ascorbic acid (6 μM), cysteine–HCl (6 μM) and sucrose (0·5 M). PVP should be added to this buffer if the seed contains much tannin.

Horizontal electrophoresis is conducted in a Shandon apparatus (Kohn type) using a 7·5% gel, prepared by the method of Lund (1965). A solution of 75 g acrylamide, 2·0 g *bis* (N,N′-methylene-*bis*-acrylamide) in 250 ml water is added to 750 ml tris-citrate-Temed (N,N,N,′N′-tetramethylethylenediamine) buffer and 100 mg ammonium persulphate added as polymerizing agent. The gel is polymerized in a flat perspex mould (150×100×6 mm) covered with a plate-glass lid from which are projected a row of thirteen celluloid slot formers (5·0×1·5×1·5 mm) approx. 25 mm from the cathode side of the gel. Each slot produced in the gel will accept 40 μl of sample. The sample slots are first filled with extracting buffer and pre-electrophoresis is conducted for 1·5 h at constant current (20 mA) and increasing voltage (90–170 V). The buffer in the sample slots is removed, the different samples of plant extracts then placed in the slots and electrophoresis continued for a further 2–2·5 h or until the borate ion boundary has migrated 50 mm towards the anode from the sample slots. After electrophoresis, the liquid is removed from the sample slots and the gel is cut horizontally using a taut-wire slicer into up to four slices, so that each slice can be stained to reveal protein and different enzyme types. The horizontal slicing of gels requires some degree of skill and the novice should begin by cutting gels into two or three thicker slices.

The protein bands are detected by immersing one slice in a solution of 0·7% amido black in 7% HOAc for 1 h. The residual dye is removed by repeatedly washing with 7% acetic acid, when the protein bands can be seen as blue-black bands on a clear background. Esterase is demonstrated by immersing a slice in 100 ml phosphate buffer (0·1 M, pH 6·3) containing α-naphthyl acetate, 100 mg, and diazo blue B (Michrome 250, E. Gurr) 50 mg, for 5–10 min. The background is clarified by washing with water and the esterases appear as red bands. Peroxidase is demonstrated by immersing a slice in 100 ml acetate buffer (0·1 M, pH 4·4) containing 100 mg *o*-dianisidine and hydrogen peroxide (30 vols) 0·5 ml. Catalase activity can be observed as areas of oxygen evolution during evaluation of peroxidase activity.

The above procedure, abbreviated as PAGE, is widely used as such or with minor modification in the study of plant proteins. An alternative protein stain to amido black which is more sensitive is Coomassie Brilliant Blue, which is used as a 0·25% solution containing 25% methanol and 10% acetic acid. Specially prepared sample slots in the gel are not essential, since it is possible to introduce the samples by making thin cuts in the gel at the start line and then inserting into them paper wicks previously soaked in the different sample solutions. Other modifications in PAGE include the incorporation of sodium dodecyl sulphate (SDS) or urea into the electrophoretic buffer, which may improve the resolving power of the system.

Electrophoresis is also conducted on gels of starch rather than poly-acrylamide. A starch support has been used, for example, for surveying plant populations for isozymic variation in leaf tissues (Gottlieb, 1981). Methods are available for staining gels for over twenty different enzyme activities. For further practical details of the gel electrophoresis of proteins, see Sargent (1969) and Deyl (1982).

(g) *Isoelectric focusing*

This technique depends on the fact that proteins vary in their isoelectric point, the pH at which there is no net charge in the molecule. Thus, if a protein mixture undergoes electrophoresis in a pH gradient, the individual proteins will become focused at different regions of that gradient and hence will separate. Application of a constant voltage to a mixture of carrier ampholytes (amphoteric compounds of closely spaced isoelectric points) in a poly-acrylamide or other gel will cause a pH gradient to form. A protein placed at any position in the gradient will acquire a charge and thus it will migrate until it reaches its isoelectric point. Proteins differing in isoelectric point by as little as 0·1 pH units can be resolved, so that the method has high resolving power. Ampholytes are expensive, but are available commercially. Procedures for isoelectric focusing are similar to those for electrophoresis and are described in Catsimpoolas (1976). The application of this technique to the separation of subunits of RuBP carboxylase (or fraction I protein) as they occur in different plants is described by Gray (1980).

(h) *Other techniques*

For separating complex mixtures of plant proteins, it is often advantageous to use isoelectric focusing in combination with electrophoresis on poly-acrylamide gel, in the presence of sodium dodecyl sulphate (SDS). The two-dimensional map so produced may distinguish as many as a 1000 indi-vidual proteins in a given plant extract (O'Farrell, 1975). A range of other two-dimensional systems have also been developed for resolving proteins. Such separations of seed proteins are illustrated in the book of Daussant *et al.* (1983).

An increasingly important method of analysing protein mixtures is that of HPLC (Catsimpoolas, 1983). For example, in a recent investigation of the allergenic proteins of grass pollen, Calam *et al.* (1983) employed two HPLC separations, among other techniques. Reversed-phase HPLC was employed on a Spherisorb S5 ODS-2 column, gradient-eluted with varying proportions of $0 \cdot 1$ M $(NH_4)_2SO_4$ and $0 \cdot 1$ M $(NH_4)_2SO_4$–acetonitrile (2:3), while size-exclusion HPLC was conducted on a TSK G 3000 SW column with a mobile phase of $0 \cdot 1$ M sodium phosphate. Good resolutions were achieved on a small scale, with excellent recovery.

7.3.3   Practical experiment

(a) *Isolation of the plant enzyme phenylalanine ammonia lyase*

Phenylalanine ammonia lyase (PAL) catalyses the de-amination of phenyl-alanine to cinnamic acid and this reaction is important in phenolic bio-synthesis. The reaction is:

$$Ph\,CH_2\,CH\,(NH_2)\,CO_2H \rightarrow PhCH = CH\,CO_2H + NH_3$$

PAL activity has been demonstrated in acetone powders of both mono- and dicotyledonous plants. The related enzyme activity, tyrase catalyses the de-amination of tyrosine to give *p*-hydroxycinnamic acid and only occurs in monocotyledons.

Both enzyme activities can be detected in acetone powders of barley seedlings by measuring the cinnamic acids formed, either chromato-graphically or spectrophotometrically. The following procedure may be used.

(b) *Preparation of acetone powder*

(1) Cut the stems from twenty 7-day-old etiolated barley seedlings into pieces approx. 3 cm long and store in the deep-freeze for 1 h.
(2) Homogenize in acetone at $-20°C$ for 1 min, filter and wash the residue with cold acetone ($\times 3$).
(3) Dry the powder between filter papers and then in a vacuum desiccator. This acetone powder may be stored at $4–6°C$ in a tightly sealed container.

(c) *Demonstration of enzyme activity*

(1) Add $0 \cdot 2–0 \cdot 4$ g acetone powder to each of two tubes containing 3 ml $0 \cdot 1\%$ phenylalanine and tyrosine in $0 \cdot 1$ M borate buffer pH $8 \cdot 8$.
(2) Add similar amounts of powder to 3 ml buffer, as a blank control. Leave at $37°C$ for 2 h, add 7 ml water and centrifuge.

(3) Acidify the supernatants and extract into ether. Evaporate the three ether extracts to dryness and take each up into 0·3 ml ethanol.

(4) Spot the three samples (using 0·05 ml) on two silica gel TLC plates, along with marker spots of cinnamic and *p*-hydroxycinnamic acids, and develop one plate in toluene–methanol–acetic acid (90:16:8) and the second one in cyclohexane–chloroform–acetic acid (4:5:1). The cinnamic acids can be seen on the dried developed plates as dark spots on a fluorescent background in UV light of 253 nm wavelength.

(5) Add 2·7 ml ethanol to the remainder of each sample in a silica cuvette and measure the spectrum of the two enzymic reaction products, using the control as the solvent blank. On adding a drop of alkali, the spectra of one of the products, cinnamic acid (λ_{max} 273 nm) undergoes a hypsochromic shift. The spectrum of the other product, *p*-hydroxycinnamic acid (λ_{max} 310 nm), should shift bathochromically in the presence of alkali.

7.4   POLYSACCHARIDES

7.4.1   Chemistry and distribution

The chemistry of polysaccharides is, in a sense, simpler than that of the other plant macromolecules mentioned in earlier sections, since these polymers contain only a few simple sugars in their structures. Indeed, the best known polysaccharides – cellulose and starch – are polymers of a single sugar, glucose. The structural complexity of polysaccharides is due to the fact that two sugar units can be linked together, through an ether linkage, in a number of different ways. The reducing end of one sugar (C1) can condense with any hydroxyl of a second sugar (at C2, C3, C4 or C6) so that during polymerization some sugars may be substituted in two positions, giving rise to branched chain structures. Furthermore, the ether linkage can have either an α- or a β-configuration, due to the stereochemistry of simple sugars, and both types of linkage can co-exist in the same molecule.

Although a few polysaccharides (e.g. cellulose) are, in fact, simple straight-chain polymers, the majority have partly branched structures. It is very difficult to determine complete sequences in such branched polysaccharides and, at present, it is only possible to define their structures in terms of a 'repeating unit' of oligosaccharide, a large number of which are linked together to produce the complete macromolecule. Some of the structural variation that can occur among polysaccharides is shown in Fig. 7.5. The structures of the individual monosaccharides from which plant polysaccharides are composed have been given in an earlier chapter (see Fig. 6.1, p. 224).

The most familiar plant polysaccharides are cellulose and starch, but many other kinds are known, which have other sugars besides, or in addition to, glucose (Table 7.3). Cellulose represents a very large percentage of the combined carbon in plants and is the most abundant organic compound of all. It is the fibrous material of the cell wall and is responsible, with lignin, for the structural rigidity of plants. Cellulose occurs in almost pure form (98%) in

Starch   Cellulose

— Glc α 1→4 Glc α 1→4 Glc —   — Glc β 1→4 Glc β 1→4 Glc —

Amylose

—Glc α 1→4 Glc α 1→4 Glc
$$6$$
$$\uparrow$$
$$1\alpha$$
— Glc α 1→4 Glc α 1→4 Glc —

Amylopectin

Xylan   Arabinogalactan

— Xyl β1→4 Xyl β1→4 Xyl —   —Gal 1→3 Gal 1→3 Gal 1→3 Gal—

   3   2   3   6   6   6   6
   ↑   ↑   ↑   ↑   ↑   ↑   ↑
   1   1α   1   1   1   1   1
  Ara   4 Me Glur   —Ara   Gal   Gal   Ara   Gal
                                     6   6
                                     ↑   ↑
                                     1   1
                                   Gal   4 Me Glur

Starch α 1→4 link   Cellulose β 1→4 link

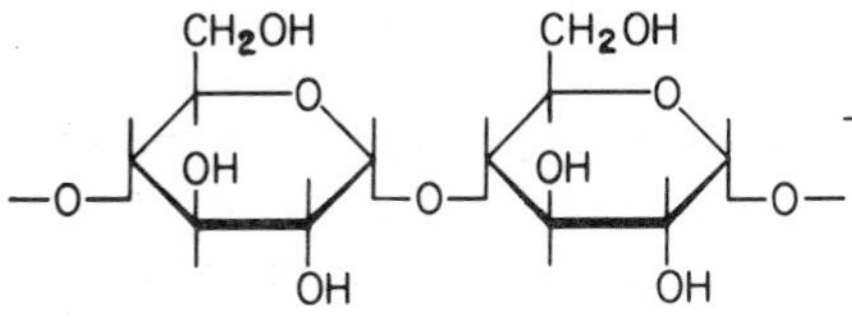

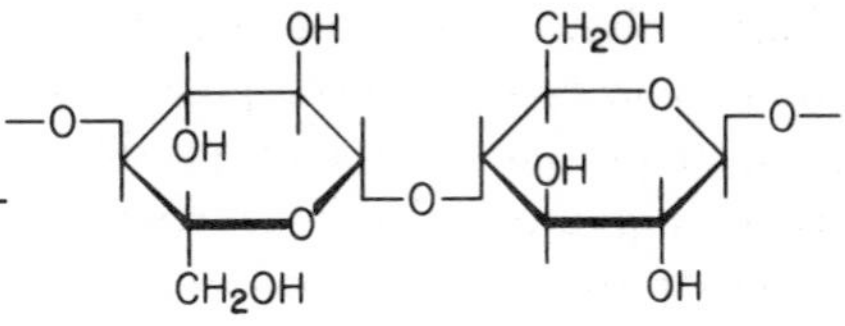

Key.
Glc = glucose,   Gal = galactose
Ara = arabinose,   Xyl = xylose
4 Me Glur = 4—methylglucuronic acid

Fig. 7.5 Structural 'repeating units' of some common polysaccharides.

cotton fibres; the wood of trees is a less abundant source (40–50%) but the most important commercially for cellulose production. Chemically, cellulose is a β-glucan and consists of long chains of $\beta1 \rightarrow 4$ linked glucose units (Fig.

Table 7.3 The main classes of polysaccharides in higher plants, algae and fungi

Class name	Sugar unit(s)	Linkage	Distribution
HIGHER PLANTS			
Cellulose	glucose	$\beta1 \rightarrow 4$	universal as cell wall material
Starch–amylose	glucose	$\alpha1 \rightarrow 4$	universal as
Starch–amylopectin	glucose	$\alpha1 \rightarrow 4, \alpha1 \rightarrow 6$	storage material
Fructan	fructose (some glucose)	$\beta2 \rightarrow 1$	in artichoke chicory, etc.
Xylan	xylose (some arabinose and uronic acid)	$\beta1 \rightarrow 4$	widespread, e.g. in grasses
Glucomannan	glucose, mannose	$\beta1 \rightarrow 4$	widespread, but especially in coniferous wood
Arabinogalactan	arabinose, galactose	$1 \rightarrow 3, 1 \rightarrow 6$	
Pectin	galacturonic acid (some others)	$\alpha1 \rightarrow 4$	widespread
Galactomannan	mannose, galactose	$\beta1 \rightarrow 4, \alpha1 \rightarrow 6$	seed mucilages
Gum	arabinose, rhamnose, galactose, glucuronic acid	highly branched	in *Acacia* and *Prunus* species
ALGAE (seaweeds)			
Laminaran	glucose	$\beta1 \rightarrow 3$	
Polysaccharide			Phaeophyceae
sulphate	fucose (and others)	—	(brown algae)
Alginic acid	mannuronic and guluronic acids	—	
Amylopectin	glucose	$\alpha1 \rightarrow 4, 1 \rightarrow 6$	Rhodophyceae
Galactan	galactose	$1 \rightarrow 3, 1 \rightarrow 4$	(red algae)
Starch	glucose	$\alpha1 \rightarrow 4, \alpha1 \rightarrow 6$	
Polysaccharide			Chlorophyceae
sulphate	rhamnose, xylose, glucuronic acid	—	(green algae)
FUNGI*			
Chitin	*N*-acetylglucosamine	$\beta1 \rightarrow 4$	widespread
Chitosan	glucosamine	$\beta1 \rightarrow 4$	Zygomycetes
β-Glucan	glucose	$\beta1 \rightarrow 3, \beta1 \rightarrow 6$	widespread
α-Glucan	glucose	$\alpha1 \rightarrow 3, 1 \rightarrow 4$	widespread
Mannan	mannose	$\alpha1 \rightarrow 2, 1 \rightarrow 6$	mainly
		$1 \rightarrow 3$	Ascomycetes

*By contrast, bacterial cell wall is particularly complex. The polysaccharide material may contain, in addition to the usual sugars, such components as glycerol, ribitol, phosphate, amino acid and *N*-acetylglucosamine.

7.5), the molecular weight varying from 100000 to 200000. Cellulose occurs in the plant cell as a crystalline lattice, in which long straight chains of polymer lie side by side linked by hydrogen bonding.

Starch differs from cellulose in having the linkage between the glucose units $\alpha 1 \rightarrow 4$ and not $\beta 1 \rightarrow 4$ and also in having some branching in the chain. Starch, in fact, comprises two components, amylose and amylopectin, which can be separated from each other. Amylose (*ca.* 20% of the total starch) contains about 300 glucose units linked in a simple chain $\alpha 1 \rightarrow 4$, which exists *in vivo* in the form of an α-helix. Amylopectin (*ca.* 80%) contains $\alpha 1 \rightarrow 4$ chains, with regular branching of the main chain by secondary $\alpha 1 \rightarrow 6$ linkages (see Fig. 7.5). Its structure is thus of a randomly multibranched type. The two forms of starch are easily distinguished by their response to iodine, amylose giving a blue and amylopectin a reddish purple colour. Starch, itself, is the essential storage form of energy in the plant and starch granules are frequently located within the chloroplast close to the site of photosynthesis.

Most plant polysaccharides, unlike starch and cellulose, are heteropolysaccharides, having more than one type of sugar unit (Table 7.3). The problem of structural determination is an acute one and it is even possible that many of them do not possess unique sequences but are collections of closely similar polymers, produced by random branching. In spite of this difficulty, these plant polysaccharides can be characterized by the amounts of different sugars they yield on acid hydrolysis and by molecular weight determination. Methylation of the polysaccharide, followed by acid hydrolysis, may indicate the length of the chain in the polymer. Again, identification of oligosaccharide fragments following partial acid or enzymic hydrolysis of the polysaccharide may provide an indication of the 'repeating unit' present.

The different classes of polysaccharide shown in Table 7.3 fall into two groups according to whether they are easily soluble in aqueous solutions or not. Those that are soluble include starch, inulin, pectin and the various gums and mucilages. The gums which are exuded by plants, sometimes in response to injury or infection, are almost pure polysaccharide. Their function in the plant is not entirely certain, although it may be a protective one.

The less soluble polysaccharides usually comprise the structural cell wall material and occur in close association with lignin. Besides cellulose, there are various hemicelluloses in this fraction. The hemicelluloses have a variety of sugar components and fall into three main types: the xylans, glucomannans and arabinogalactans. They are structurally complex and other polysaccharide types may also be found with them.

From the comparative viewpoint, polysaccharides are interesting macromolecules since they do vary in type in different groups of plant. The storage polysaccharides, for example, of a number of Composites are based on fructose rather than glucose. Thus, fructans like inulin are isolated from tubers of chicory, *Cichorium intybus* and artichoke. *Cynara scolymus* in place of the more usual starches.

Variation is also marked in the cell wall polysaccharides of different groups of organisms. Differences in cell wall composition distinguish the major classes of fungi and can be used, with other characters, as phylogenetic markers in the group (Table 7.3 and Bartnicki-Garcia, 1971). The cell wall polysaccharides of ferns and gymnosperms are distinguishable (from the angiosperms) by their frequent high mannose content (Bailey and Pain, 1971). Within the angiosperms, the differences in cell wall materials are less pronounced, but they are sufficient to be detected by simple methods. For example, the xylan fraction in the hemicelluloses of monocotyledons is usually composed of arabinoxylans, whereas in dicotyledons it consists of chains of xyloglucans interspersed with 4-*O*-methylglucuronic acid residues.

Finally, the marked variation in polysaccharide composition between different algae (seaweed) must be mentioned. Three of the main classes – brown, red and green – are clearly distinct in the types of polysaccharide they have (Table 7.3). Polysaccharides have been obtained from other classes (e.g. the blue-greens) but insufficient species have so far been examined to draw any taxonomic conclusion.

Because of the enormous commercial importance of plant polysaccharides, much has been written already about methods of their isolation and purification. Indeed, three of the five volumes in the series *Methods in Carbohydrate Chemistry* (Whistler, 1963–65) are separately devoted to cellulose, starch and general polysaccharides. The chapters covering plant polysaccharides in *Modern Methods of Plant Analysis* (Paech and Tracey, 1955a) are also very thorough and full of information. The present account of methodology is, therefore, a brief summary of the main points and reference should be made to the above major texts for greater detail.

While there are numerous general accounts of polysaccharides in biochemistry textbooks, not many deal specifically with those from plants. An excellent comparative survey of plant polysaccharides is provided in the review by Percival (1966). The same author has also published a detailed account of algal polysaccharides (Percival and McDowell, 1967). The only general introduction to polysaccharides that appears to be available is the text by Aspinall (1970). Recent accounts of storage polysaccharides in plants are contained in Loewus and Tanner (1982), while reviews of cell wall polysaccharides in different plant groups are available in Tanner and Loewus (1981). An important continuing review series is *Advances in Carbohydrate Chemistry and Biochemistry*, edited by Tipson and Horton, the latest volume of which appeared in 1983.

7.4.2 Recommended techniques

(a) *General extraction procedures*

The separation of polysaccharides from leaf or other plant tissue requires a series of extractions, by which low-molecular-weight compounds and the water-soluble polysaccharides are separately removed. The residues, after repeated extraction, will consist of more or less pure cellulose. The various procedures most commonly used are as follows.

Low-molecular-weight constituents are conveniently removed by exhaustive extraction with boiling ethanol. If the tissue is rich in lipid (e.g. seeds), it may be preferable to remove these by extraction with acetone, followed by ether–benzene (1:1).

Neutral water-soluble polysaccharides are now extracted with 1% NaCl solution or with boiling water. They are recovered from solution by pouring the extracts into several volumes of alcohol, when they are precipitated. Pectins are obtained from the above residue by extraction with 0·5% ammonium oxalate and are subsequently precipitated from solution after acidification and pouring into alcohol. Lignin is removed at this stage by extraction with 1% sodium chlorite at 70°C for 1 h. If tissues are rich in lignin, the extraction procedure must be repeated several times. Delignification can be achieved alternatively by extraction with chloramine-T (sodium *p*-toluene–sulphonchloramide) and ethanolamine (Gaillard, 1958). These extracts are discarded and the residue is washed and dried.

The hemicelluloses are now removed by extraction of the residue with 7–12% NaOH under nitrogen at room temperature for 24 h. Again, to achieve complete separation of the hemicelluloses, the extraction procedure should be repeated at least twice. The hemicelluloses are recovered by acidification of the alkaline extracts with acetic acid and precipitation with ethanol.

The remaining material is then thoroughly washed and dried and this is the pure cellulose fraction.

The various polysaccharide fractions thus obtained are then further fractionated and purified by a variety of procedures. The separation of the linear and branched polysaccharides within any one fraction is often desired. This can be achieved by reaction with iodine, as indicated below. Ion exchange and Sephadex chromatography are also applicable to polysaccharide fractionations.

Occasionally, plant polysaccharides are available naturally in a relatively pure form. This is true of the various gums and mucilages exuded from the bark of trees and also from the seeds and fruits. Such gums can often be purified simply by dissolving in water, dialysing to remove low-molecular-weight components, precipitating by pouring into ethanol and collecting and drying.

Four examples of methods for preparing plant polysaccharides will now be given. A further paragraph will mention other techniques commonly used for final purification before characterization.

(b) *Starch from potato*

Peeled potato (1 kg) is blended with 750 ml 1% NaCl and then filtered through fine muslin. The residue is re-extracted with 150 ml 1% NaCl, refiltered and this is added to the first filtrate. On standing, the starch granules, which have passed through the muslin cloth, settle out and the supernatant can then be decanted and discarded. The wet starch is washed three times with 1% NaCl, once with 0·01 M NaOH and once with water. It is then drained, dried and weighed (yield *ca.* 20 g). Procedures for isolating starch from other plant sources are given by W. J. Whelan in Paech and Tracey (1955a).

(c) *Pectin from apple*

Extract 300 g thin slices of apple with boiling ethanol for 5 min, discarding the alcohol. Mince the remaining tissue and extract it with 200 ml boiling water. Neutralize the extract with 1 M NH_4OH to pH 6·5 and evaporate *in vacuo* to 150 ml. On adding alcohol to 80% saturation, the pectic acid precipitates and it can be collected and dried. Purification is by redissolving in boiling water (20 ml), filtering, precipitating with alcohol, washing and drying.

(d) *Xylan from straw*

Powdered straw (30 g) is pre-extracted with boiling 3% sodium sulphite (one litre) and then delignified by repeated extraction first with 5% sodium hypo-chlorite at room temperature and then with 5% sodium hypochlorite containing 1% H_2SO_4 (in a fume cupboard). The residue, after washing and drying, is extracted with 400 ml 6% NaOH for 45 min at 100°C and filtered. The filtrate is treated with 200 ml Fehling's solution and the xylan–copper complex, which precipitates, is collected and washed with 80% ethanol. The copper complex is decomposed by suspending it in 96% ethanol and passing in hydrogen chloride gas at 0°C. The residue is collected by centrifugation, washed with 80% ethanol and dried.

(e) *Linear and branched hemicelluloses*

The linear and branched hemicelluloses can be separated by treatment with iodine (Gaillard, 1965). A similar procedure can also be used for fractionating amylose and amylopectin of starch. Hemicellulose (1 g) is dissolved in 100 ml $CaCl_2$ (density of solution = 1·3) and centrifuged to clarify the solution.

Iodine, 15 ml (3%) and potassium iodide (4%) are added to the supernatant. The dark blue precipitate of the linear polymer is collected. The clear brown supernatant is neutralized with sodium thiosulphate and the solution poured into 5 vols ethanol. The branched polymer which precipitates is dissolved in 5 ml 0·1 M HCl to remove calcium ion and then reprecipitated with 25 ml ethanol and collected.

The iodine complex of the linear polymer is washed with CaCl₂ solution containing iodine and potassium iodide. It is then dissolved in 100 ml hot water and the iodine neutralized with sodium thiosulphate. The polymer is precipitated after pouring the solution into 5 vols ethanol. Calcium ion is finally removed by dissolving it under nitrogen in 1 M KOH, neutralizing with 1 M HCl and reprecipitating with ethanol. The product is then washed and dried.

(f) *Other techniques*

For electrophoresis of polysaccharides, one can use cellulose acetate film and either 0·1 M $(NH_4)_2CO_3$ buffer pH 8·9 or 0·1 M acetate buffer pH 4·7, at a current of 15−20 V cm⁻¹ for 2−4 h. The polysaccharide is visualized by reaction with periodate and staining with rosaniline hydrochloride (Conacher and Rees, 1966).

Polysaccharides may be purified on a Sepharose-4B column, eluted with 1 M NaCl or on a DEAE−cellulose column, with gradient elution with NaCl solutions. The application of ion exchange and Sephadex chromatography to the separation of oat leaf polysaccharides is described by Reid and Wilkie (1969).

For fractionation of polysaccharide mixtures, the use of copper (as Fehling's solution) and iodine have already been mentioned. Barium hydroxide is also an effective reagent. For acidic polysaccharides (e.g. pectins) a useful reagent is the quaternary ammonium salt, cetyltrimethylammonium chloride. The polysaccharide salt is water-insoluble and can thus be easily separated off from the neutral polysaccharide which remains in solution.

(g) *Characterization of polysaccharides*

The most important and most easily achieved analysis is the hydrolysis and detection of the constituent sugars. Hydrolysis may be conducted in 1 M H_2SO_4 for 4 h at 100°C and the solution neutralized with $Ba(OH)_2$, filtered and the filtrate concentrated and examined for simple sugars by PC, or GLC. Procedures for doing this have already been described in an earlier chapter (see p. 225).

In the more detailed characterization of a polysaccharide, it is usual to fully methylate it. This is done with dimethyl sulphate in alkali, followed by methyl iodide and silver oxide. Methylation proceeds more rapidly if a solvent such as dimethylformamide or dimethyl sulphoxide is employed. The fully methylated carbohydrate is then normally hydrolysed with acid and the various partly methylated monosaccharides produced separated and analysed by PC or by GLC. Other procedures used for structural investigation include partial acid hydrolysis (0.1 M H_2SO_4 for 3 h at $100°C$), enzymic hydrolysis and periodate oxidation. Spectral analysis is not of great value, although IR and NMR measurements can sometimes indicate the configuration (α- or β-) of the glycosidic links.

The molecular weight or molecular size of a pure polysaccharide can be determined by gel filtration on Sephadex G-100 eluted with 0.05 M NH_4HCO_3 pH 8.0, using standard dextrins for comparison. Alternatively, high performance gel permeation chromatography may be employed, using a column of Toya Soda TSK Gel G 3000 SW eluted with 0.9% NaCl.

7.4.3　Practical experiment

(a) *Determination of sugars in polysaccharides*

A simple procedure to obtain information about plant polysaccharides is to determine the number and amounts of monosaccharides produced on acid hydrolysis. Such an analysis can be carried out directly on the total polysaccharide fraction of a leaf, following an exhaustive extraction of low-molecular-weight materials with hot ethanol (Andrews *et al.*, 1960). By its very nature, this is a relatively crude analysis and it is preferable to conduct it on polysaccharide fractions after they have been fractionated according to solubility (Bailey *et al.*, 1967).

From a comparative viewpoint, it is more meaningful to determine the carbohydrate composition of the cell wall fraction (cellulose plus hemicellulose) after removing the water-soluble polymers. There are considerable differences in sugar composition of this fraction both within the angiosperms (Gaillard, 1965) and within the ferns (Bailey and Pain, 1971). Study of the cell wall sugars is also an important technique in fungi and bacteria (Bartnicki-Garcia, 1966; 1971) and can even be employed to distinguish different species in the same genus (Wilkinson and Carby, 1971). Siegel (1962) has written a general account of the plant cell wall, and reference may be made to his book for further details.

(b) *Procedure*

(1) When isolating the total polysaccharide fraction, leaf tissue is cut into pieces, plunged in boiling alcohol and then macerated in a Waring blender for 5 min. The extraction with hot alcohol is repeated until all the

chlorophyll has been removed. The residue may be collected by filtration or by centrifugation between extractions. The final product should be a white friable powder. When isolating the cell wall fraction, extract as above and then successively extract the powder with hot water, hot 1% NaCl and hot ammonium oxalate (see p. 269) to remove all the soluble polysaccharide. The insoluble residue is finally washed and dried.

(2) The powder is heated in 1 M H_2SO_4 at 100°C for 8–16 h. After filtering and cooling, the filtrate is neutralized with $Ba(OH)_2$ or $BaCO_3$ and the precipitate of $BaSO_4$ removed by centrifugation. Add small amounts of acid and alkali exchange resins to the supernatant to remove the last traces of inorganic ion. Concentrate this clear solution to a small volume *in vacuo*.

(3) Chromatograph an aliquot of the concentrated sample on paper in two of the standard sugar solvents (see p. 225), running at the same time a number of standard marker solutions, in which the common sugars are present at known concentrations. Dry the paper, develop with aniline hydrogen phthalate, and estimate the concentrations of the different monosaccharides produced on an arbitrary scale from one to five. If further precision is required, the sugar spots can be eluted and the concentrations determined by spectrophotometric measurements (see p. 227).

Some indication of the type of results that can be obtained with the total polysaccharide fraction of plant leaves is given in Table 7.4 (data from Andrews *et al.*, 1960). As can be seen, sugars always present are galactose, glucose, arabinose, xylose and uronic acid (mainly galacturonic). Mannose and rhamnose frequently occur in traces. Occasional trace components are

Table 7.4 Monosaccharides present in polysaccharides of plant leaves

Plant species	Common name	Amount of sugar						
		Gal	Glc	Man	Ara	Xyl	Rha	Uronic
Taxus baccata	yew	2	2	tr	4	tr	—	2
Pinus sylvestris	Scots pine	3	4	4	4	2	tr	1
Cedrus atlantica	atlas cedar	tr	5	tr	2	tr	—	2
Fagus sylvatica	beech	2	2	tr	2	5	—	1
Quercus pedunculata	oak	2	2	—	4	4	—	2
Pyrus malus	apple	2	4	tr	4	3	tr	2
Ilex aquifolium	holly	4	2	—	5	5	tr	4
Aesculus hippocastanum	horsechestnut	2	5	—	2	2	tr	2
Hedera helix	ivy	2	4	tr	2	2	tr	2

Key: Gal = galactose, Glc = glucose, Man = mannose, Ara = arabinose, Xyl = xylose, Rha = rhamnose, Uronic = uronic acid, — = absent, tr = trace, 1–5 = increasing in amount (arbitrary scale).

apiose, fucose and certain monomethyl derivatives of the common sugars. Amino sugars (e.g. glucosamine) are found in cell walls of fungi and bacteria (Table 7.3).

Differences in concentration between the sugars are of considerable interest and are worth noting, though they are dependent to some extent on environmental factors. The ratio of hexose (glucose plus galactose) to pentose (arabinose plus xylose) may vary considerably; for example, this ratio is 1:2 in oak, 1:1 in pine and almost 2:1 in horse chestnut (Table 7.4).

REFERENCES

General references

Aspinall, G.O. (1970) *Polysaccharides*, Pergamon Press, Oxford.

Bailey, R.L. (1967) *Techniques in Protein Chemistry*, 2nd edn, Elsevier, Amsterdam.

Bartnicki-Garcia, S. (1971) Cell wall composition and other biochemical markers in fungal phylogeny. in *Phytochemical Phylogeny* (ed. J.B. Harborne), Academic Press, London, p. 81–104.

Boulter, D. and Parthier, B. (1982) *Nucleic Acids and Proteins in Plants, I Structure, Biochemistry and Physiology of Proteins*, Springer-Verlag, Berlin.

Catsimpoolas, N. (ed.) (1976) *Isoelectric Focusing*, Academic Press, New York.

Catsimpoolas, N. (1983) Proteins. in *Chromatography Part B Applications* (ed. E. Heftmann), Elsevier, Amsterdam, pp. 53–74.

Cohn, W.E. (1967) Column chromatography of nucleic acids and related substances. in *Chromatography* (ed. E. Heftmann), 2nd edn, Reinhold, New York, pp. 627–60.

Cowling, G.J. (1983) Nucleic acids. in *Chromatography Part B Applications* (ed. E. Heftmann), Elsevier, Amsterdam, pp. 345–76.

Croft, L.R. (1980) *Introduction to Protein Sequence Analysis*, John Wiley, Chichester.

Daussant, J., Mosse, J. and Vaughan, J. (eds) (1983) *Seed Proteins*, Academic Press, London.

Deyl, Z. (ed.) (1982) *Electrophoresis. Part B. Applications*, Elsevier, Amsterdam.

Fairbrothers, D.E. (1968) Chemosystematics with emphasis on systematic serology. in *Modern Methods in Plant Taxonomy* (ed. V.H. Heywood), Academic Press, London, pp. 141–73.

Grossman, L. and Moldave, K. (eds) (1967–8) *Nucleic Acids*, Parts A and B. in *Methods in Enzymology*, Academic Press, New York.

Hall, T.C. and Davies, J.W. (eds) (1979) *Nucleic Acids in Plants*, CRC Press, Boca Raton, Florida.

Harborne, J.B. and Van Sumere, C.F. (eds) (1975) *Chemistry and Biochemistry of Plant Proteins*, Academic Press, London.

Hindley, J. and Studen, R. (1983) *DNA Sequencing*, Elsevier Biomedical Press, Amsterdam.

Linskens, H.F. and Tracey, M.V. (eds) (1963) *Modern Methods in Plant Analysis* (Vol. VI), Springer-Verlag, Berlin.

Linskens, H.F. and Tracey, M.V. (eds) (1964) *Modern Methods in Plant Analysis* (Vol. VII), Springer-Verlag, Berlin.

Loewus, F.A. and Tanner, W. (eds) (1982) *Plant Carbohydrates I. Intracellular Carbohydrates*, Springer-Verlag, Berlin.

Marcus, A. (ed.) (1981) *Proteins and Nucleic Acids (The Biochemistry of Plants*, Vol. 6), Academic Press, New York.

Norton, G. (ed.) (1978) *Plant Proteins*, Butterworths, London.

Paech, K. and Tracey, M.V. (eds) (1955a) *Modern Methods in Plant Analysis* (Vol. II), Springer-Verlag, Berlin.

Paech, K. and Tracey, M.V. (eds) (1955b) *Modern Methods in Plant Analysis* (Vol. IV), Springer-Verlag, Berlin.

Percival, E. (1966) The natural distribution of plant polysaccharides. in *Comparative Phytochemistry* (ed. T. Swain), Academic Press, London, pp. 139–58.

Percival, E. and McDowell, R.H. (1967) *Chemistry and Enzymology of Marine Algal Polysaccharides*, Academic Press, London, p. 219.

Rhodes, M.J.C. (1977) The extraction and purification of enzymes from plant tissues. in *Regulation of Enzyme Synthesis and Activity in Higher Plants* (ed. H. Smith), Academic Press, London, pp. 245–70.

Siegel, S.M. (1962) *The Plant Cell Wall*, Pergamon Press, Oxford, p. 124.

Tanner, W. and Loewus, F.A. (eds) (1981) *Plant Carbohydrates II Extracellular Carbohydrates*, Springer-Verlag, Berlin.

Tipson, R.S. and Horton, D. (eds) (1983) *Advances in Carbohydrate Chemistry and Biochemistry* (Vol. 41), Academic Press, New York.

Whistler, R.L. (ed.) (1963–5) *Methods in Carbohydrate Chemistry* (Vols. III–V), Academic Press, New York.

Supplementary references

Akazawa, T. (1979) in *Photosynthesis II Photosynthetic Carbon Metabolism and Related Processes* (eds M. Gibbs and E. Latzko), Springer-Verlag, Berlin, pp. 208–29.

Andrews, P.G. (1964) *Biochem. J.*, **91**, 222.

Andrews, P.G., Hough, L. and Stacey, B. (1960) *Nature* (Lond.) **185**, 166.

Bailey, R.W., Haq, S. and Hassid, W.Z. (1967) *Phytochemistry*, **6**, 293.

Bailey, R.W. and Pain, V. (1971) *Phytochemistry*, **10**, 1065.

Bartnicki-Garcia, S. (1966) *J. Gen. Microbiol.*, **42**, 57.

Belford, H.S., Thompson, W.F. and Stein, D.B. (1981) in *Phytochemistry and Angiosperm Phylogeny* (eds D.A. Young and D.S. Seigler), Praeger, New York, pp. 1–18.

Biswas, S.B. and Sarkar, A.K. (1970) *Phytochemistry*, **9**, 2425.

Bolton, E.T. and Bendich, A.J. (1967) *Plant Physiol. (Lancaster)*, **42**, 959.

Bolton, E.T. and Britten, R.J. (1965) *Ann. Rept. Dept. Terrestrial Magnetism, Carnegie Inst., Wash.*, 313.

Boulter, D., Thurman, D. and Derbyshire, E. (1967) *New Phytologist*, **66**, 27, 37, 46.

Burton, K. (1956) *Biochem. J.*, **62**, 315.

Calam, D.H., Davidson, J. and Ford, A.W. (1983) *J. Chromatog.*, **266**, 293.

Cherry, J.H. and Chroboczek, H. (1966) *Phytochemistry*, **5**, 411.

Conacher, A.B.S. and Rees, D.I. (1966) *Analyst*, **91**, 55.

Crowden, R.K., Harborne, J.B. and Heywood, V.H. (1969) *Phytochemistry*, **8**, 1965.

De, G.N., Ghosh, J.J. and Bhattacharyya, K. (1972) *Phytochemistry*, **11**, 3349.

De Ley, J. (1964) *Ann. Rev. Microbiol.*, **18**, 17.

Detchou, P. and Possingham, J.V. (1972) *Phytochemistry*, **11**, 943.

Flavell, R.B., Rimpau, J. and Smith, D.B. (1977) *Chromosoma*, **63**, 205.

Gaillard, B.D.E. (1958) *J. Sci. Food Agric.*, **9**, 170, 346.

Gaillard, B.D.E. (1965) *Phytochemistry*, **4**, 631.

Gottlieb, L.D. (1981) *Prog. Phytochem.*, **7**, 1.

Gray, J.C. (1980) in *Chemosystematics, Principles and Practice* (eds F.A. Bisby, J.G. Vaughan and C.A. Wright), Academic Press, London, pp. 167–94.

Green, B. (1971) *Biochim. Biophys. Acta.*, **254**, 402.

Holdgate, D.P. and Goodwin, T.W. (1965) *Phytochemistry*, **4**, 831.

Ingle, J. (1963) *Phytochemistry*, **2**, 353.

Ikan, R. (1969) *Natural Products*, Academic Press, London, pp. 260–70.

Kirby, K.S. (1965) *Biochem. J.*, **96**, 266.

Koller, B. and Smith, H. (1972) *Phytochemistry*, **11**, 1295.

Loening, U.E. (1967) *Biochem. J.*, **102**, 251.

Loening, U.E. (1973) *Pure Appl. Chem.*, **34**, 579.

Logan, J.E., Mannell, W.A. and Rossiter, R.J. (1952) *Biochem. J.*, **51**, 470.

Lowry, O.H., Rosebrough, N.J., Farr, A.L. and Randall, R.J. (1951) *J. Biol. Chem.*, **193**, 265.

Lund, B.M. (1965) *J. Gen. Microbiol.*, **40**, 413.

McIntosh, L., Poulsen, C. and Bogorad, L. (1980) *Nature* (Lond.), **288**, 556.

McLeester, R.C., Hall, T.C., Sun, S.M. and Bliss, F.A. (1973) *Phytochemistry*, **12**, 85.

O'Farrell, P.H. (1975) *J. Biol. Chem.*, **250**, 4007.

Osborne, T.B. (1924) *The Vegetable Proteins*, 2nd edn, Longmans, Green & Co., London.

Palmer, J.D., Osorio, B., Thompson, W.F. and Nobs, M.A. (1983) *Rep. 1981–2 of the Carnegie Institute, Washington DC*, 96–97.

Reger, B.J., Ku, M.S.B., Potter, J.N. and Evans, J.J. (1983) *Phytochemistry*, **22**, 1127.

Reid, J.S.G. and Wilkie, K.C.B. (1969) *Phytochemistry*, **8**, 2053, 2059.

Sargent, J.R. (1969) *Methods in Zone Electrophoresis*, 2nd edn, BDH Chemicals Limited, Poole, Dorset.

Simpson, S.A., Lawlis, V.B. and Mueller, D.D. (1983) *Phytochemistry*, **22**, 1121.

Smith, D.B. and Flavell, R.B. (1974) *Biochem. Genet.*, **12**, 243.

Storck, R. and Alexopoulos, C.J. (1970) *Bact. Rev.*, **34**, 126.

Vaughan, J.G., Waite, A., Boulter, D. and Waiters, S. (1966) *J. Exp. Bot.*, **17**, 332.

Wade, N.L., O'Connell, P.B.H. and Brady, C.J. (1972) *Phytochemistry*, **11**, 978.

Wilkinson, S.G. and Carby, K.A. (1971) *J. Gen. Microbiol.*, **66**, 221.

Witham, F.H., Blaydes, D.F. and Devlin, R.M. (1971) *Experiments in Plant Physiology*, Van Nostrand, New York.

Wright, D.J. and Boulter, D. (1973) *Phytochemistry*, **12**, 79.

A List of Recommended TLC Systems for All Major Classes of Plant Chemical

TLC is the chromatographic system of widest application in phytochemistry (see Chapter 1, Section 1.3) since it can be applied to almost every class of compound, except to very volatile constituents. It can be applied to crude plant extracts in a preliminary survey for the presence of most compounds. In the same operation, it provides a means of separation and preliminary detection. The major reference to TLC procedures remains the book edited by E. Stahl (see p. 35), but there is also an excellent guide to the TLC of drug plants (Wagner, H., Bladt, S. and Zgainski, E.M. (1983) *Drogen Analyse*, Springer-Verlag, Berlin) which includes many colour illustrations of TLC separations of plant constituents.

A cautionary note should be added here on a common component of TLC solvent systems, namely benzene. The vapour of benzene is now recognized to be particularly harmful to health and the solvent should only be handled where adequate ventilation is available, e.g. in the fume cupboard. For many but not all purposes, benzene can be replaced by toluene, which is less harmful. As a general safety measure, prolonged exposure to any organic solvent mixture used for TLC should be avoided.

The list below of supports, solvent systems and detection methods is summarized from earlier chapters in this book and is presented here in the hope that it will provide a useful checklist for easy reference.

Abscisins	silica gel*	benzene–EtOAc–HOAc (14:6:1)	short UV
Alkaloids	silica gel	MeOH–conc.NH$_4$OH (200:3)	Dragendorff
Amino acids	silica gel	n-BuOH–HOAc–H$_2$O (4:1:1)	Ninhydrin
Amines	cellulose	n-BuOH–HOAc–H$_2$O (4:1:1)	Ninhydrin
Anthocyanidins	cellulose	conc.HCl–HOAc–H$_2$O (3:30:10)	Colour in DL
Anthocyanins	cellulose	n-BuOH–HOAc–H$_2$O (4:1:5)	Colour in DL
Ascorbic acid	silica gel*	EtOH–HOAc–H$_2$O (90:1:9)	Short UV
Aurones	cellulose	n-BuOH–HOAc–H$_2$O (4:1:5)	Colour in DL
Betacyanins	cellulose	n-BuOH–HOAc–H$_2$O (4:1:5)	Colour in DL
Biflavonyls	silica gel	toluene–HCO$_2$Et–HCO$_2$H (5:4:1)	Colour in UV

277

Cardiac glycosides	silica gel	EtOAc–pyridine–H_2O (5:1:4)	$SbCl_3$/$CHCl_3$
Carotenoids	cellulose	petrol (b.p. 40–60°C)–Me_2CO–n-PrOH (90:10:0·45)	Colour in DL
Chalcones	cellulose	n-BuOH–HOAc–H_2O (4:1:5)	Colour in DL
Chlorophylls	cellulose	petrol (b.p. 60–80°C)–Me_2CO–n-PrOH (90:10:0·45)	Colour in DL
Coumarins,			
hydroxy	cellulose	10% HOAc	Colour in UV
furano	silica gel	$CHCl_3$	Colour in UV
Cyanogenic glycosides	silica gel	$CHCl_3$–MeOH (5:1)	Ammoniacal $AgNO_3$
Cyclitols	silica gel	n-BuOH–pyridine–H_2O (10:3:1)	$Pb(OAc)_4$
Cytokinins	cellulose	n-BuOH–H_2O–conc.NH_4OH (172:18:10)	bromophenol blue/ $AgNO_3$
Depsidones	silica gel	n-hexane–Et_2O–HCO_2H (5:4:1)	H_2SO_4
Dihydrochalcones	cellulose	H_2O	Colour in UV
Diterpenoids	silica gel	n-hexane–EtOAc (17:3)	H_2SO_4
Flavanones	cellulose	5% HOAc	Colour in UV
Flavones, Flavonols			
aglycones	cellulose	HOAc–conc.HCl–H_2O (30:3:10)	Colour in UV
glycosides	cellulose	n-BuOH–HOAc–H_2O (4:1:5)	Colour in UV
Gibberellins	silica gel	benzene–n-BuOH-HOAc (14:5:1)	H_2SO_4
Glucosinolates	silica gel	$CHCl_3$–MeOH (17:3)	iodine
Glycoflavones	silica gel	EtOAc–pyridine–H_2O-MeOH (16:4:2:1)	Colour in UV
Hydrolysable tannins	cellulose	iso-BuOH–HOAc–H_2O (14:1:5)	short UV
Hydroxycinnamic acids			
free	cellulose	benzene–HOAc–H_2O (6:7:3)	Colour in UV
esterified	silica gel	toluene–HCO_2Et–HCO_2H (2:1:1)	Colour in UV
Indoles	silica gel	$CHCl_3$–EtOAc–HCO_2H (5:4:1)	p-dimethylamino-cinnamaldehyde
Iridoids	cellulose	n-BuOH–HOAc–H_2O (4:1:5)	$SbCl_3$/$CHCl_3$
Isoflavones	silica gel	$CHCl_3$–MeOH (89:11)	Folin
Isothiocyanates	silica gel	CCl_4–MeOH–H_2O (20:10:1)	ammon. $AgNO_3$
Lignans	silica gel	EtOAc–MeOH (19:1)	H_2SO_4
Lipids			
neutral	silica gel/ $AgNO_3$	iso-PrOH–$CHCl_3$ (3:197)	Rhodamine B
glyco phospho }	silica gel	$CHCl_3$–MeOH–HOAc–H_2O (170:30:20:7)	Rhodamine B
Monosaccharides	silica gel	n-BuOH–HOAc–Et_2O–H_2O (9:6:3:1)	aniline hydrogen phthalate
Monoterpenoids	silica gel	benzene–$CHCl_3$ (1:1)	vanillin/H_2SO_4

Organic acids	silica gel	MeOH–5 M NH$_4$OH (4:1)	bromothymol blue
Oligosaccharides	cellulose	*n*-BuOH–toluene–pyridine–H$_2$O (5:1:3:3)	aniline hydrogen phthalate
Phenolic acids	silica gel	HOAc–CHCl$_3$ (1:9)	folin
Phenylpropenes	silica gel	hexane–CHCl$_3$ (3:2)	vanillin/H$_2$SO$_4$
Phenols	cellulose	benzene–MeOH–HOAc (45:8:4)	Folin
Phytosterols	silica gel	hexane–EtOAc (1:1)	SbCl$_3$ in CHCl$_3$
Polyacetylenes	silica gel	CHCl$_3$–MeOH (9:1)	isatin/H$_2$SO$_4$
Proanthocyanidins	cellulose	iso-BuOH–HOAc–H$_2$O (14:1:5)	short UV
Purines Pyrimidines $\}$	cellulose*	MeOH–conc.HCl–H$_2$O (7:2:1)	short UV
Proteins	Sephadex G-100	0·02 M Na$_3$PO$_4$ buffer contg. 0·2 M NaCl	naphthalene black
Quinones			
anthra	silica gel	EtOAc–MeOH–H$_2$O (100:17:13)	Colour in DL
benzo	silica gel	hexane–EtOAc (17:3)	Colour in DL
isoprenoid	silica gel	benzene–petrol (b.p. 40–60°C) (2:3)	SbCl$_3$ in CHCl$_3$
naphtha	silica gel	petrol (b.p. 60–80°C)–EtOAc (7:3)	Colour in DL
Sapogenins	silica gel	CHCl$_3$–Me$_2$CO (4:1)	SbCl$_3$/conc.HCl
Saponins	silica gel	*n*-BuOH–H$_2$O (1:1)	SbCl$_3$/conc.HCl
Sesquiterpenes	silica gel	benzene–CHCl$_3$ (1:1)	SbCl$_3$ in CHCl$_3$
Sesquiterpene lactones	silica gel	CHCl$_3$–Et$_2$O (4:1)	Vanillin/H$_2$SO$_4$
Stilbenes	cellulose	*n*-BuOH–HOAc–H$_2$O (4:1:5)	Colour in UV
Sugar alcohols	cellulose	*n*-PrOH–EtOAc–H$_2$O (7:1:2)	alk. AgNO$_3$
Thiophenes	silica gel	benzene–chloroform (10:1)	isatin/H$_2$SO$_4$
Triterpenoids	silica gel	hexane–EtOAc (1:1)	SbCl$_3$ in CHCl$_3$
Xanthones	silica gel	CHCl$_3$–HOAc (4:1)	Colour in UV

Key: cellulose = microcrystalline cellulose plate; silica gel = standard grade plates, although activation may be beneficial in some cases; * = plate coated with fluorescent indicator; DL = daylight; UV = ultraviolet light.

Some Useful Addresses

This list of firms and addresses is not meant to be comprehensive but to provide the novice with a starting point for seeking samples of chemical markers and particular pieces of phytochemical equipment. Pure specimens of a wide range of natural plant products are now available commercially at a price, including many rare alkaloids, phenolics, terpenoids, etc. Most of the companies listed have offices or suppliers in other countries.

Rare chemicals, biochemicals

Aldrich Chemical Co.,
The Old Brickyard,
New Road, Gillingham,
Dorset, SP8 4JL,
U.K.

Fluka AG,
Chemische Fabrik,
CH-9470 Buchs,
Switzerland.

Pharmacia,
Box 175,
S 75104 Uppsala 1,
Sweden.

Sigma Chemical Co.,
P.O. Box 14508,
St. Louis,
Missouri 63178,
U.S.A.

Apin Chemicals,
Unit 1,
Milton Trading Estate,
Abingdon, Oxon., OX14 4RS,
U.K.

Koch-Light Ltd.,
37 Hollands Road,
Haverhill, Suffolk, CB9 8PU,
U.K.

Sarsyntex,
BP 100,
Av. Pdt. J.F. Kennedy,
33701 Merignac, France.

Chromatographic equipment

Buchi Lab. Techniques,
CH-9230 Flawil,
Switzerland.

Camag,
Sonnenmatt Str. 11,
CH-4132 Muttenz,
Switzerland.

E. Merck,
Darmstadt,
West Germany.

LKB Instruments,
232 Addington Road,
Selsdon, S. Croydon,
Surrey, CR2 8YD, U.K.

Schleicher and Schüll,
D-3354 Dassel,
West Germany.

Shandon Southern,
Chadwick Road,
Runcorn, Cheshire,
WA7 1PR, U.K.

Whatman Ltd.,
Springfield Mill,
Maidstone,
Kent, ME14 2LE, U.K.

Spectrophotometers, etc.

Dupont Co.,
Clinical and Instrument Systems,
Wilmington,
Delaware 19898,
U.S.A.

MSE Scientific,
Manor Royal,
Crawley, West Sussex,
RH10 1RJ, U.K.

Perkin Elmer,
Post Office Lane,
Beaconsfield, Bucks.,
HP9 1QA, U.K.

Pye Unicam,
York Street,
Cambridge, CB1 2PX,
U.K.

Techmation,
58 Edgware Way,
Edgware,
Middlesex, HA8 8JP,
U.K.

Waters Associates,
324 Chester Road,
Hartford, Northwick,
Cheshire, CW8 2AA,
U.K.

Index

Sorbitol, 237
Soxhlet extraction, 6
Spermidine, 188
Spermine, 188
Squalene, 122
Starch, 265
   from potato, 270
Stearic acid, 152
Stearolic acid, 165
Sterculic acid, 152
   NMR spectrum, 25
Steroidal alkaloids, 199, 200
Steroids
   retention times, 125
   structures, 122, 123
Sterol acetates
   separation by GLC, 13
Stigmasterol, 122
Stilbenes, 81
Strychnine, 194
Succinic acid, 145
Sucrose, 233
Sugar alcohols
   chromatography, 238
   electrophoresis, 239
   structures, 237
Sugar in floral nectars, 230
Sugar phosphates, 230
Sulphides, 170, 172
Sulpholipids, 151
Sulphur compounds
   chemistry, 169
   chromatography, 171
Syringic acid, 40

Tambuletin
   NMR spectrum, 26
Tannins
   classification, 84
   condensed, 85
   detection, 88
   estimation, 87
   hydrolysable, 86
Terpenoids
   conformation, 102
   isomerism among, 103
   main classes, 101
   path of biosynthesis, 102
α-Terpineol, 105
Terpinolene, 105
Thebaine, 195
Theobromine, 209
Thin layer chromatography, 10
   multiple elimination, 12

of simple phenols, 41
   pre-coated plates for, 11
   preparation of plates, 11
   preparative, 12
   recommended systems, 277
   solvents for, 11
Thiophenes, 170, 172
γ-Thujaplicin, 105
Thujone, 105
Tobacco alkaloids, 108
Tricetin, 70
Tricin, 70
Triglycerides, 151
   of peanuts, 159
Trimethyl-lysine, 254
Triterpenoids
   chemistry, 122, 123
   in seeds, 128
   retention times, 125
Tropane alkaloids, 198
Tropolones, 111
Tryptamine, 188, 207
Tryptophane, 178, 207
Tyramine, 188

Ubiquinones, 90, 94
Ultraviolet spectroscopy, 16
   measurement of, 17
   shifts with inorganic reagents, 18
   solvents for, 18
Umbelliferone, 47
Ursolic acid, 122

Vanillic acid, 40
Vicilin
   isolation of, 257
Vicine, 209
Violaxanthin, 130
Visible spectra of plant pigments, 19

Waxes, 160
Wyerone acid, 165

Xanthinin, 106
Xanthones, 81
   UV spectrum of, 17
Xanthophylls, 133, 136
Xylan, 265
   from straw, 270
Xylose, 224

Zeatin, 209
   MS of, 23
Zeaxanthin, 130